GEOGRAPHICAL HORIZONS

GEOGRAPHICAL HORIZONS

Edited by
John Odland
Indiana University
and
Robert N. Taaffe
Indiana University

KENDALL/HUNT PUBLISHING COMPANY
2460 Kerper Boulevard,
Dubuque, Iowa 52001

Library of Congress Catalog Card Number: 77—80460

ISBN 0—8403—1781—6

Printed in the United States of America

401781 01

Contents

Preface

On March 4-6, 1976, the Department of Geography at Indiana University held a Conference on Geography, which was supported by a grant from the Lilly Foundation. The basic purposes of this meeting were to examine recent trends in geographic research and teaching and to strengthen the professional contacts between our Department and teachers of courses in geography at private colleges. This conference was attended by 42 teachers from 26 private colleges or universities in Indiana, Illinois, and Ohio. Papers were presented by every member of our departmental faculty and six outside speakers: Reginald Golledge and Keith Semple of Ohio State University; Peter Kakela of Sangamon State; Harold Rose of the University of Wisconsin-Milwaukee; Gerald Manson of Michigan State; and Julian Wolpert of Princeton, who gave the banquet address at the conference. In addition, 15 of the private college participants served on various panels.

The sessions in the conference were devoted to five general areas of geography: economic and urban; behavioral; social; environmental and physical; and teaching aspects. Each of the thematic sessions included papers by Indiana faculty, a paper by an outside speaker, and a question-and-answer period led by a panel of participants from private colleges. The papers in this volume are grouped under the same general themes as the sessions of the conference.

In many ways, the variety of the topics and methodologies of the essays in this book reflect the diversity of the field of geography. However, this is more a source of encouragement for the future growth of multifaceted research in the discipline than a cause for alarm. Many of the papers provide a review of the state-of-the-art of some of the most rapidly expanding areas of geographic investigation and suggest some promising directions for future research. In this category are the studies treating geographic approaches to: cognitive mapping (MacKay); spatial behavior (Golledge); homicide (Rose); ethnic questions (Bennett); poverty (Wohlenberg); spatial aspects of educational phenomena (Backler); and, the energy budget of a leaf (Davis).

Another subset of studies in this book is strongly oriented toward policy issues, specifically pertaining to railroad consolidation (Black), environmental issues (Barton and Kakela), perceptions toward public housing (Jakubs), and the contingencies for emergency evacuation in areas near nuclear power plants (Wolpert).

Examples of applied and theoretical thematic research in geography are provided by the papers on urban decentralization models (Odland), growth in small and medium-sized cities (Erickson and Miller), regional growth theory (Semple), migration in centrally planned economies (Taaffe), drought in the Sahel (Hidore), and the problems of symbolization in thematic mapping (Kingsbury).

Another important theme is education in geography. This includes the application of some recent innovations in the teaching of geography at the college level, such as audio-tutorial methods (Kingsbury and Backler) and game-simulation approaches (Jakubs), and the more general problems of teaching introductory college courses in geography.

We hope that the papers in this book will convey to teachers and students at least some indication of the vitality of the field of geography and the substantial intellectual, societal, and pedagogical benefits which should be derived from a continuation of the burgeoning growth and refinement of theoretical and applied geographical analyses.

Many people and organizations have been involved in contributing to this conference and this publication. The Department of Geography would like to express its gratitude to the Lilly Foundation for their generous financial support, Dean K. Eugene Faris, the coordinator of the Lilly grant at Indiana, and his assistant, David Munger, and Dean V.J. Shiner, Jr., of the College of Arts and Sciences. Special thanks also should be given to Vice-president Robert M. O'Neil, who fortunately went far beyond the call of duty in welcoming the participants to Indiana by presenting an excellent talk on the interaction of law and geography. Many of the organizational details of the meeting were handled with great skill and many hours of overtime work by the Head Secretary of the Department of Geography, Pauline Stephans, with the able assistance of Rena Beaumont.

We would indeed be remiss if we did not also express our genuine appreciation to the participants in the conference and their institutions. We were delighted to have had an opportunity to strengthen the professional and social bonds with our colleagues in private colleges and we hope they share our enthusiasm for this type of scholarly interaction.

Urban and Economic Geography

Introduction

Urban and economic geography are traditional areas of investigation which continue to attract major portions of the total research effort in geography. The theoretical frameworks available in these areas include models for the location of production, the organization of systems of towns, and the spatial arrangement of urban land uses and these models are as general and as widely accepted as any in human geography. The construction of formal and abstract theoretical models has become a conventional procedure in urban and economic geography and the papers in this section demonstrate some of the benefits of this kind of activity, as well as the results of extensive empirical analysis.

The first paper in this section, by Odland, is concerned with urban land use patterns and shows how traditional models in this area can be broadened to yield a richer set of implications about the spatial structure of the city. Formal models of the economics of the urban land market usually incorporate conditions for residential activity and transportation but they have rarely provided detailed explanations of the spatial arrangement of land uses and population densities. This is partly due to their reliance on assumptions that the city has a single employment center which is the only focus of urban travel. Odland shows how the single center assumption may be discarded by constructing a formal model of land use patterns which includes conditions for production or employment as well as residential activity and transportation. The result is a model which provides for the full range of multiple center land use patterns, but includes the single center city as one possible solution. This model is also very well suited for analyzing the process of decentralization in the city. The formal model is logically connected to an empirical version which can be fitted to data on land use densities and Odland uses data for Chicago to test hypotheses about differences in the extent of decentralization across different urban employment sectors.

The paper by Miller and Erickson reports the results of a very extensive empirical analysis of the ecological structure of settlements. They draw on the established framework of factorial ecology but extend this method dramatically to analyze contrasts in the characteristics of settlements which differ in size and in their situation within the urban system. They utilize data for over 6,800 local areas of 4,000 to 5,000 population and perform analyses in multiple stages in which contrasts are identified between the ecological characteristics of rural vs. urban places; metropolitan areas vs. small urban places; and central cities vs. suburbs. Their study yields several results which warrant further investigation, including evidence that the differentiation of population by socioeconomic status increases in larger places; that suburbs as well as central cities include strong dimensions of poverty and family disorganization; and that black suburbanization does not necessarily reduce ghettoization.

Geographers are becoming involved in matters of public policy to an increasing degree and the paper by William Black addresses a series of crucial issues which arose in the reorganization of the northeastern and midwestern railroads. Black has wide experience in rail system planning at both state and national levels and was recently involved in the organization of the CONRAIL system. Rail system reorganization is certain to have important impacts at both local and national scales. Black points out that although many policy questions involve traditional areas of geographical inquiry such as network location, spatial interaction, and local impact analysis, geographers need to reorient their usual approaches in order to contribute effectively to decision making in this area.

The process of economic development must involve changes in the spatial organization of the economy but geographers have had relatively little success in analyzing the spatial organization of economic development. The paper by Keith Semple represents significant progress in this direction. Existing theories of spatial development imply that some stages of the development process will be characterized by inequalities in regional incomes. The association of regional income disparities with a development process; the possibility of their eventual convergence as part of that process; and the role of the public sector in regional income convergence are all issues which have importance to public policy as well as academic interest. Semple shows that inequalities in regional incomes in the United States have converged since the period of maximum disparity in the 1930s, and that convergence has occurred within each of the major employment sectors. He also demonstrates how regional income convergence is related to the influence of government spending as well as developments in the private sector such as increases in the mobility of production factors.

Economic development also involves large-scale population migrations even where development is guided by central planning. Robert Taaffe provides an extensive analysis of migration in the centrally planned economies of Eastern Europe and the Soviet Union. His conclusions have implications for understanding the relation between development and migration in general as well as in the case of centrally planned economies. Migration policies played a crucial role in the rapid industrialization and collectivization of agriculture which characterized the early stages of planned development and policies designed to direct migration in order to meet developmental goals were generally successful. Taaffe shows, however, that centralized planning agencies have had little success in controlling migration once the transition to an urban and industrial society has been accomplished. Individual locational preferences have prevailed over societal objectives despite the controls available to planners seeking to control the distribution of population.

Production Conditions and the Spatial Arrangement of Multi-center, Multi-industry Cities

John Odland
Indiana University

The spatial arrangement of land uses within cities gains much of its apparent coherence from the ways that patterns of activity focus on high density activity centers such as central business districts. Most commercial and industrial activities are located in these centers, which are recognizable as areas of high concentrations of capital and, during working hours, high concentrations of people. The convergence of most work trips on these centers is one of the major underpinnings of residential land use theory (Alonso, 1964; Senior, 1974). These concentrations of labor and capital presumably result from economic advantages gained within such concentrations: economic advantages which are usually identified as agglomeration economies or external economies of scale.

Urban land use theory is relatively weak as regards the role of these production conditions in determining land use patterns and this weakness may be a crucial hindrance to the development of at least some aspects of the theory. In particular, the variation of production conditions across different levels of spatial concentration and different production activities may play an important role in explaining the decentralization of large cities. A model of the spatial arrangement of a multi-center, multi-industry city is presented in this paper. This model allows for multiple employment centers and takes account of possible differences in production conditions across different industries and different levels of spatial concentration. The formal model is logically connected to an operational version which can be fitted to available data on land use patterns and which allows some of the elements of the theoretical argument to be examined empirically. Empirical results are presented for the Chicago metropolitan area in 1970.

PRODUCTION CONDITIONS AND LAND USE THEORY

Most formal economic models of urban land use treat the concentration economies which lead to high density employment as undifferentiated over various activities and overwhelming in comparison to economic forces which favor the dispersion of employment. These models do this by incorporating the familiar assumption that all employment is concentrated in a single central business district. The entire labor force presumably commutes to this CBD from a set of dispersed residential locations and the efficiencies gained from concentration are presumably great enough for employment activities concentrated at the CBD to absorb all the labor of the city. Most analyses then emphasize the conditions for transportation and residential occupance in the area surrounding the CBD. The single center assumption, along with an assumption of symmetry in the transportation system, simplifies these analyses considerably by making it possible to treat the two-dimensional city in terms of a single dimension: distance from the CBD.

The single-center assumption has been identified as a major theoretical weakness (Angel and Hyman, 1972; Richardson, 1976) and is at variance with readily observable urban land use patterns in which large cities ordinarily have multiple employment centers. Further, employment activities show varying responses to opportunities for high density concentration. Some activities are found only in the very high density centers. Others are apparently less responsive to opportunities to locate in high density centers and can be found in areas where the densities of capital and labor are much lower.

The concentration of employment activities within cities, in either single or multiple center patterns, apparently occurs because favorable production conditions are available when capital and labor are concentrated at high densities per unit of land. The role of such agglomeration economies or external scale economies is a major element of location theory at regional as well as urban scales. Following Hoover (1968), agglomerative factors at the urban scale can be classified into two types: those derived from internal scale economies and available to a single firm acting alone, and external scale economies associated with the joint action of several firms. The exploitation of internal scale economies can lead to spatial concentration if at least some parts of the production process must be spatially contiguous (Mills, 1972a, p. 7). Concentration of production at high densities may also occur if sets of firms seek locations in joint proximity because of direct production linkages among them. The locational effects of such linkages have been investigated in a theoretical way by Gannon (1973) and the importance of production linkages to individual urban firms has been analyzed in detail by Gilmour (1974). Lever (1975) has also investigated the role of external economies in the decentralization of manufacturing. Firms located in proximity to one another may also derive joint benefits which do not involve direct interactions among them. For example, they may be able to attract specialized services or facilities which would be unavailable to an isolated firm or their joint influence on the character of the surrounding environment may exceed that of an isolated firm. In addition, firms may simply cluster about a common attraction such as a transportation facility. When a number of firms are attracted to a particular location the resulting competition for land, and increase in land prices, is likely to alter their production decisions in the direction of greater land use intensity (Alonso, 1964, pp. 52-58).

There is clearly a varied array of reasons for urban firms to make location decisions which lead to concentration of employment and production and it is likely, given the variety of production activities in a city, that different activities will show varying responses to opportunities for concentration. Some production activities gain only modest benefits from scale economies or from location in proximity to other firms while the benefits gained from spatial concentration are vital to others. There are also negative impacts associated with concentration, including congestion, and these may also have different importance for different kinds of production.

The major focus of research on the locational effects of production conditions has been the analysis of individual firm locations but some research has been carried out on the role of production conditions in aggregative location models. In particular, Mills (1970a, 1972b), Hartwick and Hartwick (1974, 1975) and Goldstein and Moses (1975) have analyzed the location of interdependent production activities in the context of general land use models. Fales and Moses (1972) have combined the elements of the usual land rent approach with the cost-minimizing considerations of the Weber model of industrial location. Location in their model may depend on scale economies and agglomeration economies as well as transportation costs for workers and intermediate goods. Although Fales and Moses do not test a formal version of their model they provide a thorough empirical investigation of the limitations of the model for nineteenth-century Chicago.

Models which oversimplify urban land use patterns by assuming a single CBD and ignoring the effect of varied production conditions on a range of production activities will generate land use patterns which have only the most general resemblance to those observed in actual cities. They may, nonetheless, yield valuable insights regarding some current issues in land use theory (Anas and Dendrinos, 1976). In many cases, these drastic simplifications may be necessary in order to derive results formally. In some cases, however, the assumptions of a single center and simplified production conditions are untenable. In particular, the process of urban decentralization involves the dispersion of employment activities as well as the residential population and there is some evidence that the dispersion of employment may occur prior to the dispersion of residences (Mills, 1970b; Kain, 1968). Models which incorporate the assumption of a single employment center obviously do not provide a suitable basis for analyzing the dispersion of employment.

The elements which are involved in decentralization have been identified in a general way by Lave (1969) who presented a model in which a city decentralizes when agglomeration economies are outweighed by the savings in transportation costs and land rent available in a decentralized pattern. Similar arguments have been formalized in terms of a mathematical programming model by Odland (1976). Urban employment does disperse so the available concentration economies are evidently not unlimited compared to the benefits of decentralization. Further, given the variety of production activities and the varied sources of concentration economies, it is unlikely that decentralization proceeds uniformly over all categories of employment. Some activities may decentralize readily while others remain in highly concentrated patterns. Hence, decentralization may be associated with particular categories of employment activities and with an increase in the functional specialization of different areas in the city.

Some of these possibilities are investigated in the following sections. A model of the spatial arrangement of a multi-center city, which has been introduced in previous research (Odland, 1976) is extended for the multi-industry case. An empirical version of the model is used to investigate the possibility of differences in concentration economies among different sectors.

A FORMAL MODEL OF LOCATION PATTERNS

A mathematical programming model which formalizes major elements that affect the spatial arrangement of urban land uses is presented in this section. The possible effects of concentration economies on land use patterns are included by making production costs in each category of employment a function of local activity levels. Production costs for an employment activity can be lower at locations where there is a high concentration of production. This formulation leads to a complex nonlinear programming problem which has theoretical value but virtually no prospect for explicit numerical solution. Nonetheless, the programming model provides a valuable framework for empirical analysis of land use patterns. A fairly simple set of solution conditions is associated with optimum solutions and these conditions yield insights into the basis of an optimum arrangement and indicate possible sources of decentralization. The solution conditions can also be arranged to correspond with operational formulas which are the vehicles for empirical analyses of land use patterns in the following section.

The exploitation of scale economies or agglomeration economies generally requires employment to be concentrated in space but such concentration is likely to be associated with higher transportation costs for commuting, or more residential crowding, or both. The mathematical programming model is designed to identify the land use pattern which balances these opposing forces by minimizing the total of production costs, real income losses from crowding, and transportation costs for commuting. The model city consists of a finite number of contiguous and non-overlapping zones of uniform size. Several categories of employment activity are included in the model. A solution to the programming model consists of an allocation of levels of employment activity in each category to each zone along with residential activity and the number of commuters travelling between each pair of zones.

Concentration Economies and Production Costs

Scale economies or agglomeration economies are assumed to operate by affecting the per unit production costs within the zone where they originate. Production costs for activity k in zone i may depend on the level of production of the same activity in the same zone; on the level of production of one or more other activities in the same zone; or on the total of all production within the zone. Production costs within zone i are assumed, however, to be independent of activity levels in other zones in the city. That is, concentration economies are assumed to be unavailable outside the zone where they originate. The model also ignores the possible costs of transporting intermediate goods between various zones.

The per unit production costs for employment category k in zone i can be written, under these assumptions, as a function of each of the activity levels of other employment categories in the zone as well as their sum:

$$f(Q_{ik}) = f(Q_{i1}, Q_{i2}, \ldots Q_{ik}, \ldots Q_{ip}, \sum_{k=1}^{p} Q_{ik})$$

where the Q_{ik} are activity levels in p categories of employment and $f(Q_{ik})$ is the per unit production cost in activity category k. The derivatives of $f(Q_{ik})$ with respect to each of the arguments will usually be negative or zero although diseconomies of concentration (indicated by a positive sign for a derivative) are also possible. The production conditions for different activities are thus interactive within the confines of a particular zone. Since per unit production costs for any activity can depend on the levels of other activities the production costs can be altered by rearranging activities in space.

The effect of concentration within a zone on the conditions for residential activity can be handled in a similar way. That is, $g(P_{jk})$, the per capita real income loss from residential crowding for a household located in zone j with its wage-earners employed in activity k, can be a function of the total number of residents in zone j and of the number of persons in particular employment categories who also reside in zone j. The per capita real income losses could also be a function of the level of nonresidential activity in the same zone but that possibility is not examined in this research. Derivatives of $g(P_{jk})$ may be expected to be positive or zero, indicating that increases in crowding are associated with losses of per capita real income.

A Mathematical Programming Model of Spatial Arrangement

A nonlinear programming model which specifies the conditions for an optimal allocation of production, employment, and residences over a set of zones has been presented in earlier research (Odland, 1976). That model is extended here to accomodate the added complications of differing production conditions for different activities and interaction between the production conditions and production levels of different activities within a zone. The solution to the model is an optimum in the sense that the total of production costs, real income losses from residential crowding, and transportation costs for commuting is minimized.

The objective of the programming problem is to minimize for n zones and p production activities the value of

$$\sum_{i=1}^{m}\sum_{k=1}^{P} Q_{ik}f(Q_{ik}) + \sum_{i=1}^{m}\sum_{j=1}^{m}\sum_{k=1}^{P} c_{ij}N_{ijk} + \sum_{j=1}^{m}\sum_{k=1}^{P} P_{jk}g(P_{jk}) \qquad (1)$$

where N_{ijk} is the number of workers in activity k who commute from zone i to zone j and c_{ij} is the transportation cost per worker for a trip between the two zones. The minimization is subject to the following constraints:

$$\sum_{i=1}^{n} Q_{ik} - q_k W_k = 0 \qquad (2a)$$

$$\sum_{j=1}^{n}\sum_{k=1}^{P} P_{jk} - b\sum_{k=1}^{n} W_k = 0 \qquad (2b)$$

$$Q_{ik} - q_k \sum_{j=1}^{n} N_{ijk} = 0 \qquad \text{all } i, k \qquad (2c)$$

$$b^{-1} p_{ik} - \sum_{i=1}^{n} N_{ijk} = 0 \qquad \text{all } j, k \qquad (2d)$$

The first set of constraints insures full employment. The coefficient q_k is the output per worker in industry k. This value is assumed to remain constant regardless of scale or agglomeration economies. The value W_k is the number of workers in industry k. This value is assumed to be determined exogenously. The second constraint makes certain that each person has a residence. The coefficient b is the number of persons per job and is assumed to be constant over all households. The constraint (2c) states that sufficient labor reaches each zone to carry out the levels of production activity assigned to the zone. The final set of constraints states that the number of commuters leaving each zone does not exceed the number residing there.

Taking partial derivatives of the Lagrangian form of this problem yields the set of Kuhn-Tucker conditions for a solution, shown in equations (3a), (3b), and (3c). The conditions appear as pairs. The first member of each pair holds as an inequality when the associated variable has a value of zero.

$$\Phi_{ik} - \lambda_k^{(1)} + \lambda_{ik}^{(3)} \geqslant 0$$
$$Q_{ik}\,[\Phi_{ik} - \lambda_k^{(1)} + \lambda_{ik}^{(3)}] = 0 \qquad \text{all i,k} \qquad (3a)$$

$$\Gamma_{jk} - \lambda^{(2)} + \lambda_{jk}^{(4)} \geqslant 0$$
$$P_{jk}\,[\Gamma_{jk} - \lambda^{(2)} + \lambda_{jk}^{(4)}] = 0 \qquad \text{all j,k} \qquad (3b)$$

$$e_{ij} - \lambda_{ik}^{(3)} q_k - \lambda_{jk}^{(4)} \geqslant 0$$
$$N_{ijk}\,[c_{ij} - \lambda_{ik}^{(3)} q_k - \lambda_{jk}^{(4)}] = 0 \qquad \text{all i,j,k} \qquad (3c)$$

$$L'_\lambda(1) = 0; \quad L'_\lambda(2);= 0; \quad L'_\lambda(3) = 0; \quad L'_\lambda(4) = 0 \qquad \text{all i,j,k} \qquad (3d)$$

The Φ_{ik} and Γ_{jk} above are derivatives of the products $Q_{ik}f(Q_{ik})$ and $P_{jk}g(P_{jk})$ which appear in the objective function. The L' are derivatives with respect to the variables which appear as subscripts. The various λ are Lagrange multipliers which are associated with the constraints. The values of $\lambda_k^{(1)}$ and $\lambda^{(2)}$ are associated with the size of the labor force in each activity category and the size of the total population. These values are related to city size for fixed proportions of employment in the activity categories. The $\lambda_{ik}^{(3)}$ can be interpreted as location prices associated with locating activity k in zone i. The values of $\lambda_{jk}^{(4)}$ constitute prices for locating the residences of workers in activity k in zone j.

Conditions (3a) are conditions for production in each zone. In order for the optimal solution to involve location of activity k in zone i the value of $\lambda_k^{(1)}$, which is related to city size, must exceed the value of the location price for the activity and zone, $\lambda_{ik}^{(3)}$, and the level of activity is determined by their difference. Activity k is not located in zones where $\lambda_{ik}^{(3)} > \lambda_k^{(1)}$. The conditions (3b) hold in a similar way for residential activity. The conditions (3c) are conditions for commuting between pairs of zones.

The location prices vary over the zones, and, by setting the values of the Kuhn-Tucker conditions, they determine the patterns of production, residential activity, and commuting which are associated with the optimum solution. Conditions (3a) are conditions for locating production in each zone. In order for the optimum solution to involve location of activity k in zone i the value of $\lambda_{ik}^{(3)}$ must not exceed the value of $\lambda_k^{(1)}$, which is related to city size and is the same for all zones. The level of activity is determined by the difference between $\lambda_k^{(1)}$ and $\lambda_{ik}^{(3)}$. Low values of the location prices are associated with high activity levels and activity k does not locate in zones where $\lambda_{ik}^{(3)} > \lambda_k^{(1)}$. An increase in city size, experienced as an increase in the size of the labor force, would increase the value of $\lambda_k^{(1)}$ and could bring about location of activity k in zones where it previously was absent. The conditions (3b) hold in a similar way for residential activity. The spatial extent of the city is determined as the set of zones which contain production, residences, or both.

The way that location prices are related to transportation costs is shown by the conditions (3c) which are also the conditions for commuting between each pair of zones. Workers in activity k will commute from zone j to zone i as part of the optimal solution if the value of $\lambda_{ik}^{(3)} q_k + \lambda_{jk}^{(4)}$ is equal to the transportation cost between the two zones. No commuting occurs between two zones if the transportation cost exceeds this value. The appearance in these conditions of q_k, the output per worker in activity k, simply indicates that the more productive workers can be transported over greater distances as part of the optimal solution.

Transformation to an Operationally Usable Form

This programming model has a logical form which is conventional in location theory. Statements about the spatial arrangement of activities are derived from a set of antecedent conditions and the operation of a set of law-like statements. In this case, the antecedent conditions consist of the conditions for production and residential activity, the set of zone-to-zone transportation costs, and the size of the city's population or labor force. The operation of law-like statements corresponds to the formulation and solution of the nonlinear programming model. The result is an assignment of activity levels to places, or to pairs of places if the activity is commuting. A numerical solution could easily be put in the form of a map showing activity levels for each zone and, even where numerical solutions are unavailable, the solution conditions provide an abstract description of such a map.

This map-like result is very general but it fails to provide a useful framework for examining the effects of varying production conditions on the location of employment activities in actual cities. Sensitivity analyses, which would be carried out by solving the model for different sets of antecedent production conditions, are not available because of the difficulty of observing and formulating production conditions in a usable way as well as the difficulty of obtaining numerical solutions to the model. It might be possible to analyze the comparative statics of the solution by examining the behavior of the solution conditions but it would be impossible to determine if the locational shifts associated with changes in production conditions were of significant size without knowing at least relative values for some of the coefficients. Problems in relating the results of deductive models to observed location patterns are nearly universal in location theory, but this model is an exceptional case because it is possible to transform the solution conditions into a form which supports a detailed examination of actual location patterns.

The transformation to an operationally usable form is made by combining the solution conditions into a single equation. Two simplifying assumptions are necessary in order to do this. First, the model is assumed to be solved for an interior solution. This corresponds to a solution in which each activity occurs in nonzero amounts in each zone and each pair of zones exchanges some commuters. The activity levels may be very small so long as they are not zero. This assures that the first member of each pair of solution conditions holds as an equality. Second, the effects of variation in labor productivity over different activities is ignored. This amounts to setting the value of q_k in condition (3c) equal to unity for all activities.

Only the first member of each pair of solution conditions in (3a), (3b), and (3c) is relevant under these assumptions. Conditions (3a) and (3b) are both equal to zero for all zones at an interior optimum. Hence, their sum is equal to zero for each pair of zones and each activity category:

$$\Phi_{ik} - \Gamma_{jk} - \lambda_k^{(1)} - \lambda^{(2)} + \lambda_{ik}^{(3)} + \lambda_{ik}^{(4)} = 0 \qquad \text{all i, j, k} \qquad (4)$$

Condition (3c) holds as an equality for each pair of zones and, with q_k equal to unity, can be rewritten as

$$c_{ij} = \lambda_{ik}^{(3)} + \lambda_{jk}^{(4)} \qquad \text{all k} \qquad (5)$$

for any pair of zones and activity category. Substituting c_{ij} for the sum $\lambda_{ik}^{(3)} + \lambda_{jk}^{(4)}$ and rearranging terms yields a single equation which is the counterpart to the solution conditions (3a) through (3c):

$$c_{ij} = \lambda_k^{(1)} + \lambda^{(2)} - \Phi_{ik} - \Gamma_{jk} \qquad \text{all k} \qquad (6)$$

The transformation of solution conditions from the form in equations (3a), (3b), and (3c) to the form in equation (6) reverses the logical order of the statements they make and yields a functional form which can be readily applied to analyze observed location patterns. The solution conditions in the form given by (3a), (3b), and (3c) can be interpreted as rules for simultaneous assignment of equilibrium activity levels in all zones based on the relative locations of the zones. The form in equation (6) is associated with the same equilibrium distribution but the logical order of the conditions is reversed by assigning the relative locations of the zones on the basis of their activity levels. The distance between members of the pair of zones i and j, as expressed by the transportation cost c_{ij}, depends on the values of Φ_{ik} and Γ_{jk} which are functions of the levels of employment activity and residential activity. The value of the sum $\lambda_k^{(1)} + \lambda^{(2)}$ is constant for all zones in a city of given size.

AN OPERATIONAL FORM OF THE MODEL

The single-equation model given by equation (6) also makes it possible to estimate parameters for the location patterns of particular activities on the basis of actual distributions of the activities. The conditions in the alternative form of (3a), (3b), and (3c) describe the distribution of all activities over all zones simultaneously and provide no obvious basis for comparing location patterns for different activities. The equivalent conditions in equation (6), which make statements about the relative locations of pairs of zones, can be fitted to the spatial distributions of various activities by least squares methods provided that suitable functional forms are selected for Φ_{ik} and Γ_{jk}. These empirical models can then serve as a basis for comparing the locational conditions for different activity categories.

The operational models must use available data which are recorded for zones of irregular size so they are defined for densities of employment and population, in persons per square kilometer, rather than number of persons in zones of uniform size. This alters the definitions of Φ_{ik} and Γ_{jk} but does not change the logic of the model in any fundamental way. The operational models also use straight-line distance between the centers of zones instead of the transportation cost separating pairs of zones.

The definition of functional forms for Φ_{ik} and Γ_{jk} is a crucial step in making the single-equation model operational. Concentration economies may be associated with the density of activities in a particular category or set of closely related categories or they may be available because of the concentration of economic activity of all kinds. The operational form which represents Φ_{ik} should be capable of differentiating between the effects of special and general concentration economics, within the limits set by the available categorization of data. A suitable operational form to replace the Φ_{ik} of equation (6) is

$$\Phi_{ik} = b_1 \ln D^e_{ik} + b_2 \ln D^E_i \qquad \text{all k}$$

where Φ_{ik} is defined for densities of employees and:
D^e_{ik} is the density of employees in activity k in zone i;
D^E_i is the density of all employees in zone i; and
b_1, b_2 are parameters.

The operational replacement for Γ_{jk} is also capable of differentiating between the general effects of residential density and the special effects of density of employees in one category:

$$\Gamma_{jk} = b_3 \ln D^p_{jk} + b_4 \ln D^P_j \qquad \text{all k}$$

where: D^p_{jk} is the density of employees in category k residing in zone j;
D^P_j is the density of total population for zone j; and
b_3, b_4 are parameters.

Combining these operational forms yields a model which can be fitted by regression to data on distances and densities:

$$d_{ij} = b_0 + b_1 \ln D^e_{ik} + b_2 \ln D^E_i + b_3 \ln D^p_{jk} + b_4 \ln D^P_j + e_{ij} \qquad (7)$$

where: d_{ij} is the distance between zones i and j, in kilometers;
b_0 is a parameter which represents city size and replaces $\lambda_k^{(1)} + \lambda^{(2)}$ of equation (6);
and e_{ij} is an error term.

EMPIRICAL RESULTS

The model in equation (7) was fitted to data on the density of workplaces and residences for six employment categories in the Chicago area in 1970. Table 1 shows summary information on the total numbers of employees and residents and the average densities in the zones in persons per square kilometer. Data on the location of employment were obtained from employment statistics which report the location of employment for postal zones within the city of Chicago and for suburban communities outside the city in Cook county and five surrounding Illinois counties: DuPage, Will, Lake, McHenry, and Kane. The 121 zones cover this area completely. Data on the residential location of employees in each category were obtained by aggregating 1970 census tract data on numbers of employed persons by industry to the same set of 121 zones used for the employment statistics. The totals for employment and residence do not agree in each category because of differences in definition and coverage between the census and the employment statistics. Other categories of employment are available but analyses were

Table 1
Summary Data for Employment Categories—Chicago, 1970

	Total Employees	Total Residents*	Mean Density, Employment**	Mean Density, Residents*
All Employment	2,244,914	6,441,949	2,988	3,848
Durable Manufacturing	604,902	553,825	434	309
Nondurable Manufacturing	342,942	331,099	417	182
Wholesaling	219,709	136,171	344	77
Retailing	385,132	431,369	423	245
Finance, Ins., Real Estate	147,908	164,210	424	103
Services	272,431	630,188	535	394

*Figures for "All Employment" are all persons. Figures for each category are numbers of employees.
**Densities are in persons per square kilometer.

performed only for categories which apparently fulfilled the assumption of an interior solution for the 121 zones. Categories which have zero employees in some zones were not analyzed.

The results of fitting the model by ordinary least squares are shown in Table 2. These equations include only those regression coefficients which were shown to be significantly different from zero on the basis of a t-test. The equations are based on 14,641 observations, the square of the number of zones. The correlation coefficients are not large but the standard errors, shown in parenthesis beneath each coefficient, are small.

The regression equations reveal some important differences among location patterns for the various activity categories. These differences reflect the varying attraction of concentrations of activity in the same category as well as concentrations of total employment and residential activity. The signs of the coefficients for total employment and total residential density are, as expected, negative for each activity. This indicates that the city is arranged so that high density concentrations of employment and high density concentrations of residential activity are near one another. The distance separating a pair of zones increases as the densities of employment and residence increase.

Differences in the dispersion between different activity categories are indicated by the relative values of coefficients associated with D_i^e and D_j^r . The values of the coefficients for D_i^e are positive for the two categories of manufacturing activity. This indicates that concentrations of manufacturing are more dispersed with respect to residential density and the density of total employment than are the other activity categories. That is, the members of a pair of zones with given densities of total employment and total residence would be likely to be located farther apart if a substantial proportion of the employment were in manufacturing.

Table 2

Regression Equations for Six Categories of Employment

Durable Manufacturing	103.74238	$+ 0.51595 \ln D_i^e$	$- 4.60371 \ln D_i^E$	$- 5.79607 \ln D_j^R$	
R^2 = .2201	(1.1951)	(.1418)	(.1758)	(.1223)	
Nondurable Manufacturing	104.87690	$+ 1.81810 \ln D_i^e$	$- 5.80455 \ln D_i^E$	$- 5.79607 \ln D_j^R$	
R^2 = .2294	(1.1454)	(.1320)	(.1569)	(.1216)	
Wholesaling	80.32883	$- 2.82209 \ln D_i^e$	$- 0.54441 \ln D_i^E$	$- 2.50378 \ln D_j^r$	$- 3.36894 \ln D_j^R$
R^2 = .2510	(1.4303)	(.1433)	(.2011)	(.1654)	(.2002)
Retailing	100.75404	$- 1.49232 \ln D_i^e$	$- 2.89080 \ln D_i^E$	$- 0.41125 \ln D_j^r$	$- 5.39766 \ln D_j^R$
R^2 = .2241	(1.2336)	(.1630)	(.1589)	(.1822)	(.2146)
Finance, Ins., Real Estate	94.93420	$- 0.31676 \ln D_i^e$	$- 3.75954 \ln D_i^E$	$- 1.70041 \ln D_j^r$	$- 4.11185 \ln D_j^R$
R^2 = .2257	(1.3278)	(.0893)	(.1274)	(.1639)	(.2303)
Services	101.36137	$- 0.47770 \ln D_i^e$	$- 3.60407 \ln D_i^E$	$- 5.79607 \ln D_j^R$	
R^2 = .2199	(1.1875)	(.1531)	(.1755)	(.1234)	

The negative coefficients for D_i^e in the equations for wholesaling and retailing indicate that these activities are less dispersed than total employment. In the case of wholesaling, this occurs because the activity is strongly oriented to the center of the region. Density of employment in wholesaling declines more rapidly with distance from the center than other employment categories. Retailing is less strongly oriented to central locations and the lower dispersion in this category probably reflects an orientation to residential locations rather than the city center. Coefficients for employment densities in services and financial activities are also negative but are relatively small.

SUMMARY AND CONCLUSIONS

The application of formal models of urban land use as models of the spatial arrangement of cities has typically been restricted by oversimplification of the production or employment sector of the urban economy. These simplifications, which include prior specification of the location of production as well as treatment of production as homogeneous activity, limit the array of spatial arrangements which the associated city may assume. The spatial possibilities for most such models are limited to a range of monocentric residential density functions. The model presented here relaxes the restrictive assumptions and, by doing so, greatly enriches the range of spatial configurations which are possible for the model city. This enrichment is accomplished by formulating the problem of arranging the city as a problem of allocating employment, residences, and commuting over a set of discrete zones. Solution conditions for the resulting nonlinear programming model provide a basis for qualitative statements about the spatial arrangement of the city.

The solution conditions also provide a basis for estimating the parameters associated with the observed spatial arrangement of particular cities. Combining solution conditions for pairs of zones yields a single equation and operational forms of the equation can be fitted to data on distances and densities by regression. This equation predicts the spatial separation between members of a pair of zones based on their internal characteristics and can provide appropriate descriptions for a broader array of spatial arrangements than the traditional monocentric density functions. The parameter estimates are not independent of the sizes and shapes of the zones, however, and the sizes of the zones are especially crucial for this model. External scale economies and other density-related effects of production conditions are localized and are unavailable outside some unknown spatial range centered on the locations where they are generated. The capacity of a set of parameter estimates to detect evidence of these effects depends on the relation between this unknown range and the sizes of the zones. Parameter estimates based on zones which are substantially larger or substantially smaller than the spatial range of external economies will fail to detect influences of those economies on spatial arrangements.

Empirical results for the Chicago area indicate that there are significant differences between the spatial arrangement of different categories of employment and the differences are probably associated with differences in production conditions. The employment categories are highly aggregated and the zone sizes may be incongruent with the spatial range of density-related effects on production conditions. The results indicate, however, that this approach has considerable potential for empirical investigation as well as theoretical anlysis. Complete exploitation of the potential for empirical investigation will require more detailed investigations. In particular, it may be important to identify, on the basis of a detailed categorization of land uses, the sets of land uses which gain benefits from mutual proximity, as well as the spatial range within which those benefits are available.

REFERENCES

Alonso, W. *Location and Land Use.* Cambridge: Harvard University Press, 1964.

Anas, A., and Dendrinos, D. "The New Urban Economics: A Brief Survey." In G.J. Papageorgiou (ed.) *Mathematical Land Use Theory.* Lexington: D.C. Heath & Co., 1976, pp. 23-51.

Angel, S., and Hyman, G. "Urban Transport Expenditure." *Papers of the Regional Science Association* 29 (1972):105-123.

Chicago Area Labor Market Analysis Unit. "Where Workers Work: A Survey of Employment Covered Under the Illinois Unemployment Compensation Act." Chicago: Illinois Department of Labor, 1970.

Fales, R., and Moses, L. "Land Use Theory and the Spatial Structure of the Nineteenth Century City." *Papers of the Regional Science Association* 28 (1972):49-80.

Gannon, C.A. "Intra-Urban Industrial Location Theory and Inter-Establishment Linkages." *Geographical Analysis* 5 (1972):214-244.

Gilmour, J.M. "External Economies of Scale, Inter-Industrial Linkages and Decision-Making in Manufacturing." In F.E.I. Hamilton (ed.) *Spatial Perspectives on Industrial Organization and Decision Making,* New York: John Wiley, 1974, pp. 335-362.

Goldstein, G.S., and Moses, L. "Interdependence and the Location of Urban Activities." *Journal of Urban Economics* 2 (1975):63-84.
Hartwick, J.M., and Hartwick, P.G. "An Activity Analysis Approach to Urban Model Building." *Papers of the Regional Science Association* 35 (1975):76-85.
———. "Efficient Resource Allocation in a Multi-nucleated City with Intermediate Goods." *Quarterly Journal of Economics* 74 (1972):100-117.
Hoover, E.M. "The Evolving Form and Organization of the Metropolis." In Perloff, H.S. (ed.) *Issues in Urban Economics.* Baltimore: The Johns Hopkins Press, 1968, pp. 237-284.
Kain, J.F. "The Distribution and Movement of Jobs and Industry." In J.Q. Wilson (ed.) *The Metropolitan Enigma.* Cambridge: Harvard University Press, 1968, pp. 1-44.
Kemper, P., and Schemner, R. "The Density Gradient for Manufacturing Industry." *Journal of Urban Economics* 1 (1974):410-427.
Lave, L.B. "Congestion and Urban Location." *Papers of the Regional Science Association* 25 (1970):133-150.
Lever, W.F. "Manufacturing Decentralization and Shifts in Factor Costs and External Economies." In Collins, L. and Walker, D.F. (eds.) *Locational Dynamics of Manufacturing Industry.* New York: John Wiley, 1975, pp. 295-324.
Mills, E.S. "The Efficiency of Spatial Competition." *Papers of the Regional Science Association* 25 (1970):71-82.
———. "Urban Density Functions." *Urban Studies* 7 (1970):6-20.
———. *Studies in the Structure of the Urban Economy.* Baltimore: The John Hopkins Press, 1972.
———. "Markets and Efficient Resource Allocation in Urban Areas." *Swedish Journal of Economics* 74 (1972):100-117.
Odland, J. "The Spatial Arrangement of Urban Activities: A Simultaneous Location Model." *Environment and Planning* 8 (1976):779-792.
Richardson, H. "Relevance of Mathematical Land Use Theory to Applications." In G.J. Papageorgiou, (ed.) *Mathematical Land Use Theory.* Lexington: D.C. Heath & Co., 1976, pp. 9-22.
Senior, M.L. "Approaches to Residential Location Modelling 2: Urban Economic Models and some Recent Developments." *Environment and Planning* 6 (1974):369-409.

Situational Contrasts in the Characteristics of Settlement System Places

Theodore K. Miller
Indiana University
and
Rodney A. Erickson
Pennsylvania State University

The structural properties of socioeconomic traits which characterize human populations have provided a topic of considerable interest to geographers and other social scientists. This interest has ranged from an academic curiosity focused on human ecology to a pragmatic approach for classifying population groups. Research has frequently emphasized the comparative aspects of socioeconomic structure for a variety of observational units.

Pioneering research in social area analysis[1] and factorial ecology[2] which included a wide array of socioeconomic variables has generated numerous case studies and cross-sectional analyses of the structural dimensions of populations. Although such studies have included the analysis of many types of observation units, none has provided a systematic examination of characteristics covering the entire array of places in the settlement pattern as a whole including non-urban areas.

In the following analysis, situational contrasts in the characteristics of local areas in the United States are examined. A multistage factor analytic methodology is applied to a large set of population data on local areas recently made available by the U.S. Bureau of the Census.[3] The primary purpose of the analysis is to develop a systematic and integrative comparison of the characteristics of local areas which comprise various settlement system places. A secondary and related purpose is to examine the effects of successive spatial disaggregation on ecological analysis employing a factor analytic methodology. Such analyses can facilitate more insightful theoretical investigations of socioeconomic structure (and change, using subsequent samples) and a comparative basis from which to evaluate analyses more limited in geographic scope and population size.

FRAMEWORK AND APPROACH

The multivariate analytic techniques of factor analysis and principal components analysis have been used extensively to isolate the underlying structure of data sets by reducing many specific characteristics to a smaller number of dimensions or patterns.[4] Factorial ecology is the application of these techniques to problems of areal differentiation. Past research has attempted to discover the basis for areal differentiation through several different scales of study:

(1) partitioned geographic space, such as smaller units of a nation or region; (2) city systems, and (3) intraurban/city studies.[5]

Geographers and economists have typically utilized sets of national or regional population characteristics to examine spatial variations in indices of socioeconomic well-being focusing on differences in income, occupation, and education.[6] County observation units have been used in most studies of the United States. While the structural dimensions of development and poverty revealed by these studies were, in part, conditioned by the variables which were included in each data, set, the county-level surveys yielded a number of similar results. The common structural patterns identified were: (1) a general level of economic health, well-being or poverty dimension; (2-3) an urban-rural dichotomy reflecting urbanization-industrialization characteristics and agriculture-resource development, respectively; (4) labor force participation rates; and (5) a racial gap in socioeconomic well-being. Other national studies which have utilized states as observation units have reported similar results.[7]

Numerous studies have also focused on cities as observation units within an urban system.[8] Substantive emphasis in this research has been of two types: functional or economic specialization theories; and, socioeconomic characteristics analysis. Functional studies of the urban system have been used primarily to develop taxonomies or classifications relating to dominant economic activities such as industrial, commercial, service, and special function. The latter type of study has reduced large sets of socioeconomic data to a limited number of dimensions which are used to develop urban profiles. Among those factors that have typified urban socioeconomic characteristics are the following: (1) socioeconomic status; (2) nonwhite population; (3) age composition; (4) residential mobility; (5) foreign-born concentration; and (6) population size. City size has emerged as a dominant feature in the ecological structure of cities.

City-metropolitan studies or urban ecologies have been completed for dozens of cities in the United States using census tracts, wards, and municipalities as observation units.[9] The objective of this research has typically been to determine how subareas can form the basis for theoretical constructs regarding socioeconomic space and the associated patterns of urban development. However, it has proven difficult to derive generalizations concerning a real differentiation from urban ecologies because of variations in study areas, observation units, variable selections, time frames, and methodologies.

The following analysis combines many of the geographic aspects of previous factorial ecologies, inasmuch as many elements of the rural-urban dichotomy, urban system structure, and city studies are present. A multistage approach is utilized to proceed from an aggregated to a progressively more disaggregated analysis of characteristics in various settlement system places thereby permitting a comparison of population attributes based on urban size and/or intrametropolitan location (Figure 1). The data set contains 6,861 local area observation units with associated socioeconomic characteristics.

The analysis begins with an examination of characteristics in those areas that are located within any urban places as distinguished from those that are not (rural places). The contrast in Stage 2 is between urban local areas that are classified as metropolitan urbanized areas[10] and those that are not. Stage 3 focuses on local areas in small metropolitan urban places (50,000 to 499,999 population in the urbanized area) as opposed to those in large urban places (500,000 or more population in the urbanized area). Finally, in Stage 4, the characteristics of local areas

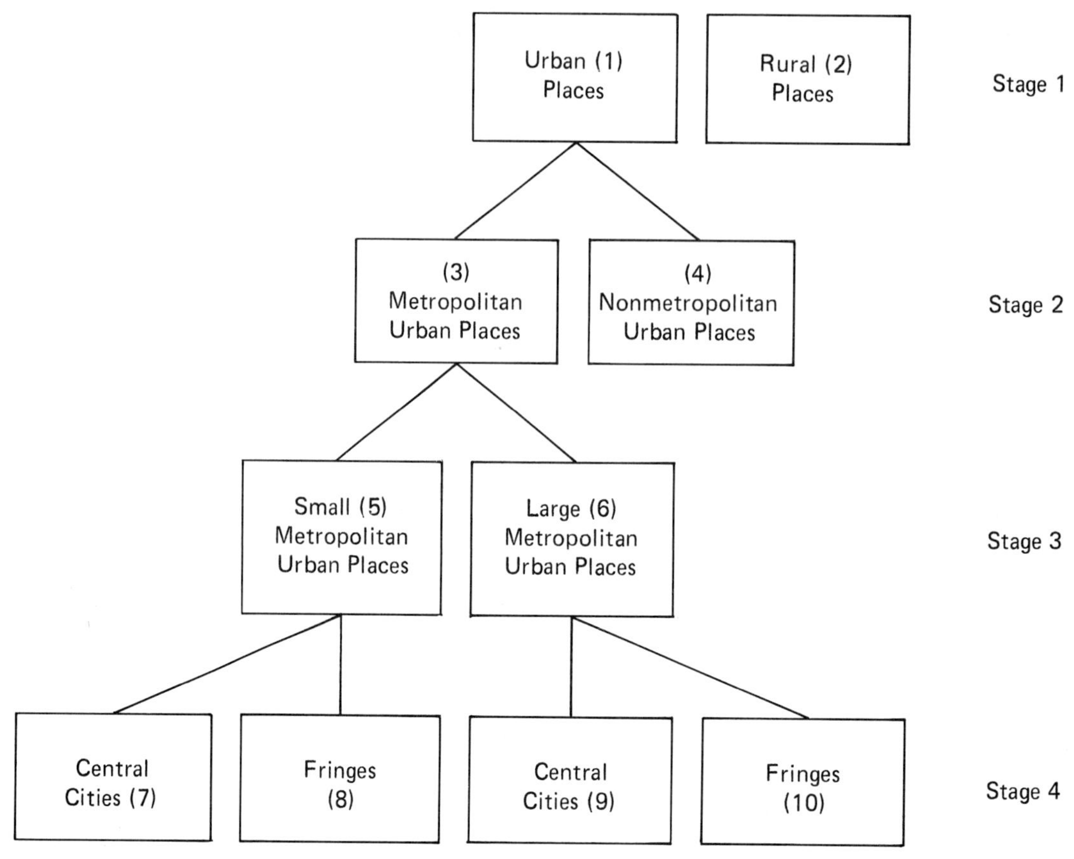

Figure 1. Multi-stage Analytical Approach.

that are situated within the central city of a metropolitan urban place are compared with those that are in the fringe of the urbanized area. The effect of this approach is to allow a comparative analysis of many distinct situational types including both within and between-stage contrasts. The analytical stages identified above will also facilitate a consideration of the effects of underlying heterogeneity in ecological analysis.

DATA AND METHODOLOGICAL PROCEDURES

Since 1960, the U.S. Bureau of the Census has provided access to basic census records through the Public Use Sample.[11] For 1970 one of the data sets available within the Public Use Sample is the Neighborhood Characteristics Sample. In this sample, associated with each randomly selected household record, is a set of information descriptive of the area surrounding the base housing unit. The geographic size of this area varies with population density and contains a residential population of approximately 4,000 to 5,000 persons.

Obviously, this local area will not correspond to a "neighborhood" in the sociological sense of the term. Indeed, there is no reference to any kind of homogeneity criterion in the definition of these areas. The lack of any particular homogeneity bias sets this data apart from others based on areal units such as census tracts and provides a better indication of the nature of various local areas than would data based on areas defined in terms of significant social or land use divides.

The analysis utilizes observations of local areas randomly selected as a 1 in 10,000 sample drawn from a national data set. The population characteristics are represented by 34 variables which include elements of age structure, family structure, ethnic composition, education, mobility, employment occupation, and income (Appendix A). The values for individual variables in local areas are expressed as absolute values or percentages of total area population or other aggregates.

The method of multivariate analysis employed in the following study is generally referred to as "common factor analysis."[12] Because factor analysis is a procedure that can be used in a variety of ways and so as to minimize the confusion that often surrounds reports of its use, a concise description of the criteria employed in applying the method of the data set is given here.

All possible common factors were extracted using the principal axis procedure employing squared multiple correlation coefficients as communality estimates. Subsequently, all factors were rotated to a varimax solution. In assessing the factor model, distinction is made between primary and non-primary factors. Primary factors are those that were associated with more than one primary variable. Primary variables were defined as those with loadings greater than $\sqrt{.5}$, since these variables seen unambiguously associated with a particular factor (i.e., they could never have a higher loading with any other factor).

Given the massive nature of the data set and the multistage analytical framework, the factors which emerged from the entire set of analytical steps are first identified. Subsequently, the results of the various analytical stages are discussed focusing on those aspects which differentiate the settlement system places.

EMPIRICAL ANALYSIS

The analytical design outlined in Figure 1 requires ten separate factor analyses. The results for primary factors, following the procedure discussed above, are given in Appendix B (Tables B1-B10). While at least four primary factors emerge from each analysis, in the majority of cases there are five, which is also the maximum number found. Examination of the primary variables associated with each primary factor leads to the identification of nine essentially different factors which are listed in Table 1.

Many of the terms in the factor labels are used in a conventional way. Socioeconomic Status references a complex of variables including educational attainment, type of employment and income level. Life Cycle Stage includes age characteristics and size and structure of households. Race represents the proportion of the population that is Negro. Poverty includes the proportion of families with annual incomes less than $5,000, the proportion of families and persons with income below the poverty level, and a Gini index[13] of income concentration. Aging and Mobility are largely self-explanatory. The former refers to the proportion of elderly population living alone, while the latter indexes the rate at which people change residential locations.

Table 1
Primary Factors in Situational Contrasts

1. Socioeconomic Status
2. Life Cycle Stage
3. Race, Poverty and Family Organization
4. College
5. Mobility
6. Poverty
7. Family Organization
8. Aging
9. Poverty and Family Organization

Other terms, however, are not as frequently employed in the context of this type of research. In particular, the term "Family Organization" refers to a variable complex including proportions of separated and divorced population, families with female heads and persons aged 17 or under in husband-wife families. The College factor includes a measure of the concentration of population in the 16 to 21 age cohort living in group quarters.

Given the relatively small number of different primary factors, it is clear that particular factors are likely associated with more than one settlement situation type. A cross-classification of factors with situation type, which appears as Table 2, indicates the extent of this multiple association.

Consideration of Table 2 suggests that there are clear differences between situation types with regard to primary factor structure. These differences can be most fruitfully explored in comparison with what appears to be a relatively common pattern, that of the following situation types: urban (1), metropolitan urban (3), small metropolitan urban (5), large metropolitan urban (6), and small metropolitan central city (7). Table 2 indicates that these situation types are all identical in terms of factor composition. Tables B-1, B-3, B-5, and B-6 in Appendix B provide a detailed summary of primary factors and loadings for major aggregates of urban settlement places. It indicates a factor structure very similar to what has become associated with the factorial ecology of urban areas, where factors such as socioeconomic status, life cycle stage and race-poverty-family organization have been found to constitute the primary metropolitan dimensions.[14]

Five situation types deviate from the common pattern of the previous situation types. Four situation types—rural (2), nonmetropolitan urban (4), small metropolitan fringe (8) and large metropolitan central city (9) deviate in a major way, since their factor composition does not include one or more of the three primary metropolitan dimensions. The fifth deviate situation type, large metropolitan fringe (10) differs in a somewhat less significant way from the common pattern because it is different only with respect to the "college" factor. A comparison of the five deviate situation types shows that they are mutually different, at least using the

Table 2
Primary Factors in Situational Contrasts

Factor \ Settlement Place	Urban	Rural	Metropolitan Urban	Nonmetropolitan Urban	Small Metropolitan Urban	Large Metropolitan Urban	Small Metropolitan Central City	Small Metropolitan Fringe	Large Metropolitan Central City	Large Metropolitan Fringe
Socioeconomic Status	X	X	X	X	X	X	X	X	X	X
Life Cycle Stage	X	X	X	X	X	X	X	–	X	X
Race, Poverty and Family Organization	X	–	X	–	X	X	X	–	–	X
College	X	–	X	X	X	X	X	–	X	–
Mobility	X	X	X	–	X	X	X	X	X	X
Poverty	–	X	–	X	–	–	–	–	–	–
Family Organization	–	–	–	X	–	–	–	X	–	–
Aging	–	–	–	–	–	–	–	X	–	–
Poverty and Family Organization	–	–	–	–	–	–	–	–	X	–

definitions employed here. Thus, the analytical results indicate that there are six different situation types in the United States, which include five deviate types and the common urban-metropolitan type.

The distinction outlined above can also be cast into the framework of Figure 1, which can be viewed as a network-type structure with 10 nodes. The nodes are divisible into an interior set (situation types 1, 3, 5, and 6) and an exterior set (situation types 2, 4, 7, 8, 9, and 10). The interior nodes consist of aggregations of the more fundamental exterior nodes and are also homogeneous with respect to factor structure. The exterior nodes, alternatively, consist of a set of situation types that have mutually different factor structures.

Thus, insofar as investigating the impact of aggregation in ecological analysis is concerned, the results of this analysis are striking. All aggregations of the exterior situation types emerge with a similar structure, even when the component parts are mutually different. This is most clearly illustrated in the cases of situation types 7 (small metropolitan central city) and 8 (small metropolitan fringe) which are aggregated to form type 6 (small metropolitan urban), and types 9 (large metropolitan central city) and 10 (large metropolitan fringe), which form type 6 (large metropolitan urban). Even though types 7, 8, 9, and 10 were found to be mutually different, types 5 and 6 have a homogeneous factor structure. At the very least, then, there is an aggregation problem when working at a scale included in the interior nodal set, for example at an urbanized area-wide scale. It remains to be seen whether further disaggregations of the exterior settlement situation types would yield still different factor structures.

EXTERIOR SITUATION TYPES: SETTLEMENT PLACE CONTRASTS

It has been demonstrated that there are six mutually different situation types located on the exterior of the settlement place network of Figure 1. It is apparent from Table 1 and Appendix B that these six are not different in every detail. Many of the specific differences that do exist, moreover, appear to be rather minor in nature. Consideration in this section will be restricted to those aspects of difference that appear to be significant. In particular, the analysis will focus on Socioeconomic Status, Race, Life Cycle Stage, Family Organization, and Poverty, all of which are characteristics of populations that emerge as being important here, as well as in much of the previously cited literature.

Without doubt, the most obvious differences that exist among the various settlement situation types concern race, poverty, and family organization concepts. As the list of primary factors demonstrates (Table 1), these concepts emerge in a variety of combinations. These range from small metropolitan central cities and large metropolitan fringes where all are combined into a single factor, to rural places, where a strictly Poverty factor emerges and Race and Family Organization are not associated with any primary factor. These results suggest that there is a relationship between the settlement size characteristic and the extent to which these concepts are organized into complexes. In rural areas (situation type 2), only a poverty factor exists; in nonmetropolitan urban areas (situation type 4), poverty and family organization occur individually in separate factors; at the level of metropolitan-urban areas more complex combinations are the norm. However, the four constituents of the metropolitan urban level (situation types 7-10) are hardly identical in terms of these combinations and therein lies the most interesting aspect of these results.

The nature of the results for metropolitan urban areas can be most fruitfully elaborated against the hypothesis of an expectation derived from previous analyses of such places. The results of these analyses have generally been similar to those found here for situation types 5 and 6 where Race, Poverty, and Family Organization have come together on a single factor.[15] Prevailing conceptions of metropolitan areas seem to suggest that central city and fringe areas have different properties in terms of Race, Poverty, and Family Organization. In particular, it is in central cities that one would expect to find the most intense manifestation of the Race, Poverty, and Family Organization complex. In fringe areas, one would expect the ties that bind this complex to be weaker, particularly with respect to race. The rationale for this hypothesis is that the black population living in suburban areas has generally achieved a higher socioeconomic status relative to central city black populations.[16] Thus, to the extent that there is poverty in suburban areas, one would expect this to be less associated with race than is the case in central cities. Our results conform to this expectation in the case of small metropolitan areas but fail to do so for large metropolitan areas.[17]

For central cities of metropolitan areas with less than 500,000 population, there is a well-defined factor associated with the Race, Poverty, and Family Organization complex. The loadings for the primary variables associated with that factor are given in Table 3. With regard

Table 3
Race, Poverty, and Family Organization:
Primary Factor Loadings
in Small Metropolitan Urbanized Areas[1]

	Central Cities
% Families with Female Heads	.840
% ≤ 17 in Husband-Wife Families	−.763
% Separated and Divorced	(.676)
% Negro	.715
% Families with Income < $5,000	.852
% Families with Income Below Poverty	.906

Parenthesis Indicate a Nonprimary Variable.

1. Small metropolitan urbanized areas are defined as having 50,000 to 499,999 population in 1970.

to the fringe areas in such places, however, race and poverty variables are not associated with any primary factor. Rather, they generally exhibit weaker loadings in the .4 to .7 range. Also appearing as a weakly loading variable is the proportion of families with female heads. The remaining Family Organization variables combine with two associated Life Cycle Stage variables to form a factor that seems to epitomize the familism that is so central to many conceptions of the suburban life-style.

The results for metropolitan areas with greater than 500,000 population shown in Table 4 stand in marked contrast. Specifically, as regards the three poverty-related variables, all have loadings greater than .8 in the central city analysis, while in fringe areas, one of these variables, the proportion of families with less than a $5,000 income, falls below .8 to a level of .696. Thus, the poverty portion of the factor is somewhat less coherent in the fringe but still impressively strong. A somewhat greater difference exists with regard to the family composition variables. In the central city, all three variables again have loadings of .8 or more, while in the fringe areas all fall below .8 to .763 (proportion of families with female head), −.677 (proportion aged 0-17 in husband-wife families, and .581 (proportion of population separated and divorced). Family composition, then, is clearly a less well-integrated complex of characteristics in fringe areas. It is with regard to the race variable that expectation is not met. As stated previously, the prevailing image is one of central cities in large urbanized areas where race and poverty are closely intertwined; while to the extent that blacks have suburbanized, there has been a concomitant social mobility, implying a weakening of that bond.

The results of this analysis indicate that the hypothesis drawn from previous studies is not supported. Race is less closely associated with the poverty complex factor in central cities (.661 loading) than in the fringe (.724 loading). Although the difference in loading is not extraordinarily large, it suggests the need for a reexamination of the race-resources issue in large

Table 4
Race, Poverty, and Family Organization:
Primary Factor Loadings in Large Metropolitan Urbanized Areas[1]

	Central City	Fringe
% Families with Female Heads	.886	.763
% $\leqslant$ 17 in Husband-Wife Families	–.800	(–.677)
% Separated and Divorced	.801	(.581)
% Negro	(.661)	.724
% Families with Income $<$ \$5,000	.823	(.696)
% Families with Income Below Poverty	.848	.818
% Persons with Income Below Poverty	.851	.801

Parenthesis Indicate Nonprimary Variable.

1. Large metropolitan urbanized areas are defined as having greater than 500,000 population in 1970.

American cities.[18] For example, it appears that the commonly accepted idea that suburbanized blacks have escaped the situation implied by the race-resources factor is an illusion created by the general weakening of the bonds between the nonracial components of the resources complex in suburban areas. Race remains, however, a central element in that somewhat less coherent pattern. In contrast, it appears that while poverty and family (dis)organization form a very well-defined complex in central city areas, the race element has tended to disassociate from it. These findings suggest that the race and resources issue is becoming one most meaningfully conceived of in the suburban context, while in the central cities it is being transformed into one more purely focused on resources.

Differences between the various situation types with regard to Socioeconomic Status and Life Cycle Stage are less dramatic but nevertheless apparent. For Socioeconomic Status, there is a fairly consistent decrease in the coherence of the factor with settlement size. This can be seen in Table 5, which shows the loadings for primary variables on the various Socioeconomic Status factors. For large metropolitan areas, the same seven variables have loadings greater than .7 for both central city and fringe situation types. In the next size category, that of small metropolitan areas, there are five and three primary variables associated with Socioeconomic Status in central city and fringe areas respectively. With regard to the remaining situation types, there are two primary variables on the factor for nonmetropolitan urban areas and three for rural areas. Thus, it seems clear that with increasing size and complexity comes a sharper delineation of population along Socioeconomic Status lines, a finding that is certainly consistent with the original hypothesis of social area analysis.[19]

With respect to Life Cycle Stage, the differences between settlement situation types are much less apparent than for Socioeconomic Status. Indeed, only in the case of the small metropolitan area fringe is there a significant deviation from the typical case (e.g., small

Table 5
Socioeconomic Status
Factor Loadings (Primary Variables)

Situation Types / Variables	Large[1] Metropolitan Urban Place		Small[2] Metropolitan Urban Place		Nonmetro-politan Urban Place	Rural Place
	Central City	Suburb	Central City	Suburb		
% Population 25-54 with 12+ School Years Completed	.766	.819	.730			
% Population 25-54 with 4+ College Years Completed	.880	.922	.864	.851	.763	.857
Median School Years Completed for Population Aged 25-54	.774	.812				
% Managerial and Professional Workers	.898	.932	.909	.897	.883	.902
% Blue-collar Workers	–.796	–.854	–.849	–.777		
% Families with $15,000+ Annual Income	.836	.861	.711			.752
Median Family Income	.727	.801				

1. Large Metropolitan Urban Places are defined as having 500,000 or more population in their urbanized area.
2. Small Metropolitan Urban Places are defined as having 50,000-499,999 population in their urbanized area.

metropolitan area central cities) of a factor with four primary variables: (1) proportion of population 0-17 years old, (2) proportion aged 65 and older, (3) mean household size, and (4) proportion of primary individual type households. In such fringe areas we find the second and fourth variables as primary variables on a factor termed Aging, while, as mentioned previously, the first and third join with two other variables, proportion of population aged 0-17 in husband-wife families and proportion of ever-married population separated and divorced, to form a factor that clearly references the type of familism that is so central to many conceptions of the suburban life-style. It appears from these results that many suburban portions of large metropolitan areas did not conform to that idealization in 1970.

CONCLUSIONS

The analysis has extended the basic concepts of social area analysis and factorial ecology to consider a spectrum of places in the settlement system. The analysis has thus permitted an initial systematic comparison of the structural characteristics of local areas in places that range in size and situation from rural areas to large central cities. The multistage analytical approach has also made it possible to observe the effects of aggregation in ecological analysis.

One implication of the analysis is that the typical metropolitan area ecological result, with localities differentiated by Socioeconomic Status, Life Cycle Stage, and Race-Resources, is in no small part a product of aggregation effects. A remarkably similar factor structure characterizes all aggregates of urban places, regardless of size differentiation. However, localities within central city and suburban components of metropolitan urbanized areas, when isolated, are different from that of the metropolitan norm in many important respects. This finding is also valid with respect to localities in the nonmetropolitan settlement system.

The results of the analysis indicate that, aside from the common factor structure of basic urban aggregates, there are also notable contrasts between component intrametropolitan and other nonmetropolitan settlement place types. With regard to Race, Poverty, and Family Organization variables, there is a definite relationship between the settlement size characteristic and the extent to which the variables are associated in complexes. In general, the smaller the settlement place, the less complex are the associations of Race, Poverty, and Family Organization within a single factor and the more likely are these variables to be observed loading on separate, distinct factors. Similarly, the results indicate that with increasing settlement place size comes a sharper delineation of population along a Socioeconomic Status dimension.

In larger metropolitan urban areas, the findings appear to be of particular interest. In general, the Socioeconomic Status variables form the most clearly-defined complex in suburban areas. However, the manifestation of the poverty-family organization complex is not restricted solely to central city areas. Suburban areas, while exhibiting a weaker degree of family (dis)organization, nonetheless possess a rather well-defined poverty complex.

An examination of the association of race with the poverty-family organization factor indicates that the "conventional wisdom" in which black resource patterns mirror white patterns with rising resources in suburban areas may be an illusion created by the general weakening of the family (dis)organization complex in suburban areas. The findings provide support for other related research suggesting that blacks re-ghettoize in suburban areas, and evidence that open housing policies and black suburbanization are not in themselves sufficient conditions for enhanced economic opportunity and social welfare.

NOTES

1. E. Shevsky, and M. Williams, *The Social Areas of Los Angeles: Analysis and Typology* (Berkeley, Cal.: University of California Press, 1949); W. Bell, "The Social Areas of the San Francisco Bay Region," *American Sociological Review* 18 (1953), pp. 29-47.
2. D.O. Price, "Factor Analysis in the Study of Metropolitan Centers," *Social Forces* 20 (1942), pp. 449-455; P.R. Hofstaetter, " 'Your City'—Revisited: A Factorial Study of Cultural Patterns," *American Catholic Sociological Review* 13 (1952), pp. 159-168; H.B. Kaplan, "An Empirical Typology for Urban Description," unpublished Ph.D. dissertation (Department of Sociology, New York University, 1958).
3. U.S. Department of Commerce, Social and Economic Statistics Administration, Bureau of the Census, *Public Use Samples of Basic Records from the 1970 Census* (Washington, D.C.: Government Printing Office, 1972).
4. The reader unfamiliar with factor analysis and principal components analysis is directed to the following references: R. Rummel, *Applied Factor Analysis* (Evanston, Ill.: Northwestern University Press, 1970); H. Harman, *Modern Factor Analysis* (Chicago, Ill.: University of Chicago Press, 1966).
5. P.H. Rees, "Factorial Ecology: An Extended Definition, Survey, and Critique of the Field," *Economic Geography* 47 (Supplement, 1971), pp. 220-233.
6. J.H. Thompson, et al., "Toward a Geography of Economic Health: The Case of New York State," *Annals of the Association of American Geographers* 52 (1962), pp. 1-20; G. Malcolm Lewis, "Levels of Living in the Northeastern United States © 1960: A New Approach to Regional Geography," *Transactions, The Institute of British Geographers* 45 (September 1968); S. Brunn and J.O. Wheeler, "Spatial Dimensions of Poverty in the United States," *Geografiska Annaler* 53, Series B (1971), pp. 6-15; R.A. Barnett, "Economic Health in Oregon: A Factor Analysis," *Growth and Change* 1 (1970), pp. 19-26.
7. D.M. Smith, *The Geography of Social Well-Being in the United States* (New York, N.Y.: McGraw-Hill Book Company, 1973), pp. 79-103.
8. B.J.L. Berry, "Latent Structure of the American Urban System," in B.J.L. Berrry, ed. *City Classification Handbook* (New York, N.Y.: Wiley-Interscience, 1972), pp. 11-60; J.K. Hadden and E.F. Borgatta, *American Cities: Their Social Characteristics* (Chicago, Ill.: Rand McNally & Company, 1965); L. King, "Cross-Sectional Analysis of Canadian Urban Dimensions, 1951 and 1961," *Canadian Geographer* 10 (1966), pp. 205-224.
9. G.W. Carey, "The Regional Interpretation of Population and Housing Patterns in Manhattan through Factor Analysis," *Geographical Review* 56 (1966), pp. 551-569; G.W. Carey, et. al., "Educational and Demographic Factors in the Urban Geography of Washington, D.C.," *Geographical Review* 58 (1968), pp. 515-537; P.H. Rees, "The Factorial Ecology of Metropolitan Chicago, 1960," in B.J.L. Berry and F. Horton, eds., *Geographic Perspectives on Urban Systems* (Englewood Cliffs, N.J.: Prentice-Hall, 1970), pp. 306-394; C.F. Schmid and K. Tagashira, "Ecological and Demographic Indices: A Methodological Analysis," *Demography* 1 (1964), pp. 194-211.
10. Suburban areas are more adequately represented by using data for urbanized areas rather than SMSA boundaries. The urbanized area includes only those enumeration districts meeting criteria of population density as well as adjacent incorporated places.
11. *Public Use Sample, op. cit.*
12. R. Rummel, *op. cit.*, pp. 104-112.
13. The Gini Index of income concentration ranges from .01 to .99. As the index approaches 1.00, the inequality of the income distribution increases.
14. P. Rees (1970), *op. cit.*; R. Murdie, *Factorial Ecology of Metropolitan Toronto 1951-1961,* Research Paper no. 116 (Department of Geography Research Series, University of Chicago, 1969).
15. ——— (1970), *op. cit.*,, pp. 333-339; J.K. Hadden and E.F. Borgatta, *op. cit.*, pp. 42-44.
16. J.J. Palen and L.F. Schnore, "Color Composition and City-Suburban Status Differences: A Replication and Extension," *Land Economics* 41 (1965), pp. 87-91; L.F. Schnore, *Class and Race in Cities and Suburbs* (Chicago, Ill.: Markham Publishing Co., 1972).
17. Palen and Schnore also found that small metropolitan areas did not exhibit the expected status differences among either blacks or whites in areas outside the South.
18. The "conventional wisdom" concerning the spatial pattern of economic opportunity for nonwhites has also been challenged in an analysis of the twelve largest SMSA's using U.S. Office of Economic Opportunity data for 1965: B. Harrison, "The Intrametropolitan Distribution of Minority Economic Welfare," *Journal of Regional Science* 12 (1972), pp. 23-43.
19. R.A. Murdie, "The Social Geography of the City: Theoretical and Empirical Background," in L.S. Bourne, ed., *Internal Structure of the City* (London: Oxford University Press, 1971), pp. 279-290.

Appendix A

Definition of Variables

1	POP17	Population 0-17 years old/Total population
2	POP65	Population 65 years and over/Total population
3	POPHH	Population in households/Total households (Range: 0.0 to 9.9)
4	PRIND	Primary individuals/Total households
5	POPGQ	Population in group quarters (including inmates)/Total population
6	FAMFH	Families with female head/Total families
7	P17HWF	Persons 0-17 in husband-wife families/Total persons 0-17 years old
8	SDPOP	Separated and divorced population/Total ever-married population
9	CEBEMF	Children ever born to ever-married females 35-44 years old/Total ever-married females 35-44 years old
10	NEGPOP	Negro population/Total population
11	PSPHER	Persons of Spanish heritage/Total population
12	PFORST	Persons of foreign stock/Total population
13	P16NES	Persons 16-21 not enrolled in school and not high school graduates/Total population 16-21 years old
14	P16T21	Population 16-21 years old/Total population
15	P3T34C	Persons 3-34 enrolled in college/Persons 18-21 years old
16	POT7SC	Persons 25-54 with 0-7 years of school completed/Total persons 25-54 years old
17	P12MSC	Persons 25-54 with 12 or more years of school completed/Total persons 25-54 years old
18	P4MYC	Persons 25-54 with 4 or more years of college completed/Total persons 25-54 years old
19	MEDSYC	Median years of school completed for persons 25-54 years old
20	MOBSH5	Persons 5 and over living in same house as 5 years ago/Total persons 5 years old and over
21	MOBDC5	Persons 5 and over living in different county 5 years ago/Total persons 5 years old and over
22	MCLF16	Male civilian labor force 16 years old and over/Total civilian males 16 years old and over
23	FCLF16	Female civilian labor force 16 years old and over/Total civilian females 16 years old and over
24	YPUNLF	Persons 16-21 years old not enrolled in school, unemployed or not in labor force/Total persons 16-21 years old not enrolled in school

25 UNEM16 Unemployed persons 16 years old and over/Total civilian labor force 16 years old and over

26 EPTKWM Employed professional, technical, and kindred workers; and managers and administrators except farm managers, 16 years old and over/Total employed persons 16 years old and over in ECLF

27 EFMLFF Employed farmers, farm managers, farm laborers, and farm foremen 16 years old and over/Total employed persons 16 years old and over

28 EBCW16 Employed blue collar workers (craftsmen and kindred workers, operatives including transport equipment operatives, and laborers except farm) 16 years old and over/Total employed persons 16 years old and over

29 F<5KY Families with less than 5,000 family income/Total families

30 F>15KY Families with $15,000 or more family income/Total families

31 MFY Median family income (in thousands of dollars)

32 GINIY Gini Index of income concentration

33 FYBPL Families with income below the poverty level/Total families

34 PYBPL Persons with income below the poverty level/Total population

Appendix B

Table B-1
All Urban Places

Variable	Primary Factors and Loadings				
	I	II	III	IV	V
1. POP17			.835		
2. POP65			–.770		
3. POPHH			.909		
4. PRIND			–.769		
5. POPGQ				.820	
6. FAMFH		.846			
7. P17HWF		–.770			
8. SDPOP		.729			
9. CEBEMF					
10. NEGPOP		.753			
11. PSPHER					
12. PFORST					
13. P16NES					
14. P16T21				.814	
15. P3T34C					
16. POT7SC					
17. P12MSC	.770				
18. P4MYC	.884				
19. MEDSYC	.740				
20. MOBSH5					.828
21. MOBDC5					–.798
22. MCLF16					
23. FCLF16					
24. YPUNLF					
25. UNEM16					
26. EPTKWM	.910				
27. EFMLFF					
28. EMCW16	–.813				
29. F<5KY		.708			
30. F>15KY	.798				
31. MFY					
32. GINIY					
33. FYBPL		.802			
34. PYBPL		.778			

Table B-2

All Rural Places

Variable	Primary Factors and Loadings				
	I	II	III	IV	V
1. POP17			.832		
2. POP65			−.755		
3. POPHH			.919		
4. PRIND					
5. POPGQ					
6. FAMFH					
7. P17HWF					
8. SDPOP					
9. CEBEMF					
10. NEGPOP					
11. PSPHER					
12. PFORST					
13. P16NES					
14. P16T21					
15. P3T34C					
16. POT7SC					
17. P12MSC					
18. P4MYC		.857			
19. MEDSYC					
20. MOBSH5				.814	
21. MOBDC5				−.787	
22. MCLF16					
23. FCLF16					
24. YPUNLF					
25. UNEM16					
26. EPTKWM		.902			
27. EFMLFF					
28. EMCW16					
29. F<5KY	.850				
30. F>15KY		.752			
31. MFY					
32. GINIY	.740				
33. FYBPL	.905				
34. PYBPL	.895				

Table B-3
Metropolitan Urban Places

Variable	Primary Factors and Loadings				
	I	II	III	IV	V
1. POP17			.875		
2. POP65			−.805		
3. POPHH			.905		
4. PRIND			−.762		
5. POPGQ				.790	
6. FAMFH	.872				
7. P17HWF	−.782				
8. SDPOP	.746				
9. CEBEMF					
10. NEGPOP	.727				
11. PSPHER					
12. PFORST					
13. P16NES					
14. P16T21				.830	
15. P3T34C					
16. POT7SC					
17. P12MSC		.783			
18. P4MYC		.898			
19. MEDSYC		.762			
20. MOBSH5					.814
21. MOBDC5					−.787
22. MCLF16					
23. FCLF16					
24. YPUNLF					
25. UNEM16					
26. EPTKWM		.913			
27. EFMLFF					
28. EMCW16		−.831			
29. F<5KY	.807				
30. F>15KY		.809			
31. MFY		.707			
32. GINIY					
33. FYBPL	.868				
34. PYBPL	.850				

Table B-4
Nonmetropolitan Urban Places

Variable	Primary Factors and Loadings				
	I	II	III	IV	V
1. POP17		−.751			
2. POP65					.709
3. POPHH					−.764
4. PRIND					.739
5. POPGQ		.903			
6. FAMFH					
7. P17HWF				−.761	
8. SDPOP				.772	
9. CEBEMF					
10. NEGPOP					
11. PSPHER					
12. PFORST					
13. P16NES					
14. P16T21		.902			
15. P3T34C					
16. POT7SC					
17. P12MSC					
18. P4MYC			.763		
19. MEDSYC					
20. MOBSH5					
21. MOBDC5					
22. MCLF16					
23. FCLF16					
24. YPUNLF					
25. UNEM16					
26. EPTKWM			.883		
27. EFMLFF					
28. EMCW16					
29. F<5KY	.882				
30. F>15KY					
31. MFY					
32. GINIY					
33. FYBPL	.919				
34. PYBPL	.926				

Table B-5
Metropolitan Urban Places
with 50,000-499,999 Population

Variable	Primary Factors and Loadings				
	I	II	III	IV	V
1. POP17			.740		
2. POP65			–.783		
3. POPHH			.870		
4. PRIND			–.774		
5. POPGQ				.847	
6. FAMFH	.821				
7. P17HWF	–.714				
8. SDPOP					
9. CEBEMF					
10. NEGPOP	.722				
11. PSPHER					
12. PFORST					
13. P16NES					
14. P16T21				.805	
15. P3T34C					
16. POT7SC					
17. P12MSC		.706			
18. P4MYC		.856			
19. MEDSYC					
20. MOBSH5					.847
21. MOBDC5					–.800
22. MCLF16					
23. FCLF16					
24. YPUNLF					
25. UNEM16					
26. EPTKWM		.908			
27. EFMLFF					
28. EMCW16		–.828			
29. F<5KY	.815				
30. F>15KY		.729			
31. MFY					
32. GINIY					
33. FYBPL	.890				
34. PYBPL	.859				

Table B-6
Metropolitan Urban Places
with 500,000 or More Population

Variable	Primary Factors and Loadings				
	I	II	III	IV	V
1. POP17			.908		
2. POP65			–.819		
3. POPHH			.901		
4. PRIND			–.755		
5. POPGQ				.772	
6. FAMFH	.886				
7. P17HWF	–.810				
8. SDPOP	.782				
9. CEBEMF					
10. NEGPOP	.719				
11. PSPHER					
12. PFORST					
13. P16NES					
14. P16T21				.838	
15. P3T34C					
16. POT7SC					
17. P12MSC		.789			
18. P4MYC		.906			
19. MEDSYC		.791			
20. MOBSH5					.791
21. MOBDC5					–.769
22. MCLF16					
23. FCLF16					
24. YPUNLF					
25. UNEM16					
26. EPTKWM		.915			
27. EFMLFF					
28. EMCW16		–.825			
29. F<5KY	.829				
30. F>15KY		.837			
31. MFY		.746			
32. GINIY					
33. FYBPL	.874				
34. PYBPL	.872				

Table B-7
Central Cities of Metropolitan
Urban Places with 50,000-499,999 Population

Variable	Primary Factors and Loadings				
	I	II	III	IV	V
1. POP17			.752		
2. POP65			–.788		
3. POPHH			.921		
4. PRIND			–.812		
5. POPGQ				.864	
6. FAMFH	.840				
7. P17HWF	–.763				
8. SDPOP					
9. CEBEMF					
10. NEGPOP	.715				
11. PSPHER					
12. PFORST					
13. P16NES					
14. P16T21				.864	
15. P3T34C					
16. POT7SC					
17. P12MSC		.730			
18. P4MYC		.864			
19. MEDSYC					
20. MOBSH5					.823
21. MOBDC5					–.752
22. MCLF16					
23. FCLF16					
24. YPUNLF					
25. UNEM16					
26. EPTKWM		.909			
27. EFMLFF					
28. EMCW16		–.849			
29. F<5KY	.852				
30. F>15KY		.711			
31. MFY					
32. GINIY					
33. FYBPL	.906				
34. PYBPL	.882				

Table B-8
Remainder of Urbanized Area in Metropolitan
Urban Places with 50,000 to 499,999 Population

Variable	Primary Factors and Loadings				
	I	II	III	IV	V
1. POP17		.751			
2. POP65			.782		
3. POPHH		.844			
4. PRIND			.786		
5. POPGQ					
6. FAMFH					
7. P17HWF		.793			
8. SDPOP		−.712			
9. CEBEMF					
10. NEGPOP					
11. PSPHER					
12. PFORST					
13. P16NES					
14. P16T21					
15. P3T34C					
16. POT7SC					
17. P12MSC					
18. P4MYC	.851				
19. MEDSYC					
20. MOBSH5				.827	
21. MOBDC5				−.799	
22. MCLF16					
23. FCLF16					
24. YPUNLF					
25. UNEM16					
26. EPTKWM	.897				
27. EFMLFF					
28. EMCW16	−.777				
29. F<5KY					
30. F>15KY					
31. MFY					
32. GINIY					
33. FYBPL					
34. PYBPL					

Table B-9
Central City of Metropolitan Urban Places
with 500,000 or More Population

Variable	Primary Factors and Loadings				
	I	II	III	IV	V
1. POP17			.848		
2. POP65			−.816		
3. POPHH			.893		
4. PRIND			−.780		
5. POPGQ				.826	
6. FAMFH	.886				
7. P17HWF	−.800				
8. SDPOP	.801				
9. CEBEMF					
10. NEGPOP					
11. PSPHER					
12. PFORST					
13. P16NES					
14. P16T21				.853	
15. P3T34C					
16. POT7SC					
17. P12MSC		.766			
18. P4MYC		.880			
19. MEDSYC		.774			
20. MOBSH5					.760
21. MOBDC5					−.713
22. MCLF16					
23. FCLF16					
24. YPUNLF					
25. UNEM16					
26. EPTKWM		.898			
27. EFMLFF					
28. EMCW16		−.796			
29. F<5KY	.823				
30. F>15KY		.836			
31. MFY		.727			
32. GINIY					
33. FYBPL	.848				
34. PYBPL	.851				

Table B-10

Remainder of Urbanized Area in Metropolitan Urban Places with 500,000 or More Population

Variable	Primary Factors and Loadings				
	I	II	III	IV	V
1. POP17			.903		
2. POP65			–.707		
3. POPHH			.925		
4. PRIND			–.788		
5. POPGQ					
6. FAMFH		.763			
7. P17HWF					
8. SDPOP					
9. CEBEMF					
10. NEGPOP		.724			
11. PSPHER					
12. PFORST					
13. P16NES					
14. P16T21					
15. P3T34C					
16. POT7SC					
17. P12MSC	.819				
18. P4MYC	.922				
19. MEDSYC	.812				
20. MOBSH5				.838	
21. MOBDC5				–.828	
22. MCLF16					
23. FCLF16					
24. YPUNLF					
25. UNEM16					
26. EPTKWM	.932				
27. EFMLFF					
28. EMCW16	–.854				
29. F<5KY					
30. F>15KY	.861				
31. MFY	.801				
32. GINIY					
33. FYBPL		.818			
34. PYBPL		.801			

Geographic Research and Regional Rail Transport Policy

William R. Black
Indiana University

When I was first asked to participate in this conference, I had just completed two years of directing three studies on rail abandonment and state rail planning.[1] It was my intent, at the time, to present a paper suggesting areas where geographic research could assist in the development and formulation of regional rail transport policy by the public sector. However, since that time I have been involved in the private railroad sector and as a result I have altered the general approach and will address the subject of geographic research which should be undertaken that would be of some utility in the formulation of rail transport policy by the public sector as well as the private sector.

As an academic who has spent time in both the public and private sector, I am bothered by the number of times I have heard the phrase, "It's academic." You may have a sophisticated notion of what the phrase means, but if you are honest with yourself you must admit that the phrase often means unpractical and pedantic. Certainly when members of a university community discuss academic research, this is not the meaning they are applying to that term. However, when staff members of state and federal agencies and corporations use the phrase, I am not at all certain whether they are giving credit to the origin of the research, or commenting on its utility.

There is absolutely no reason to avoid research that has societal utility. Indeed, Richard Morrill has made a plea for geographers to become more involved in this area which he refers to as policy analysis. This request has not generated a significant response from the geographic profession, and this may be due to some misconceptions held by the members of the academic community.

It is common for university and college researchers to assume that, although it is not public information, government knows the answers to questions relevant to, or important in, the formulation of policy. Generally speaking, this is false. Federal agency policy is derived from legislation and appropriations, while state policy is more often than not a reaction to federal programs and entitlements (e.g., urban transit, interstate highways, speed limits, outdoor advertising, rail subsidies, and so forth).

Another misconception of researchers is that all of the research necessary or worth doing in the private sector has been done internally and has simply never been released for public consumption due to its confidential nature. Unfortunately, the private sector is not quite so fortunate. Although there is considerable research in this sector, most of it is of a problem solving nature and the outcome is a policy statement specific to the problem addressed. In other words, scientific research that would lead to generalizations is absent.

The bankruptcies of nine railroads in the Midwest and Northeast have been attributed on more than one occasion to unresponsive management.[2] Was management unresponsive or was it fully occupied with minor policy decisions due to the lack of a body of generalizations for such decision-making?

It is within this context that I would like to identify a series of major research problems which researchers currently have the necessary methodologies and capabilities to analyze. If such examinations are to have utility, they must be rigorous and at the same time capable of application. Each research design assumption decreases the usefulness of the research. Methodologies which result in different outcomes based on the number of iterations, factors, method of rotation, and so forth are the subject of theoretical analysis, and should not be used in this context.

The problems which follow are, once again, actual problems facing government and/or industry. Although several of these have been investigated in the past, the general areas noted are, for the most part, devoid of generalizations.

TERMINAL LOCATION AND RAIL RELOCATION

The first problem where substantive research is necessary is in an area which has been recognized by urban geographers as an uneconomical situation for the better part of this century.[3] This is the location of large rail terminal facilities and yards within congested parts of major metropolitan areas. Such land use decreases development options for city government and increases existing congestion. This congestion itself is a problem for the rail industry since it interferes with the car routing and yard efficiencies which the industry has developed. At the same time the industry is faced with monumental property tax bills that are increasing at a greater rate than savings from new techniques.[4]

A study to evaluate the utility of major terminal and yard relocation is clearly necessary. The problem can be reduced to a comparative cost analysis where the tax savings must be weighed against construction costs. It seems apparent that there are a series of critical land taxation values which support the relocation decision for yards and facilities of different sizes. A related problem is that of relocating rail lines in urban areas.[5]

JOINT GOVERNMENT-INDUSTRY POLICY ON ABANDONMENT

A second problem is in the area of rail abandonment and concerns government-industry policy toward rail abandonment. The industry is concerned with access to markets and it has no desire to retreat from areas where a potential market exists. Government, on the other hand, is concerned with access to rail and it has no desire to see rail retreat from an area where a "political" market exists. The multitude of hearings over the past two to three years has seen the rail sector in conflict with government, but they need not be adversaries.[6] There is reason to believe that if the industry decides a rail line should be abandoned, there is no projectable market. If government believes a market exists, they should meet with the railroad and present their case. Lines without potential growth that are uneconomical will remain so, whether government or industry subsidizes them.

The greater the economic loss to the railroad in keeping a line the less the political loss to government in its abandonment. Similarly, the lower the economic loss to the railroad, the greater the political impact. Determining the position and function of this relationship would yield far more than arguments for line retention by government or rail abandonment by industry.

COSTS OF RAIL ABANDONMENT

While this problem is under analysis, there is a need in the interim for a comprehensive study of the economic costs to government of abandoning lines. It is clear that we do not know the range of such impacts or their flow throughout the economic system.

An input-output study which examines the principle sectors which use rail transport and their interactions with other sectors of the economy is clearly necessary. Such a study should focus on the impact of transport costs—increases or decreases—on the economy.

RAILROAD LABOR COSTS

The problem of labor costs is still another area. Labor costs are viewed by railroad management and the public sector as a major problem inhibiting the efficiency of railroads.[7] How significant are labor costs as a component of total transport costs? What are the benefits to the railroad of reducing crew size? What are the costs to the public sector in terms of income transfers and decreases in safety?

IMPACT OF RATE FLEXIBILITY ON RAIL VIABILITY

Another area where we will very shortly begin to see results is in the area of rate flexibility by railroads. Under recent legislation, railroads have the flexibility to increase or decrease freight rates by seven percent.[8] The problem has not been adequately researched and, as a result, the Interstate Commerce Commission has initiated a trial period for such flexibility. There is clearly a need for solid research by universities on this subject as opposed to the rhetoric which typifies much of the research to date.

STANDARDS OF PERFORMANCE

Another problem is in the area of service standards. Whenever the public sector becomes involved in the private sector, as they are in the rail sector now, there is the new to measure performance and require that it meet certain standards. In the next year such standards will be developed. At the present time, however, this seems like an impossible task without major legislative changes.

DISTRIBUTION CRITERIA FOR RAIL PROGRAMS

If the public sector were to subsidize or assist in covering the costs of the rail sector, it would be necessary to determine the most equitable criteria for the allocation of such assistance. For example, if it were necessary to insure rail lines of the private sector and subsidized rail lines of the public sector against liability risks, should such costs be spatially allocated on a train-mile, ton-mile, track-mile, or car-mile basis? Related to this is the question of the most equitable costing rationale for subsidized service. Should such costs be avoidable costs, marginal costs, or others?[9]

TRANSCONTINENTAL RAILROADS

There is a growing belief in some sectors of the rail industry as well as some states that the United States needs transcontinental railroads. Aside from the general attractiveness of the concept, what would be the general benefits to shippers and industries? What would a few transcontinental railroads do to the competitive balance between railroads in the United States today? Has areal specialization developed to the point that existing commodity flows would no longer justify such linkages?

SHORT LINE RAILROADS

The last few years have seen a rapid growth in the number of short line railroads that have been formed to prevent abandonment of local rail services. It is ironic that such small operations can operate effectively without the scale economies of the larger railroads, which have viewed such operations as unprofitable. What are the economics of these situations? The first response is that labor costs are significantly lower on short lines. However, this is not a sufficient response by itself and more research is clearly necessary.

TRAFFIC RETENTION

A major methodological assumption in the procedures currently utilized to determine subsidies is that all revenues generated on a line proposed for abandonment are attributable to that line.[10] This creates two problems. First, it assumes that the railroad would not retain any traffic if the line were abandoned and traffic would not move to another point which continues to be served by the railroad. Second, it might result in potential double counting of revenues if the traffic were to originate *and terminate* on lines being considered for abandonment. Obviously, more research is necessary on empirical traffic retention estimation.

ENVIRONMENTAL IMPACTS OF ALTERNATE MODES

There are many unknowns in the environmental area as it applies to the ecological costs of alternate transport modes. Standards currently in use are not accurate for the majority of the problems being researched.[11] For example, emittant standards are based on trucks on the Interstate Highway system, not on remote state or county roads. Railroad data suggest that mode is more desirable from an environmental perspective. However, there is undoubtedly a break-even point where the modes would be equal polluters of the environment. This is a critical question that should be reviewed and analyzed before meaningful comparisons can be made.

There are many other research problems that need to be examined. Among these are: the forecasting of markets and market shares by two or more railroads, as well as by rail and truck; the market size necessary to retain economical rail service; the impact of inflation on water-rail transfers; comparative costs of alternative routes; and many others.

The research questions noted above clearly have a spatial dimension. However, some might argue that these are not sufficiently geographic to merit analysis by the profession. Nothing could be further from the truth. These questions and problems may be viewed as fundamental inquiries into network location, spatial interaction and flows, and the costs of interaction. Admittedly, they will not be answered by pure abstract geographic research utilizing only spatial variables, but all disciplines utilize each others' variables and indicators as the problem merits.

These are research questions which we will hear much more about during the next several years. Both government and industry are too crisis or reaction oriented, or too biased to undertake many of these studies. It is at this point that the university research community could make a substantive contribution by undertaking their analysis.

NOTES

1. These studies were: Governor's Rail Task Force, *USRA Segments in Indiana: State Analysis and Recommendations, Vols. 1, 2, 3, 4,* (Bloomington, Indiana: The Center for Urban and Regional Analysis, Indiana University), 1974; Public Service Commission of Indiana, *Indiana State Rail Plan, Preliminary* (Bloomington, Indiana: The Center for Urban and Regional Analysis, Indiana University), 1975; and, Council of State Governments, *The States and Rural Rail Preservation: Alternative Strategies* (Lexington, Kentucky: Prepared by the Council of State Governments under a contract from the Economic Development Administration), November 1975.
2. See, for example: Joseph R. Daughen and Peter Binzen, *The Wreck of the Penn Central* (Boston: Little, Brown, and Co., Inc.) 1971.
3. An excellent example of this geographic research is: Harold M. Mayer, "Localization of Railway Facilities in Metropolitan Centers as Typified by Chicago," *Land Economics* XX (1944) 299-315.
4. William E. Lahner, "Regulation and the Plight of the Railroad Industry," *Proceedings* of the Annual Conference on Taxation, National Tax Association—Tax Institute of America, Houston, Texas, 1976 (in press).
5. This area has seen considerable activity lately, see: JoAnne McGowan and Hoy A. Richards (eds.) *Proceedings of the 1975 National Rail Planning Conference,* (Washington, D.C.: Federal Railroad Administration, U.S. Department of Transportation), 1976.
6. These hearings are summarized in: Rail Services Planning Office, *The Public Response to the Secretary of Transportation's Rail Service Report, Vols. 1, 2, 3,* (Washington, D.C.: Interstate Commerce Commission), 1974 and 1975.
7. See: William L. Thornton, *How to Deal with the Railroad Crisis: An Open Letter to Congress,* St. Augustine, Florida: Florida East Coast Railway Company, June 1975.
8. This is provided for in the Rail Revitalization and Regulatory Reform Act of 1976, Public Law 94-210 enacted February 5, 1976.
9. Under existing legislation the subsidies are paid on an avoidable cost basis, i.e., costs that wouldn't be incurred if the line was abandoned.
10. This is true in the Rail Services Planning Office of the Interstate Commerce Commission approach as well as in the procedure utilized by the United States Railway Association. The latter is presented in detail in: United States Railway Association, *Viability of Light-Density Rail Lines,* (Washington, D.C.: US), March 1976.
11. See: William R. Black, "Energy and Other Trade-offs in Rail and Truck Shipment and Their Impact on Industrial Location." Presented at the Annual Meeting of the American Association for the advancement of Science, New York City, January 1975.

Regional Development Theory and Sectoral Income Inequalities

R. Keith Semple
The Ohio State University

Income is an important indicator of human well-being which in turn is an important correlate of regional development. Researchers investigating the intricacies of regional development sooner or later must concern themselves with the problems associated with personal income: its generation, its growth, its level, and its disparity. The concern manifests itself in the availability of a rich literature seeking answers to a variety of questions. How is income generated? How does it grow? What is a satisfactory level of income? Why do income disparities exist? Are income disparities a necessary feature of development?

Of particular importance in the present study are the problems of regional income disparities and trends in these disparities with development. Disparities have traditionally been investigated across classes of individuals and among groups of regions. The present research goes one step further and investigates the regional income disparity problem for the various sectors of the economy. Specifically it disaggregates regional incomes into such sectors as manufacturing, wholesaling, retailing, government expenditures and agriculture. Before this is accomplished, however, the study reviews both the theoretical and empirical evidence suggesting a relationship between development and income trends. The investigation covers the period 1919-1974 and disaggregates regional per capita incomes of the United States.

NATIONAL EFFICIENCY AND REGIONAL EQUITY

During the process of development, a policy of fastest economic growth may generate distinct geographic inequalities, concentrate wealth and power in a few advanced regions, and condemn lagging areas to lengthy periods of poverty. Conversely, policies of regional equalization may slow down the growth of the total economy, (Alonso, 1968). If this were true there would appear to be a conflict between the goals of national efficiency and regional equity. The case for aggregate efficiency, a free market process and noninterference with private enterprise is best expressed by Cameron (1968) as the theory of national demand. The case for a national policy that attempts to intervene in the market process in order to assist lagging regions is summarized by Cumberland (1971) as the theory of planned regional adjustment. The argument for some sort of middle ground appears to be the path of least resistance for modern regional development policies (Sakashita and Kamoike, 1974).

The Theory of National Demand

The theory of national demand stresses the axiom that there can be no long-run problem of regional economic distress because competitive market forces will result in an optimal spatial distribution of economic activity. According to this theory, a decline in economic activity in lagging areas is inevitable and should be encouraged since this would strengthen the national or aggregate economy. Cumberland (1971) points out that short run regional unemployment and declining incomes may result from changing demand factors for regional goods or from declining regional competitiveness. However, according to the theory, (Sakashita and Kamoike, 1973) local unemployment and low incomes in the long run will either attract new investment to take advantage of profit opportunities made available by low cost factors of production or, if this fails to happen because the region has other competitive disadvantages, the unemployed will emigrate to areas of greater opportunity and the distressed area will adjust to a lower level of economic activity.

This national demand theory has been formalized by Fukuchi and Nobukuni (1970). They develop an interesting discussion on the tradeoff relation between national growth and regional income equality. Utilizing a difference equation approach, the authors develop a measure of *absolute* variance of per capita output as an index of regional income inequality. Sakashita and Kamoike (1973) find the F-N model less than satisfying and reformulate it using a differential equation model with a *relative* variance measure of per capita output as an index of regional income. They find that the reformulated model produces somewhat more realistic long-run regional stability conditions.[1] The authors conclude that there is no conflict or trade-off relation between high national economic growth and regional-income equality as far as the S-K model is concerned. In other words the free market economy generally does not augment regional disparities over time. The theory maintains that while local subsidies to raise levels of economic activity might be justified on the basis of income redistribution, such subsidies are not likely to raise levels of productivity substantially and would probably be required on a permanent basis.

The Heckscher-Ohlin School and the School of Social Physics also adhere to the theory of national demand or the principle of long-run stable equilibrium which maintains that the equilibrating forces in the economy far out weigh the disequilibrating ones, (Warntz, 1965). The members of both schools conclude that regional income differences have a tendency to disappear over time in an economy which is left to itself. On the other hand, Myrdal (1957) adheres to the principle of circular and cumulative causation which states that the disequilibrating forces of the economy far outweigh the equilibrating ones. This leads him to the conclusion that regional income differences will have a tendency to widen over time in an economy which is left to itself. Regional income differences as seen by Olsen (1967) are the result of a complicated set of interactions between both equilibrating and disequilibrating forces. The implication is that some inequality always exists with the inequalities becoming greater or less depending upon the net magnitude of the two forces.

The Theory of Planned Regional Adjustment

The theory of planned regional adjustment opposes the theory of national demand with respect to the determination of regional incomes. The planned adjustment theory assumes that market processes cannot be depended upon automatically to provide optimal spatial distributions of income nor to guarantee maximum aggregate efficiency. Whereas the national demand theory explains the tendency of regional income disparities to diminish, the planned regional adjustment theory explains their tendency to enlarge. The latter theory emphasizes the role of the regional allocation of social overhead capital and maintains that it determines the relative productivity of private capital in each region. It also assumes (Cumberland, 1971) that unnecessary structural employment results from such factors as immobility of factors of production, failure to achieve economies of agglomeration and scale, and incomplete information and misallocation of public and private investment. By implication, aid to lagging regions can be justified not only on the grounds of equity but also, if properly administered, on the grounds of aggregate efficiency.

The logic of the theory rests upon three basic assumptions. The most important assumption maintains that a disproportionately small share of public and private capital is being invested in lagging areas as compared to advanced regions. The second assumption maintains that some types of private business activities could reduce costs and increase profits by locating in low cost rural areas rather than in high cost regions. The last assumption provides that modest but timely aid to lagging regions can assist them to arrest their decline and thus avoid continuing large scale public and private costs.

In spite of arguments for maximizing national efficiency or minimizing regional equity, a middle of the road policy that mixes both strategies appears to be most useful since neither the national demand nor the planned regional adjustment is tenable in its extreme form.

A Practical Mixed Strategy

The national demand theory of universal long run adjustment is refuted by the many regions where low incomes, unemployment, and excess capacity persist. As an end result, the income differences that emerge tend to become self-perpetuating (Hughes, 1961) and lack any prospect of convergence (Booth, 1964). While the national demand theory of regional economic adjustment has not been sufficiently supported by empirical evidence to result in a pattern of regional prosperity acceptable to the nation, it nevertheless provides insight into the nature of the adjustment problem (Cumberland, 1971). Similarily, the theory of planned regional adjustment is not acceptable in its extreme form, which would insist that economic decline could be arrested in every region. Changing patterns of national demand and industrial location patterns make it highly improbable that every region can remain sufficiently adaptable and responsive to avoid decline. Nor is it possible that a nation could allocate sufficient resources to regional aid programs to underwrite the prosperity of every region, if regions are defined as small local areas. However, the theory of planned regional adjustment provides useful insight into the problem of how to assist those regions which appear to have a reasonable opportunity for sustained growth. Mera (1967) agrees with this approach and identifies the cause of regional income inequality by associating disparity with regional differences in production functions. Sakashita and Kamoike (1974) also appear to favor the practical mixed strategy over their previous national demand theory. By 1974 they extend their earlier work to include public as well as private capital. By utilizing a two-region Cobb-Douglas production function in which room for public policy is introduced by the inclusion of public capital and public investment, they show, by phase analysis, that regional inequality of per capita income may disappear at a convergence point if enough public capital is available, or will equalize at some practical level related to policy variables.

Regional economic policies at the national level are often directed toward improving the interregional distribution of output and income even though a reduction of aggregate output results (Holmes and Monro, 1970). Mera (1975) also suggests that there is a cost associated with this type of policy and he demonstrates that the cost of efficiency increases when the regional disparity of per capita income increases at the efficiency distribution. He concludes that there is a persistent market force that works for creating concentrations of development at a few locations where higher per capita products are produced and that the spatial allocation of private factors of production cannot be fully controlled by the allocation of social capital. As a result he maintains that a proper course of action to be taken by political decision makers is gradualism. That is to say, they should not attempt to equalize fully the per capita incomes of all regions within a short period of time but try to lessen income disparities among them gradually. The degree of loss in the aggregate efficiency associated with any specific degree of income equalization ought be the same regardless of how long a time it has taken to reach the distributional level, but certainly political controversies should be less if the equalization policy is implemented gradually.

INCOME DISPARITY

A considerable literature has grown up dealing with the secular impact of economic development on the regions making up a national economy. Hirschman (1958) indicates that positive spread effects occur as the result of economic development. Kuznets (1955) also makes it clear that income is more equally divided in mature than in developing economies. Both researchers agree that as development takes place income equity increases. Kuznets stresses the impact development has on classes of people, Hirschman stresses the regional impact. In both cases convergence rather than divergence is the rule. Empirical research provides abundant evidence for the verification of each hypothesis.

Income Inequality by Class

Class income inequality tends to diminish as level of income rises. In order to understand the development process it is necessary to understand why there are distributional differences in income by class. Aigner and Heins (1967) ask a fundamental question: "Is there anything about the process of capital accumulation and hence economic development that tends to make the personal distribution of income become more equal?" Soltow (1960) indicates that educational level plays an important role in support of such a relationship, as shown by the existence of greater equality among more highly skilled or educated groups of people. Kravis (1960) expects that social and cultural attributes such as barriers to mobility, human characteristics, racial discrimination and political organizations impinge on the income distribution. Aigner and Heins (1967) report that the Kuznets hypothesis is supported in situations where a high degree of control is exercised over other population attributes. The median age of the population is also important because older populations have a high degree of skill rigidity which in itself causes an enforcement and proliferation of already established equality barriers. Kuznets' observations are interesting in this context because they suggest that economic development per se may be a harbinger of social justice. Aigner and Heins support such an hypothesis and favor some sorts of social and economic policies designed to redistribute resources.

Gunther and Leathers (1974) test the Kuznets hypothesis for depressed regions, pointing out that if it holds true for lagging regions then income inequality by class should be somewhat greater in these areas than for the nation as a whole and should tend to decrease as these regions achieve higher levels of development. They find that the hypothesis is only partially applicable in depressed areas although there is a tendency for greater inequality where high rates of unemployment and relatively large amounts of poverty exist. Smolensky (1961) asks a most significant question: If regional per capita income inequalities diminish over time does this imply greater equality in the income distribution of people? This does not necessarily follow, but Smolensky does find that the degree of income inequality among persons in the United States as a whole has been accompanied by parallel changes in the degree of per capita income inequality among individual states. In particular, income inequality among persons and states increased in the 1920s and declined through the thirties and forties with the rates of convergence accelerating after the Second World War. Brimmer (1971) reports that during the 1960s class incomes continued to converge with the proportion of income going to the lowest groups doing so at the expense of those in the highest quintile. He concluded that the distribution of income has continued to move toward greater equality among families despite rather wide divergences in the performance of the economy. He attributes this phenomena to the upgrading of the skills of the labor force and a reduction in the fraction of the population which derives its income from the lower paying occupations. Brimmer also notes a definite decline in the share of income in the form of proprietary income, for example; dividend and rental income to upper income groups, and a general increase in the median income of blacks relative to whites.

Chiswick and Mincer (1972) find that for adult males the most important force creating income parity has been the business cycle through its effects on the dispersion of weeks of employment. Al-Samarrie and Miller (1967) indicate that the agricultural impact on income inequalities is declining with its decreased share of the national economy. Chiswick (1971) adds that effective compulsory schooling accompanies development and produces more efficient capital markets and more equal rates of return. Development is also associated with the exit of the very young and the aged from the work force and the effective entry of females to positions of equity. Finally, Ahluwalia (1974) notes that the extent of inequality varies widely among countries but the following patterns can be identified. The socialist countries have the highest degree of overall equality in the distribution of income. This is as one would expect since income from the ownership of capital does not accrue as income to individuals. The observed inequality is due mainly to inequality in wages between sectors and skill classes. Most underdeveloped countries show markedly greater relative inequality than developed countries. Ahluwalia confirms that class income inequality first increases and then decreases with development. He notes however that nations with large absolute growth rates tend to have larger class income inequalities. Finally he once again confirms the Kuznets hypothesis that the rate of economic growth and equality are positively related suggesting that the objectives of growth and equity are compatible.

Income Inequality by Region

Hirschman (1958) notes that lagging regions tend to catch up with the leading regions of a nation as economic development progresses. During the early stages of economic development, regional growth within a developing nation may be characterized by polarization (Hale, 1973). The growth of the economy may be oriented or polarized, toward specific urban centers (Semple, Gauthier, Youngmann, 1972). The impetus for the emergence of a growth pole may originate within the pole itself and the pole may then provide growth stimuli to a surrounding hinterland. The chain of causality, however, may operate in the reverse direction with growth in the hinterland providing the primary stimulus for the emergence of the growth pole (Casetti and Semple, 1969; Parr, 1973). The point remains however that for an economy to attain higher income levels it must first develop within itself one or more regional centers of economic strength (Hansen, 1967). Polarization may also be in the form of propulsive industries or lead firms with technological linkages to all sectors of the economy. Growth may be traced through an elaborate regional input-output matrix in an abstract, Perrouxian space (Thomas, 1975).

In the early stages of development, the advantage lies with the growth poles or developed centers, which benefit from the existence of overhead facilities, external economies, political power, spatial preferences of decision-makers, immigration of the more vigorous and educated elements from the lagging regions, flows of funds from the land-wealthy in the hinterlands to the financial markets in the cities, and a variety of other factors. These factors lead to polarization and increases in the differences of regional incomes. After a certain point, however trickling-down effects come into prominence (Hirshman, 1958). The spread of literacy and bureaucratic practices improves knowledge in the lagging areas. The opening of transportation routes to reach these areas as markets for the developed centers also opens them as possible locations for productive activities. In the early stages of development there will be increasing disparity between developed and underdeveloped regions, but later there will be a tendency toward equalization or uniformity as the economy reaches maturity, (Semple and Golledge, 1970; Semple and Wang, 1971). Recent empirical studies support this view and refute the conclusions reached by Myrdal (1957) concerning the inequity that results from the vicious circle of backwash effects.

Williamson (1965) for example, found that regional disparities are greater in less developed countries and smaller in the more developed while over time regional disparities increased in the less developed countries and decreased in the more developed. These findings suggest that regional inequality, if plotted against economic development, would result in a bell-shaped curve, with a maximum being reached at the transition from the take-off to the mature stage (Alonso, 1968). The peak of inequality may be delayed until a later stage in more complex regional economies where structural underemployment exists. Partial verifications of the bell-shaped curve have been provided by Williamson (1965), Easterlin (1958, 1960), Hanna (1959), Theil (1967), and Semple and Gauthier (1972). Perin and Semple (1976) explain why the theoretical regional income inequality curve tends to be bell-shaped (Figure 1). Figure 2 provides the most recent and complete verification of the bell-shaped curve for the U.S. The trend shows inequalities in per capita state income for the period 1919-1974. Information statistics provide the basis for the index of relative inequality. A mathematical derivation of the partitioned inequality statistic may be found in Semple and Griffin, (1971), Semple and Gauthier (1972) and Perin and Semple (1976). The analysis calculates in an iterative fashion an inequality index for selected years throughout the interval 1919-1974[2]. Figure 2 provides, in addition to the national inequality trend, a between regional trend, a between subregional trend, and a within subregional trend. The between regional analysis partitions the United States into the traditional North, South, and West. The between and within subregional analysis partitions the three major regions in turn into three census geographical subdivisions[3].

As yet there has been little theoretical evidence to suggest the nature of these subnational income inequality trends, but since it is evident that they reflect national conditions and by

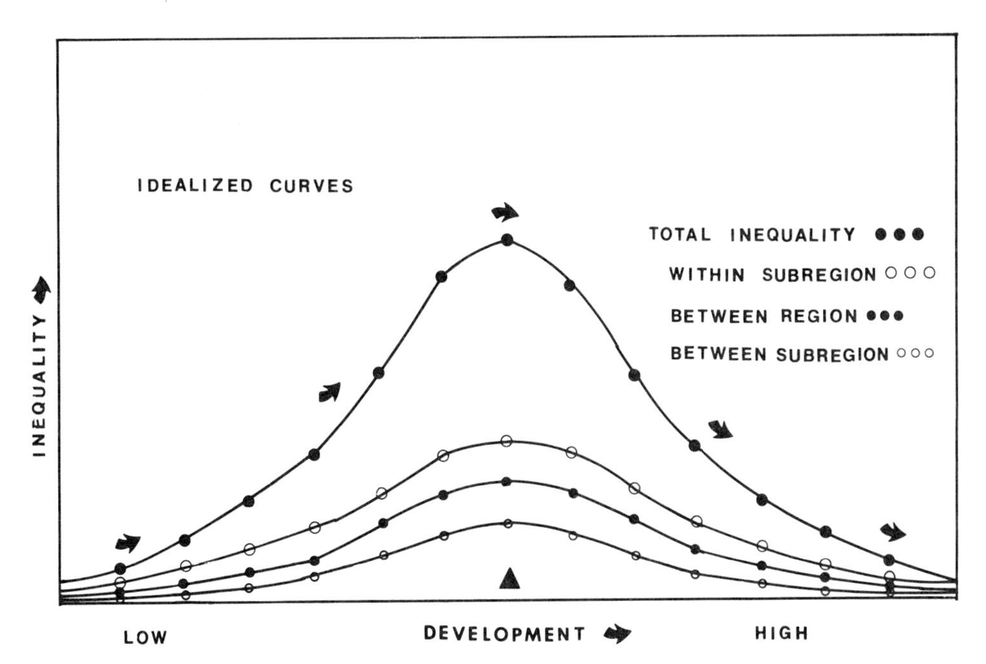

Figure 1. Theoretical per Capita Income Inequality Curve with Regional and Subregional Partitioning.

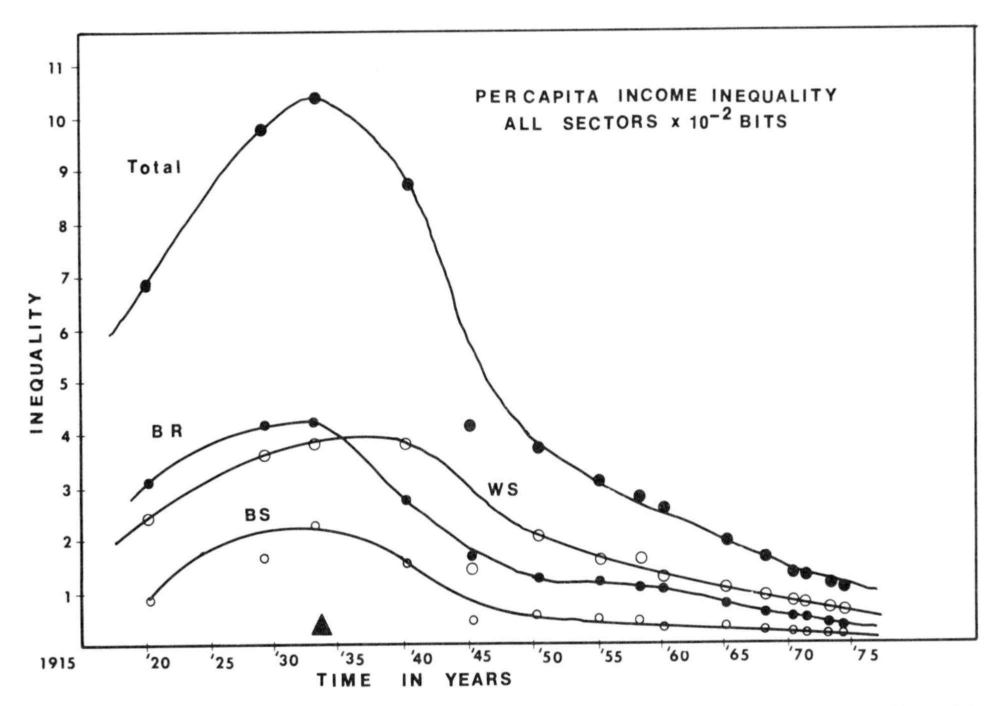

Figure 2. U.S. per Capita Income Inequality with Regional and Subregional Partitioning 1920-1974.

definition sum to the national inequality levels then there ought to be great similarity. The empirical evidence indicated by the regional and subregional trends in Figure 2 confirms this similarity. The bell-shaped income inequality trend with a 1933 peak confirms the partial findings of previous writers. The between region and between subregion curves parallel each other with greater inequality between the major regions than between the minor subregions within these major regions. After 1935 the largest amount of inequality occurs for the states within subregions. This suggests that inequalities at the local level are now a more significant problem than the traditional North-South inequality pattern.

Since the nature and behavior of national and regional income trends appears to be well established, one might ask about the trends in the sectoral subcomponents of both the national and regional incomes in the United States. For example, do manufacturing, service, and agricultural per capita income inequality trends exhibit the same properties individually as when they are aggregated? The next section provides answers to this question.

SECTORAL INCOME INEQUALITY BY REGION

Participatory and Nonparticipatory Regional Income

For analytical purposes total income may be divided into two categories: income received by persons for their participation in current production, known as participatory income, and that received from investment and transfer payments, known as nonparticipatory income. Participatory income in the United States accounts for around eighty percent of all income (Graham, 1964). This form of income is distributed by state and industry and consists of wages, salaries, supplemental income, and earnings of employed persons. The nonparticipatory income consists of two components: property income (rents, dividends, and interest) and transfer payments. This category of income is not analyzed in the work that follows since it cannot be allocated by industry or sector. Nevertheless a few general remarks must be made concerning its impact on regional inequality before turning to participatory income.

Changes in property income in the United States during the postwar era generally contributed to the relative shift of income share from the Northeast and Central regions to the South and West. Since savings from which investments are made are accumulated out of current income, and since the geographical trend of income change to the South and Southwest has been underway for many years, the relative shift in property income is to be expected. On the other hand, the amount paid out in the form of transfer payments depends mainly on population and public policy. Because there was much less variation in the rates of change in transfer payments in the postwar era than in income from current production; transfers tended to retard the geographical redistribution of income, (Graham, 1964). The net effect of the twenty percent nonparticipatory portion of regional income appears to range from neutral to slightly positive in its influence on regional income inequality patterns. With this discussion as background information, the subsequent findings deal only with sectoral participatory income.

The analysis follows the lead of the U.S. Department of Commerce, Office of Business Economics and breaks participatory income into a number of categories or sectors. Trends in the income inequality of wage earners by sector and state over time provide the basis for the subsequent findings. The sectors selected for analysis include manufacturing, wholesaling, retailing, agriculture, service, federal government, state and local government, transportation and utilities, and finance. Comparable data are available for manufacturing back to 1919, for the next three sectors from 1929 and for the remaining sectors from the late thirties.[4]

The Analysis

State per capita incomes for each sector were calculated for selected time intervals 1919-1974. The inequality inherent in these data was calculated in the same way as inequality for the aggregate income data. Figures 3 through 11 display the partitioned regional income inequality trends for each sector. A number of conclusions emerge from the analysis. The first three sectors, manufacturing, wholesaling, and retailing comply with the overall income trends, (Figure 2). That is, they display a symmetrical bell-shaped curve with inequalities peaking in the early or late 1930s. For manufacturing there is an interesting cyclical increase in inequality for the within subregional inequality during the 1960s.

For the remainder of the sectors, with the notable exception of agriculture, behavior is as expected. That is from the late 1930s to the present there is a systematic trend toward greater regional and subregional equality of participatory income. Interesting cycles are noted for within subregional inequality for the service sector, for federal government during the war, and for finance during the 1960s.

Agriculture is an exceptional sector. From the 1930s to the present, erratic and changing trends occur. Greater inequality exists today than at any time in the past forty years. It is to be noted, however, that the role of agriculture in the overall economy greatly diminished during this period.

The Explanation of Regional Trends

For the past fifty years and especially since 1940 dramatic changes have been taking place in the spatial economy of the United States. The period has witnessed vast migrations of rural workers to urban centers; the automation of manufacturing; the rise of the white-collar worker and the service industries; a postwar renewal of immigration; a drastic decline in the agricultural employment; the rise of big government and the associated inroads of the welfare state; the growth of the civil rights movement and the associated increase in mobility for minorities; the increasing participation of women in the labor force; the rise of minimum wages with the associated removal of marginal jobs; the removal of the very young and the very old from active participation in the work force; the introduction of the space age and the rise in educational and technical awareness; and the achievement of a more uniform set of wage and fringe benefits (Graham, 1964).

The Role of the Private Sector

It is evident that the most substantial reductions in sectoral income inequalities on a regional basis were accomplished during the war years and since then convergence has continued at a much slower rate. Batra and Scully (1972) suggest technical progress as a basis for the long run convergence of regional incomes. Scully (1969, 1971) however, notes the persistence of the North-South income gap and concludes that the social-political milieu as well as industrial mix, quality of education, wage discrimination against nonwhites, and lowkeyed unionization have worked against equalization in regional incomes. Countering these aspects of the North-South disparity is the fact that industry expanded at greater rates in southern areas at the expense of the North.

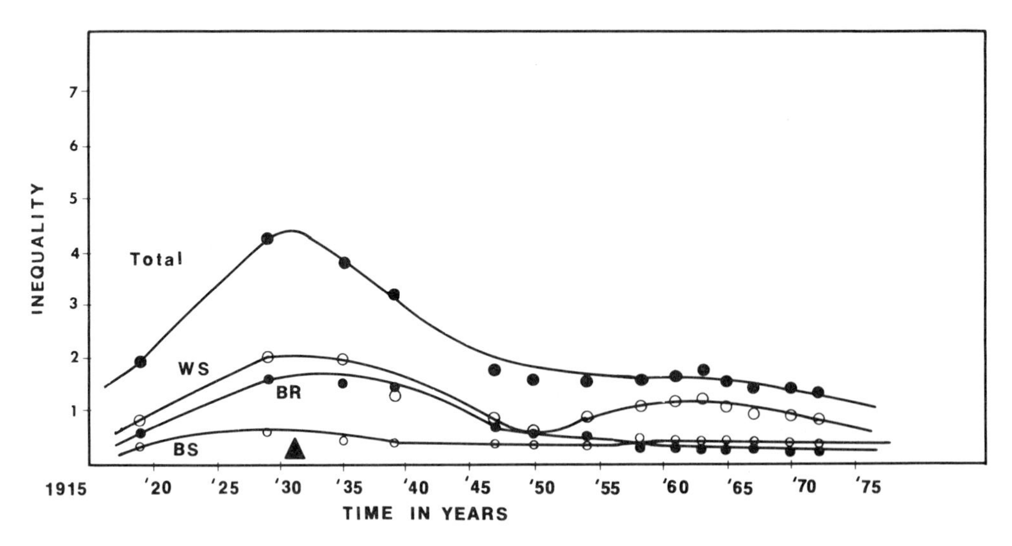

Figure 3. Per Capita Manufacturing Regional Income Inequality 1919-1972.

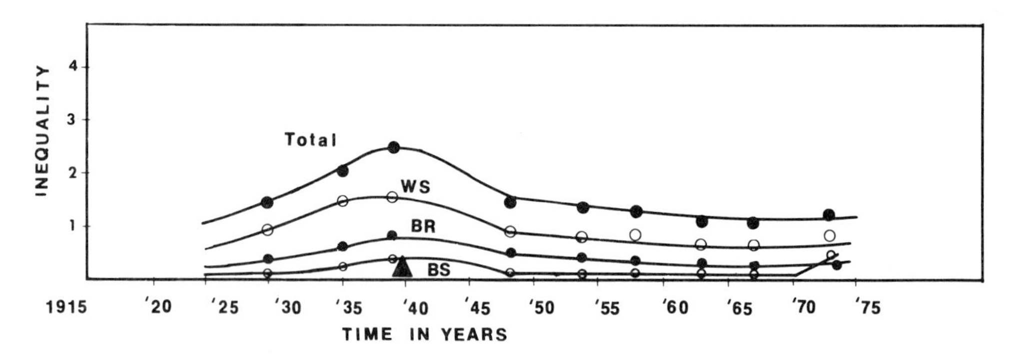

Figure 4. Per Capita Wholesale Regional Income Inequality 1929-1973.

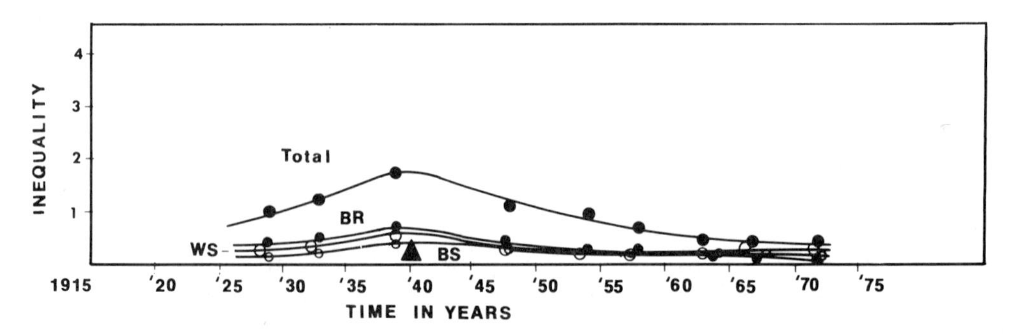

Figure 5. Per Capita Retail Regional Income Inequality 1929-1973.

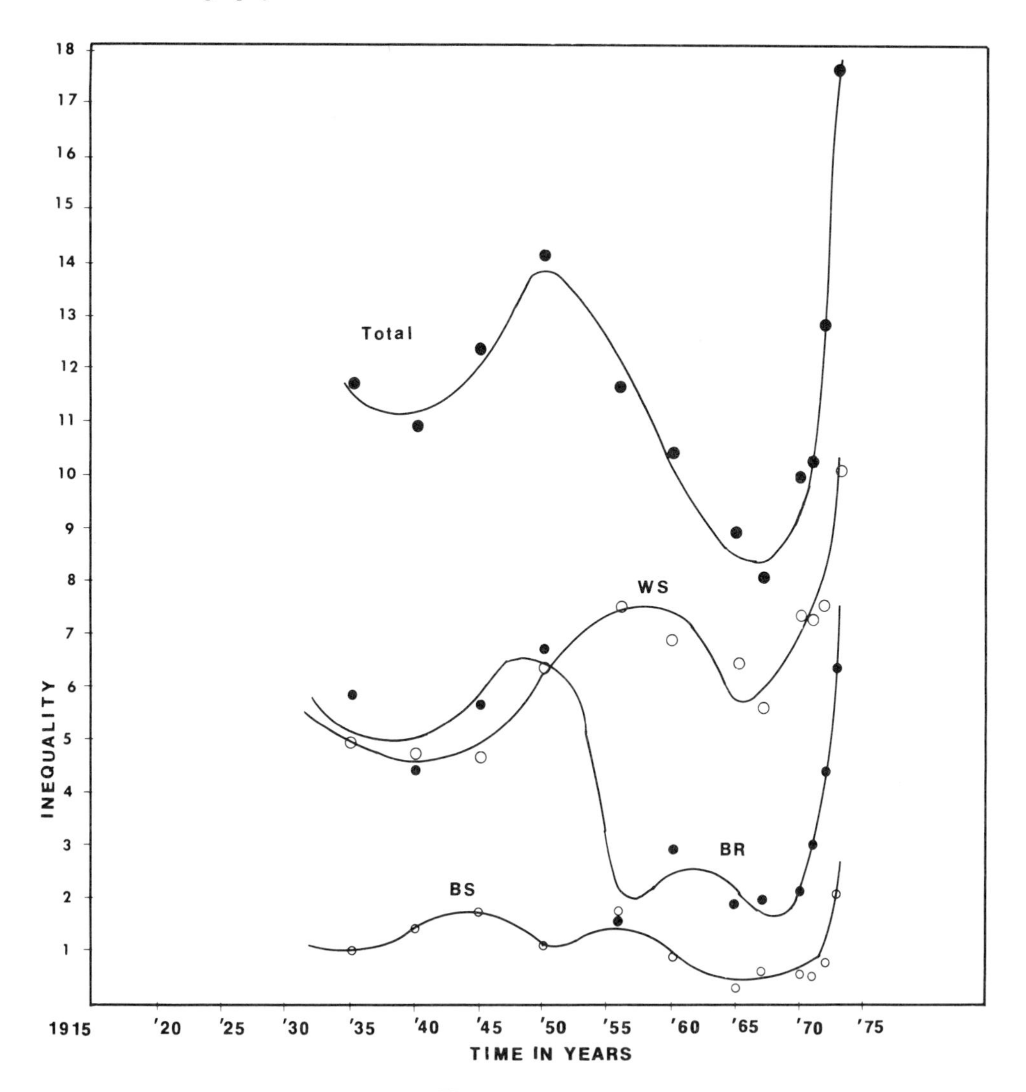

Figure 6. Per Capita Agriculture Regional Income Inequality 1935-1973.

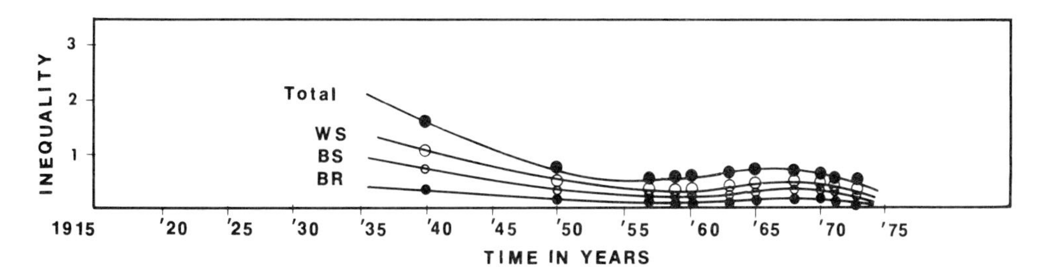

Figure 7. Per Capita Transportation Regional Income Inequality 1940-1973.

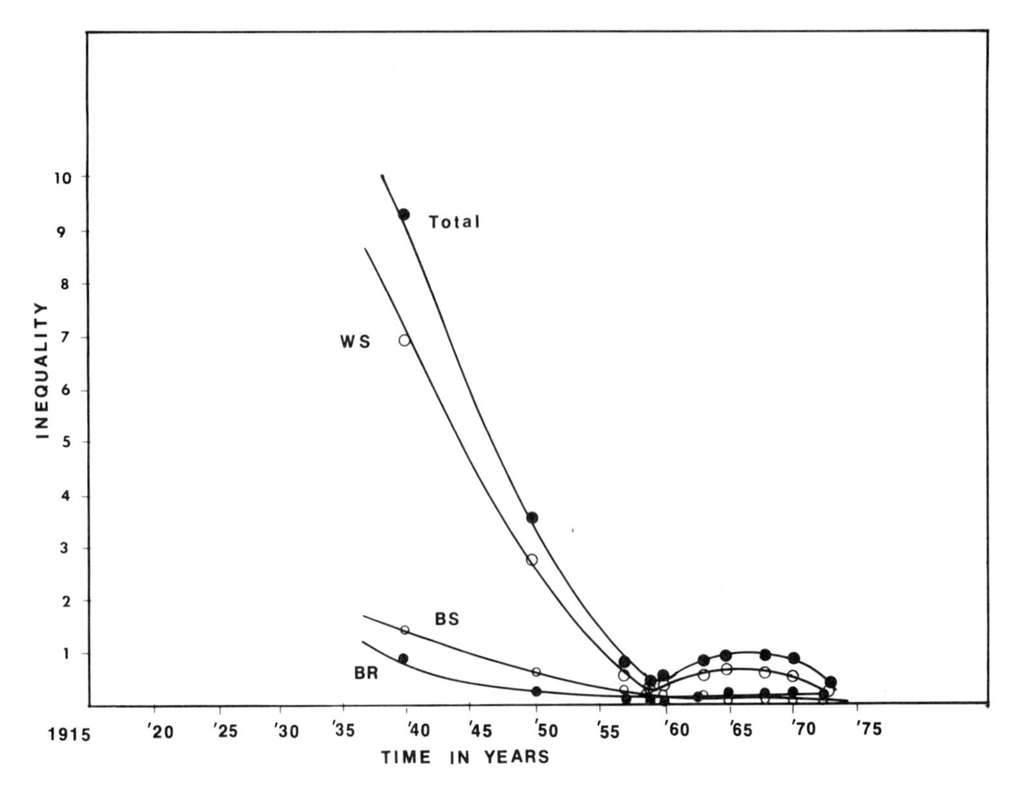

Figure 8. Per Capita Finance Regional Income Inequalities 1940-1973.

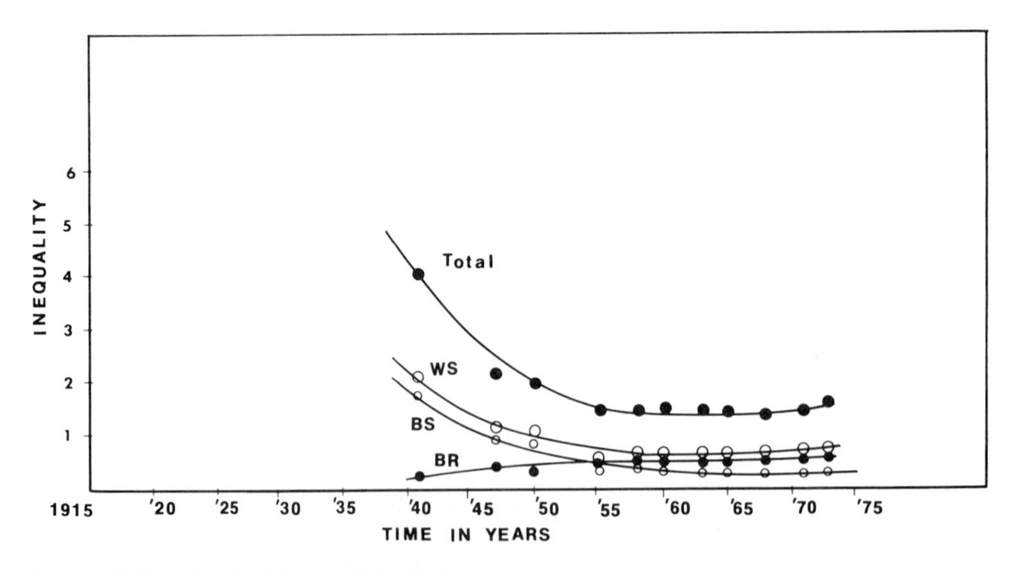

Figure 9. Per Capita State and Local Government Regional Income Inequality 1940-1973.

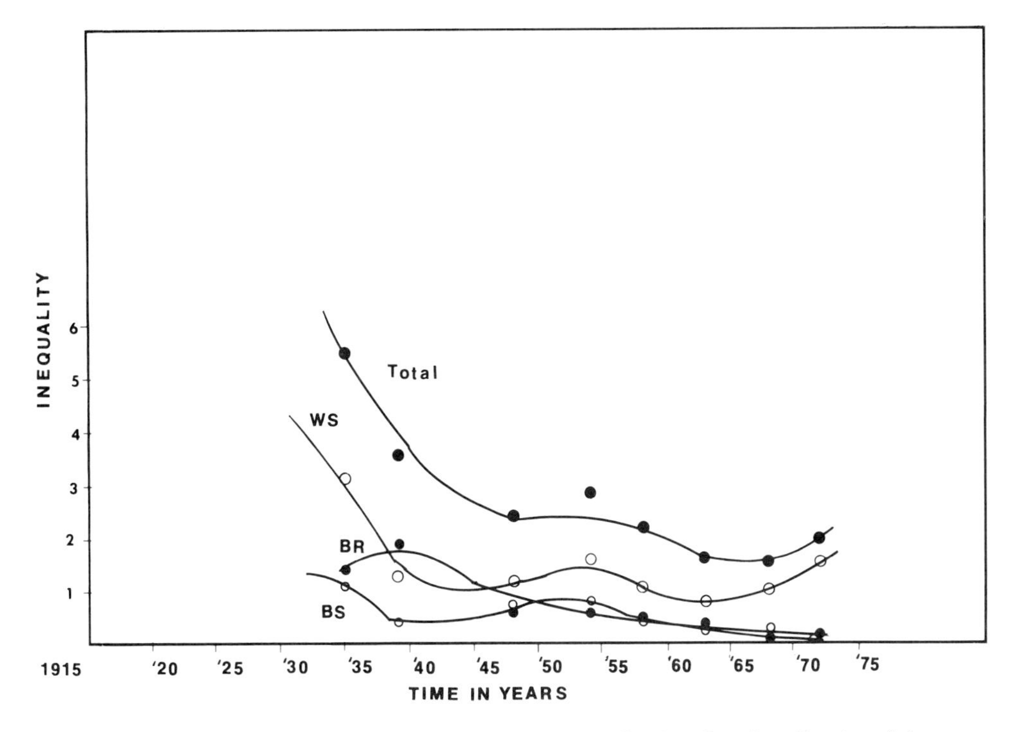

Figure 10. Per Capita Service Regional Income Inequality 1935-1972.

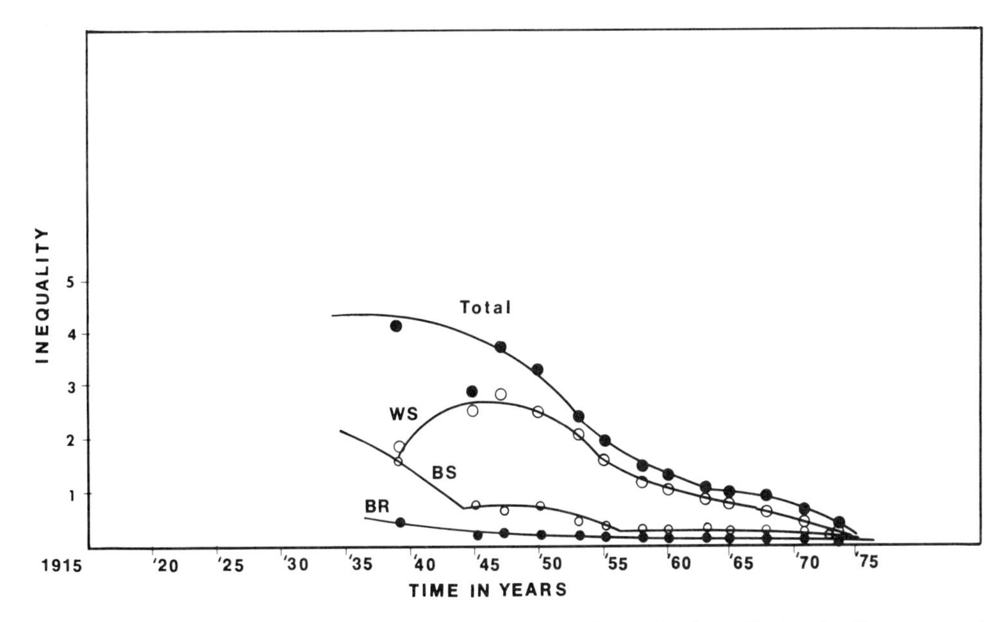

Figure 11. Per Capita Federal Government Regional Income Inequality 1939-1973.

The postwar period also saw a sharp increase in the utilization of underemployed labor resources. The huge demand occasioned by the war and postwar period for nearly all types of goods and services stimulated production in the South more than elsewhere. Both employment and productivity rose but there was little change in population. The unusually large military presence in the South also boosted demand for local goods and services. Agriculture, also rose from the depression years to full utilization of capacity with associated relative income gains. The record high volume of farm income reached in the 1947-1950 period pushed per capita incomes unusually high in the Plains and Rocky Mountains. Thereafter, however, the trend was downward until 1968 when incomes bottomed out and began their upward spiral into the mid-1970s. The effect of this income cycle can be seen on the agricultural income inequality trends in Figure 6.

Since the war the nation has seen the dramatic rise of the West, especially California. This rise, along with the associated relative decline of the North, has had an effect on the decision making and power structure of the country. This period, for example, has witnessed a shift in corporate headquarters away from the Northeast to other parts of the nation and a faster growth rate for those headquarters located outside the North. The net result is a trend toward a more dispersed regional pattern of headquarter locations in the postwar period, (Semple, 1973).[5]

Associated with the improvements of the private sector in industry, agriculture, and commerce is the growing role that government has begun to play in the process of regional and national equity.

The Role of Government

Government spending to equalize the economic well-being of all regions appears to be an adequate and appropriate policy goal within the constraints outlined earlier. However, the exact spatial strategy that ought to be taken is not clear. That is, should governmental agencies stress a growth centers development policy that seeks a more efficient utilization of regional resources by concentrating various programs in the central cities of multi-county labor market areas or should aid be spread around in a more uniform manner? Lewis and Prescott (1972) are not sure what policies are best but suggest that many elements of growth center strategies warrant reconsideration. Nichols (1969) suggests that it is probably advisable to concentrate investment in the area that has the strongest linkages with the surrounding region in order to stimulate strong income multiplier effects in higher order and higher linked centers. At another level entirely Ehrenberg and Goldstein (1975) emphasize the interrelationships among various wage categories and income sectors. They show that there is a tendency for the wages of governmental employees to be both occupationally and geographically interdependent.

Wiedenbaum (1966) provides one of the most interesting studies concerning governmental spending and regional equity. He specifically studies the allocation of federal expenditures in relation to the regional distribution of income in the United States by investigating two types of spending: federal grants-in-aid and defense and nondefense expenditures. He reports that the influence of the geographic distribution of federal programs on regional income differentials can be considered progressive, proportional or regressive. The progressive programs are those that tend to reduce the inequality in the distribution of personal income among regions. The proportional programs are those that have little or no effect on regional income distributions and the regressive programs are those that tend to accentuate inequality in the geographic distribution of income.

He concludes that the current expansion of domestic civilian programs in the federal government is resulting in a shift in the geographic distribution of federal expenditures. This should work toward greater equality in the regional distribution of income. There are, however, some interesting comparisons to be made. The regional distribution of spending for defense and space programs differs quite significantly from the pattern of federal nondefense programs. Low income states tend to get a greater per capita share of expenditures for the nondefense public programs. In contrast the high income states tend to receive a larger than proportional share of expenditures for defense and space programs. This reflects the dependence on industrial areas for the design and production of weapon and space systems. As a result a shift to defense spending, all other things being equal, would create greater income inequalities and vise-versa.

Significant differences exist among the geographical allocations of various types of nondefense spending of the federal government according to Wiedenbaum. Farm price support subsidies, for example, favor low income areas to a greater extent that any other governmental program. Also aid to elementary and secondary education is more heavily oriented toward low income regions.

SUMMARY AND CONCLUSIONS

The foregoing analysis has dealt with the problem of inequalities within a free market society. Peet (1975) provides a most interesting paper dealing with the Marxist notions of inequality and poverty. Specifically the paper combines a theoretical explanation of the origins of inequality with some empirically derived generalizations about who is poor and exactly how inequality persists. The Marxist view is that income inequality is inherent in the capitalist mode of production. As different types of labor require different levels of education and skills, so wages must differ between various categories of workers. Peet shows that income inequality is necessary to produce the variety of labor needed by the various levels of a multitude of economic activities. Also important is the notion that capital is simply historical labor power accumulated by the capitalist class because it has been able to pay labor a sum less than the value of the goods produced by the workers.

It follows that a private enterprise economy will inevitably have great income inequalities between the capitalist class, which controls the use of accumulated past labor and receives part of the production of many workers in the form of profit, and the workers who only receive income in the form of wages (Peet, 1975). The key to the analysis, however, lies in the fact that as capital accumulates interclass inequalities will grow and, although some improvements may occur for the worker, his relative social position will decline as the accumulation of capital places an increasing share of income into the hands of the owners of the means of production.

Although the theory is interesting and appears to be a consistent way of explaining class income inequalities its theoretical conclusions cannot be borne out by the present empirical findings. Nor does it agree with the findings of non-Marxist income theory. In fact, not only do class and regional incomes converge in advanced Western economies but sectoral regional incomes converge as well. Even proprietary income is declining as a source of class inequality in the United States. If Marxist theory has any value at all in income theory and the study of inequality then it only appears to be valid for the stage of increasing inequality or that period of inequality up until 1930 for the United States (Figure 2). It appears that the theory has failed to allow for the self-adjusting mechanisms within advanced Western societies which counter rising inequality with tendencies toward equality. Marxist theory also appears to have failed to appreciate the increased role that government would ultimately play in national development and welfare. Today it is clear that the trend toward equality of income in the United States is in part the result of shifts of capital and labor to lagging regions and in part due to expenditures by government. This government-private sector partnership appears to have succeeded extremely well in the United States in relieving income disparity. Marx also failed to foresee the day of universal education and a national development policy that stressed equal economic opportunity for all, rising income levels, and full employment. He wrote in an era when capitalism stressed economic freedom and opportunity at the expense of social justice, security, redress of inequalities, and the equality of human life. Today attitudes have changed (Cumberland, 1971).

It must never be forgotten, however, that achieving full equalization of per capita incomes, especially incomes in all regions, requires strong determination by the government and a willingness to sacrifice a substantial proportion of the productive efficiency of the economy (Mera, 1975). Nevertheless, Mera adds, it is important to recognize the role of mobility in improving the distribution of income and increasing the aggregate output. When the mobility of labor is high, regional wage differentials are rapidly reduced. In addition, mobility of factors help to increase the aggregate efficiency of production. Therefore if efforts are directed toward removing obstacles against interregional mobility of population and capital, then regional income differentials can be achieved without sacrificing substantial aggregate production efficiency.

NOTES

1. For the sake of interregional comparison a measure of relative variance appears to be more appropriate than an absolute one (Sakashita, 1967 and Theil, 1967). For empirical studies, additional refinements are suggested by Paglin (1975). If per capita income in every region is doubled then few people would think that regional income disparity is increased although an absolute variance index is exactly four times larger than before.
2. Data for state per capita income are available back to the 1880s and earlier. This writer feels however that they are not robust enough to be compared with more modern data sets. There are, for example, indications that inequalities were larger after the period of the Civil War then at the turn of the century.
3. The North consists of New England, Middle Atlantic States, and Great Lakes; the South, South Atlantic, East South Central, and West South Central; and the West, West North Central, Rocky Mountain, and Pacific.
4. Data were collected from the following sources:
   a. U.S. Dept. of Commerce, Bureau of the Census *Statistical Abstract*. See various yearly volumes for income and employment data by state and all sectors other than agriculture.
   b. U.S. Dept. of Agriculture, *Agricultural Statistics*. See various yearly volumes for income and employment data by state and region for agriculture.
   c. U.S. Dept. of Commerce, Bureau of Economic Analysis, *Survey of Current Business*. See the August issue of the annual volumes for personal income by state and sector. For a summary of all data 1929-1956 see the *Special Report* of the Survey of Current Business, 1956.
5. Pred (1974, page 22) makes the comment that the recent analysis by Semple (1973) is carried out at an unnecessarily and unrealistically high level of spatial aggregation and questions the use of a system entropy analogy (footnote 15). Professor Pred's comments are unfounded and misleading. The paper by Semple was written so as to come to grips with the notion that as national economies grow and mature there appears to be a tendency for a more equitable regional distribution of wealth and power. National development studies dealing with such topics often partition a nation into "have and have-not" regions or analyze the "North-South" problem within that country. The present paper investigates the links between development (the growth and maturation) of the economy and the associated dispersion of corporate power. For this class of problem the level of spatial aggregation is both realistic and necessary if results are to be applied to other works in the development literature carried out at a similar scale of analysis. It is agreed that for Pred's purpose (the study of major job providing organizations in urban centers) this level of aggregation is inappropriate. What Pred needs is a study dealing with the dispersion of employment controlled from corporate headquarters located in urban centers. Semple's paper dealt with the dispersion of wealth (assets, revenues, sales) at a regional level, not jobs or employment at the urban level. Surely it is a vacuous

research practice to utilize primary research findings to reinforce introductory review expositions when the review deals with a different substantive field, at a different level of spatial aggregation and has a different purpose from that of the primary research.

Professor Pred also questions the use of the systems entropy analogy. He states "Semple suggests that the trend toward increased dispersion of major organizational headquarters represents increased 'systems entropy'," and points out that this is questionable. I agree but this is not what I suggest. I suggest that the entropy equations developed to measure the dispersion of a system of corporate headquarters provide maximum measures when complete equality holds within a region (all regions have equal amounts of corporate activity present) and minimum measures when concentration exists. The use of information statistics and entropy analogies is quite appropriate for investigations of such corporate systems.

BIBLIOGRAPHY

Ahluwalia, Montek S. "Income Inequality: Some Dimensions of the Problem." *Finance and Development* 11 (1974):2-8, 41-42.

Aigner, D.J., and Heins, A.J. "On the Determinants of Income Inequality." *American Economic Review* 57 (1967):175-181.

Alonso, Wm. "Urban and Regional Imbalances in Economic Development." *Economic Development and Cultural Change* 17 (1968):1-13.

Al-Samarrie, A., and Miller, Herman P. "State Differentials in Income Concentration." *The American Economic Review* 57 (1967):59-72.

Batra, R., and Scully, G.W. "Technical Progress, Economic Growth and the North-South Wage Differential." *Journal of Regional Science* 12 (1972):375-386.

Brimmer, A.F. "Inflation and Income Distribution in the U.S." *The Review of Economics and Statistics* 53 (1971):37-48.

Booth, E.J.R. "Interregional Income Differences." *Southern Economic Journal* 31 (1964):44-51.

Cameron, G.C. "The Regional Problem in the United States; Some Reflections on Available Federal Strategy." *Regional Studies* 2 (1968):207-220.

Casetti, E., and Semple, Keith R. "Concerning the Testing of Spatial Diffusion Hypotheses." *Geographical Analysis* 1 (1969):254-259.

Chiswick, B.R. "Earnings Inequality and Economic Development." *Quarterly Journal of Economics* 85 (1971):21-39.

Chiswick, B.R., and Mincer, J. "Time-Series Changes in Personal Income Inequality in the United States from 1939, with Projections to 1985." *Journal of Political Economy, Supplement*, 1972, pp. 534-566.

Cumberland, John H. *Regional Development: Experiences and Prospects in the United States of America*. The Hague: Mouton, 1971.

Easterlin, R.A. "Long Term Regional Income Changes: Some Suggested Factors." *Papers and Proceedings of the Regional Science Association* 2 (1958):313-325.

Easterlin, R.A. "Interregional Differences in Per Capita Income, Population and Total Income, 1840-1950." *Trends in The American Economy in the Nineteenth Century*. Princeton: Princeton University Press, 1960, pp. 73-140.

Ehrenberg, R.G., and Goldstein, Gerald S. "A Model of Public Sector Wage Determination." *Journal of Urban Economics* 2 (1975):223-245.

Fukuchi, T., and Nobukuni, M. "An Econometric Analysis of National Growth and Regional Income Inequality." *International Economic Review* 11 (1970):84-100.

Graham, Robert E. "Factors Underlying Changes in the Geographic Distribution of Income." *Survey of Current Business* 44 (1964):16-32.

Gunther, William D., and Leathers, Charles D. "Income Inequality in Depressed Regions: Some Empirical Evidence." *Land Economics* 50 (1974):176-180.

Hale, Carl W. "Growth Centers, Regional Spread Effects and National Economic Growth." *Journal of Economics and Business* 26 (1973):10-18.

Hanna, F.A. *State Income Differentials, 1919-1954*. Durham: Duke University Press, 1959.

Hansen, N.M. "Development Pole Theory in A Regional Context." *Kyklos* 20 (1967):709-727.

Hirschman, A.O. *The Strategy of Economic Development.* New Haven, Conn.: Yale University Press, 1958.

Holmes, R.A., and Monro, J.M. "Regional Nonfarm Income Differences in Canada: An Econometric Study." *Journal of Regional Science* 10 (1970):65-74.

Hughes, R.B. "Interregional Income Differences: Self-Perpetuation." *Southern Economic Journal* 28 (1961):41-45.

Kravis, I.B. "International Differences in the Distribution of Income." *Review of Economics and Statistics* 42 (1960):408-416.

Kuznets, S. "Economic Growth and Income Inequality." *American Economic Review* 45 (1955):1-28.

Lewis, W.C., and Prescott, James R. "Urban-Regional Development and Growth Centers: An Econometric Study." *Journal of Regional Science* 12 (1972):57-70.

Mera, Koichi. "Tradeoff Between Aggregate Efficiency and Interregional Equity: A Static Analysis." *Quarterly Journal of Economics* 81 (1967):658-674.

———. *Income Distribution and Regional Development.* Tokyo: University of Tokyo Press, 1975.

Myrdal, Gunnar. *Economic Theory and Underdeveloped Regions.* London: Gerald Duckworth, 1957.

Nichols, Vida. "Growth Poles: An Evaluation of Their Propulsive Effect." *Environment and Planning* (1969):193-208.

Olsen, Erling. "Regional Income Differences: A Simulation Approach." *Papers of the Regional Science Association* 20 (1967):7-17.

Paglin, Morton. "The Measurement and Trend of Inequality: A Basic Revision." *The American Economic Review* 65 (1975):598-609.

Parr, John B. "Growth Poles, Regional Development, and Central Place Theory." *Papers of the Regional Science Association* 31 (1973):173-212.

Peet, Richard. "Inequality and Poverty: A Marxist-Geographic Theory." *Annals of the Association of American Geographers* 65 (1975):564-571.

Perin, Dan E., and Semple, Keith R. "Recent Trends in Regional Income Inequalities in the United States." *Regional Science Perspectives* 6 (1976):65-85.

Pred, Allan R. *Major Job-Providing Organizations and Systems of Cities.* AAG Commission on College Geography, Resource Paper, no. 27, 1974.

Sakashita, Noboru. "Regional Allocation of Public Investment." *Papers of the Regional Science Association* 19 (1967):161-182.

Sakashita, Noboru, and Kamoike, Osamu. "National Growth and Regional Income Inequality: A Consistent Model." *International Economic Review* 14 (1973):372-382.

———. "National Growth and Regional Income Inequality: Further Results." *Journal of Regional Science* 14 (1974):81-87.

Scully, G.W. "Interstate Wage Differentials: A Cross Section Analysis." *American Economic Review* 59 (1969):757-773.

———. "The North-South Manufacturing Wage Differentials, 1869-1919." *Journal of Regional Science* 11 (1971):235-252.

Semple, R. Keith, and Golledge, R.G. "An Analysis of Entropy Changes in A Settlement Pattern Over Time." *Economic Geography* 46 (1970):157-160.

Semple, R. Keith, and Wang, L.H. "A Geographical Analysis of Changing Redundancy in Interurban Transportation Links." *Geografiska Annaler* 53, Series B (1971):1-5.

Semple, R. Keith, and Griffin, J.M. *An Information Analysis of Trends in Urban Growth Inequality in Canada.* Dept. of Geography, Discussion Paper no. 19, Ohio State University, Columbus, Ohio, 1971.

Semple, R. Keith, and Gauthier, H.L. "Spatial-Temporal Trends in Income Inequalities in Brazil." *Geographical Analysis* 4 (1972):169-179.

Semple, R. Keith; Gauthier, H.L.; and Youngmann, Carl E. "Growth Poles in Sao Paulo, Brazil." *Annals Association of American Geographers* 62 (1972):591-598.

Semple, R. Keith. "Recent Trends in the Spatial Concentration of Corporate Headquarters." *Economic Geography* 49 (1973):309-318.

Semple, R. Keith, and Scorrar, D.A. "Canadian International Trade." *Canadian Geographer* 19 (1975):135-148.

Semple, R. Keith, and Brown, L.A. "Scales of Resolution in Spatial Diffusion Studies: A Perspective." *Professional Geographer* 29 (1976):8-16.
Smolensky, E. "Industrialization and Income Inequality, Recent United States Experience." *Papers of the Regional Science Association* 7 (1961):67-88.
Soltow, Lee. "The Distribution of Income Related to Changes in the Distribution of Education, Age, and Occupation." *Review of Economics and Statistics* 42 (1960):450-453.
Tabb, W.K. "Decreasing Black-White Income Differentials: Evaluating The Evidence and a Linear Programming Framework for Urban Policy Choice." *Journal of Regional Science* 12 (1972):443-456.
Theil, Henri. *Economics and Information Theory*. Chicago: Rand McNally, 1967.
Thomas, Morgan D. "Growth Pole Theory, Technological Change and Regional Economic Growth." *Papers of the Regional Science Association* 34 (1975):3-26.
Warntz, Wm. *Macrogeography and Income Fronts*. Philadelphia: Regional Science Research Institute, Monograph no. 3, 1965.
Weidenbaum, Murray L. "Shifting the Composition of Government Spending: Implications for the Regional Distribution of Income." *Papers of the Regional Science Association* 17 (1966):163-178.
Williamson, J.G. "Regional Inequality and the Process of National Development: A Description of the Patterns." *Economic Development and Cultural Change* 13 (1965):3-45.

The Migrational Process in Centrally Planned Economies

Robert N. Taaffe
Indiana University

In recent years, Western scholars have placed increasing emphasis on the study of migration. An earlier preoccupation with the examination of massive population transfers, often international in scope, has been replaced by a growing concern with macro-scale and micro-scale migrational analysis.[1] Macro-scale studies have involved the investigation of statistical materials, usually governmental, to derive insights into aggregative causes and effects of different types of migrational flow patterns. In Western societies, interurban migration is the dominant focus of this type of analysis whereas rural-urban movement is the preeminent theme of migrational studies of lesser-developed countries. By contrast, micro-scale analysis has relied primarily on interviews and scaling measures to ascertain migrational attitudes and behavior. Many of these studies have examined movement within metropolitan areas.

A specific approach to the investigation of migration which has relevance for this paper concerns the interaction of economic development and the geographic mobility of the labor force. This includes a substantial body of literature devoted to migration in relation to regional labor markets, industrialization, urbanization, and agricultural change. Much of this work has assumed that economic motives are the sole basis for interpreting the aggregative migrational behavior of labor. For example, the macro-level studies of labor-force migration by Schultz and Sjaastad rest on the assumption that individual decisions to migrate are based on the present discounted value of future earnings in alternative locations.[2] However, this abstract and intuitive notion appears to be quite narrow and, moreover, does not lend itself to testing or measurement. It is much less satisfactory than broader formulations of migrational decision making, such as those employing the concept of place utility.[3]

A type of research which is related directly to the aims of this paper involves the migrational implications of regional growth theory and the policies of regional planning in Western societies.[4] Unfortunately, this work usually has treated migration implicitly and has not examined it directly and comprehensively. The retardation of out-migration from lagging regions has been assumed to be a desirable and inevitable consequence of programs designed to reduce regional developmental inequities through investment in new industries and social overhead capital in regional growth centers. In general, studies in the field of regional growth theory have provided relatively few insights into the complex interrelationships between programs of regional development and aggregative or individual migrational behavior.

Even though the migrational process in centrally planned economies represents the only example of a comprehensive effort to incorporate migrational and developmental variables into a unified national plan, remarkably little attention has been paid to this experience in the West. In his review and extensive bibliography of migrational research published by the Regional Science Institute in 1975, Shaw listed approximately 600 references pertaining to about 45 countries.[5] However, he was unable to cite a single work describing migration in a centrally planned economy.

This paper will attempt to trace in a tentative and exploratory manner, some of the more important aspects of the migrational experience in economies which are planned centrally in an effort to obtain some findings which will have broader relevance for an understanding of the complexities of the migrational component in regional economic planning. The results should point out the hazards of efforts to derive facile or comprehensive solutions to these problems in virtually any society.

Most of the examples will be drawn from the experience of the U.S.S.R., which has served as the basic model of migrational planning for Eastern Europe, and from the example of Socialist Bulgaria, which provides a useful illustration of how the Soviet model has been modified to respond to specific national conditions. To a much lesser extent, the migrational record of the other Communist nations in Europe will be cited.[6] No explicit reference will be made to other countries with central planning and predominantly agrarian economies.

In order to place the changing and interrelated problems of migration and development in perspective, an evolutionary model of migration in centrally planned economies will be outlined. This will be followed by a summary of the basic methods of migrational and settlement planning. Finally, some of the apparent reasons for the persistence of migrational problems in the planned economies and possible remedies will be examined.

THE EVOLUTION OF MIGRATION IN CENTRALLY PLANNED ECONOMIES

As the nations of Eastern Europe and the U.S.S.R. have undergone a dramatic transformation from agrarian states to complex industrial societies, major changes have occurred in the patterns of migration which have given rise to different sets of problems in different phases of development. In addition, the divergent levels of industrialization and urbanization among these countries have resulted in cross-sectional variations in migrational characteristics. Many of these can be attributed to temporal developmental disparities even though migrational differences also reflect varying national conditions. For example, the German Democratic Republic and Western Czechoslovakia were highly urbanized before the introduction of central planning and have since encountered the type of migrational problems of mature industrialized societies rather than those of the predominantly agrarian states in the Balkans or the U.S.S.R.

Although several alternative sets of migrational stages could be defined and defended, the most useful categorization for the purposes of this study includes three phases: migration in agrarian societies before the advent of central planning; the initial industrial transformation and collectivization; and the migrational process in industrialized societies.

MIGRATION IN THE AGRARIAN SOCIETIES PRIOR TO THE INTRODUCTION OF CENTRAL PLANNING AND COLLECTIVIZATION

Most of the Communist nations of Europe (and all the Communist states of Asia) were predominantly rural before the establishment of Communist rule and in many regions, such as the Balkans and the U.S.S.R., over three-fourths of the population resided in villages. Even today several East European nations have a majority of their population living in the countryside.

Although rural-urban migration was a persistent characteristic of these societies, the magnitude of these flows was constrained seriously by the relatively modest expansion of urban employment opportunities in the heavily agrarian countries or regions. The outflow of rural migrants was substantially less than the major population gains of villages resulting from demographic factors. The degree of fertility in the countryside was much higher than even the substantial rate of mortality. The retention of a large share of the rural population and the high rates of natural population increase led to the creation of a pervasive problem of rural overpopulation and underemployment which was a major obstacle to development in most of Eastern Europe and the Russian Empire.[7] A common regional manifestation of this problem was the low income of densely settled rural areas and, in a more comprehensive sense, the widespread existence of lagging regions in which the level of livelihood, health, and sociocultural characteristics was considerably below that of the most developed regions. During this early phase, most migrants travelled from villages to national or regional urban capitals or between villages. The intensity of interurban migration was relatively low. To encourage settlement in frontier regions, as in Tsarist efforts to populate Siberia, material incentives and landowning opportunities were employed.

As an aftermath of the Second World War, an enormous volume of migrational movement in the East European countries involved international population transfers, including the expulsion of approximately 14 million Germans from this region. In many cases, new regions of settlement were opened in areas evacuated by the Germans. The classic example of this is the Polish effort to settle the northern and western territories acquired from the Germans with Polish occupants, including two million Poles repatriated from Polish lands annexed by the Soviet Union.[8] The Czech movement into the mountainous fringes of Bohemia evacuated by Germans was a more modest example of ethnically related new settlement. The East Germans also had to contend with a massive net loss of population to West Germany, estimated at three million, which counterbalanced the in-migration of some of the expellees from East Europe. Some large-scale international population transfers have occurred since the start of central planning and collectivization of agriculture, including: the movement of a large number of Turks from Bulgaria; the temporary international labor flows within the COMECON Bloc; and the employment of one million Yugoslavs in Western Europe. However, the basic migrational concern in the centrally planned economies is on the internal movement of people in relation to the objectives of regional economic development.

INDUSTRIAL TRANSFORMATION AND THE COLLECTIVIZATION OF AGRICULTURE

The strategy of development adopted by the centrally planned economies has resulted in the assignment of the highest priority to rapid capital-goods industrialization and also has led to sweeping organizational and socioeconomic changes in the agricultural sector. Migration has played a crucial role in the attainment of the objectives of this developmental path and, in turn, has posed increasingly complex problems for diverse aspects of the planned process of development.

Clearly, the most important contribution of migration has been the supply of workers to meet the growing labor demands of the industrial enterprises in cities. The combination of an accelerated pace of industrialization and the liberation of a substantial part of the underemployed agricultural labor force after collectivization and farm mechanization resulted in a massive rural outflow to the burgeoning cities and this has continued to the present. In general, the combination of rural migration, the legal redefinition of villages into towns and the urban annexation of nearby villages has accounted for two-thirds to three-fourths of the total growth of the urban population in most of the centrally planned economies of Europe.

In addition to the better jobs and the higher salaries offered by urban industries, the exodus of villagers also seems to have been motivated by the perceived sociocultural amenities of cities and the psychological impact of the glorification of the industrial worker. The negative aspects of village life associated with the turbulence of the first waves of collectivization and the disparities in the level of rural life compared to that of cities also were important components of these decisions. This movement was not planned in an explicit manner and within a relatively short time, the rate of the urban influx from the countryside seriously taxed the labor-absorbing capabilities of many of the leading cities and induced a wide range of urban problems resulting from overcrowding. In an effort to control the imbalance between urban demands for labor and rural supplies, restrictions on migration to large cities were widely imposed.

An evaluation of the migrational effects during this phase of development can be enhanced by an examination of the impact of migrational change on different types of settlements or regions. In the villages, the large-scale exodus, primarily of young males, was roughly similar in magnitude to the level of natural demographic increases. During the early years of collectivized agriculture the total rural population remained relatively stagnant and substantial surpluses of farm workers persisted in most regions. The harmful effects of the departure of rural youth did not become a major planning issue until a subsequent stage of the migrational process.

Rural-urban migration has been marked by an extremely pronounced distance-decay effect resulting from migrant flows to local small towns when employment opportunities were available and to nearby large cities even if they were not. In this context, distance has been a surrogate indicator of information availability, proximity to friends and relatives, and the advantages of a known environment, rather than a measure of transport accessibility. The early patterns of movement are reinforced on a long-term basis by the establishment of a network of personal communication between the initial urban migrants and potential movers in the countryside, with information flowing essentially toward the villages followed by the reverse flow of urban-bound villagers along the same channels. Nearness to friends and relatives plays a major role in the migrational structure of any society. This factor has been intensified in centrally planned economies where quite often the only possibility of moving to a large city involves sharing an apartment or house with an established urban resident.

The small towns became learning centers in a stage-migration process which started in the villages and eventually terminated in large cities. The acquisition of nonagricultural labor skills in small towns enabled former peasants to find positions in larger urban areas and break through the administrative barriers to large city migration. This learning function, however, was limited in scope by the low level of industrialization in small towns.

The primate cities and other heavily populated urban settlements strengthened their role as the preeminent growth centers for the nation as a whole and its leading regions after the start of centralized industrial development. Despite legal restrictions on their growth, they have continued to dominate the urban hierarchy. In this role, they serve as polarized centers of attraction for migrants from villages and small towns. These large cities have a particularly high percentage of their population in the labor-force age bracket because of the dominance of young migrants moving to cities for work and education. As noted above, the heavy influx of migrants responding to the economic and cultural opportunities in large urban centers occasionally exceeded the availability of jobs and usually created extraordinary strains on the housing supply, communal economy and other urban services. In the early 1960s, 40% of the new migrants to Bulgarian towns, primarily village women, were unable to find employment.[9] To deal with this problem, migrational controls were imposed on every Bulgarian town regardless of size. But even these extensive restrictions have had only limited success in controlling urban in-migration.

On a regional level, the phase of the migrational process associated with industrialization and collectivization made major strides toward relieving the problem of large-scale rural overpopulation in densely settled agrarian regions by diverting a large number of villagers to cities. However, the strong deterrence of distance resulted in a rate of progress toward the resolution of the problem of agricultural underemployment which was sharply differentiated regionally. Major gains in this respect occurred in regions near expanding growth centers, such as the Moscow area in which approximately 50% to 60% or urban in-migrants since 1926 have originated locally, and only modest changes were realized in peripheral regions.[10] Many areas in Eastern Europe and the Soviet Union have retained their agrarian nature because of a limited level of local urban employment opportunities. This low degree of rural mobility also has been characteristic of a rural population which is ethnically distinct from the majority of the local urban population. For example, the villages of Soviet Central Asia are inhabited by Turkic (or Iranian) peoples and the largest cities are predominantly Slavic. A similar situation occurs in the rural Turkish regions of Bulgaria. In these cases, the concept of distance should be expanded to encompass an ethnic dimension.

State intervention in the migrational process after the start of centralized planning was most evident in the settlement of frontier regions. This movement involved mass resettlement programs or major investments designed to develop the resources of these regions. The U.S.S.R. relied on patriotic exhortations to encourage the short-term migration of Komsomol construction workers to Soviet Siberia. The most successful example of permanent frontier-region settlement, however, was the occupation of the farmlands and industrial towns evacuated by the Germans in the northern and western borderlands of Poland.[11]

Although problems of excessive crowding in large cities, housing shortages, and low rural incomes persisted or emerged during the migrational phase associated with early industrialization and collectivization, the positive role of internal migration was far more evident than the negative aspects. The major attainments were the achievement of important developmental goals resulting from the transfer of part of the rural population to cities. These include: (1) the expansion of the industrial work force in cities by mass out-migration from villages and the impressive increases subsequently in the technical skills of this labor; (2) the striking gains in the levels of education, medical care, and cultural attainments associated with urban life (and despite urban overcrowding); and, (3) an easing of the pressing socioeconomic problems of rural overpopulation which had been major deterrents to development in the past.

MIGRATION IN CENTRALLY PLANNED INDUSTRIALIZED SOCIETIES

During the most recent phase of the migrational process most of the centrally planned economies in Europe have become highly industrialized and interurban movement gradually has become a leading type of migrant flow. For the most part, the positive (and negative) features of the preceding phase have continued. However, a new set of interrelated migrational problems has emerged as the developmental tasks embarked upon by the planned economies have become more complex and as the attainment of individual preferences has gained preeminence over the fulfillment of centrally determined migrational objectives.

The migrational changes during this period have had a major impact on the development of different types of settlements and problem regions. In the countryside, the continuation of massive out-migration, amounting to an annual average of 1.6 million villagers in the U.S.S.R., has resulted in the virtual evacuation of some rural areas by the young.[12] Sample surveys of actual and potential rural migrants provide a profile of the typical migrants indicating they are the brightest and best of rural youth.[13] They are the most ambitious, the most highly educated and technically competent, and the most active in the village Komsomol and social organizations. In contrast to one of the common assumptions of mover-stayer differences, the least likely migrants among village youth are those who do not participate in any social or political organizations. Even when young villagers have received special training in farm machinery use or are employed in non-agricultural village activities in industry, administration or service, they have a particularly high propensity to migrate. In some regions of the Soviet Union, over 90% of the graduates of tractor schools have moved to cities where their rural training has enabled them to obtain industrial jobs requiring mechanical skills.[14]

The surveys reveal that the attractiveness or pull of socioeconomic and cultural opportunities in cities provides the dominant motive for rural out-migration and purely economic incentives play a progressively smaller role in these decisions. The single most important reason inspiring village youth to leave, uncovered in a Bulgarian survey of 10,000 young villagers, was the greater opportunity for marriage in cities, which reflects, in part, village sex imbalances derived from an earlier preponderance of male out-migration.[15] In addition, frequent mention is made of the push factors based on the inadequate level of social and cultural life in villages, the hard work on farms with lower remuneration than industrial employment, and the relatively limited number of opportunities for women to acquire the most attractive village jobs.

This continued exodus is quite beneficial to the lagging agrarian regions which still have a substantial surplus of rural population. However, a pronounced quantitative deficit in the agricultural labor force has emerged in some of the regions which have experienced heavy migrational losses, such as most of the Slavic farming areas of the U.S.S.R. and Bulgaria. The problems induced by the volume of rural labor losses in these regions have been intensified by an apparent decline in the quality of agricultural labor stemming from the sharp increase in the average age of this work force. The mechanization, modernization, and reorganization of agriculture necessary for the attainment of economic goals could best be implemented by the technically proficient and ambitious young people who have left the countryside.[16]

Another serious problem has been the demographic consequences of the rural migrational losses. The villages are aging at a rapid rate and partially as a result of this, the number of rural births has declined precipitously. A widespread trend is that net fertility rates have declined most sharply for women in the older brackets of the child-bearing ages and women in this category have become increasingly dominant in the village age structure. In some countries such as Bulgaria, the crude birth rate in the villages already is below that of cities even though rural net fertility rates are higher than urban ones in nearly all age groups.[17] The aging of the rural population also has led to a recent upsurge in crude mortality rates which had declined earlier because of major health improvements in villages and a greatly reduced rate of infant mortality.

The decline in the natural population increases of villages derived from these demographic changes has been an increasingly common characteristic of the centrally planned economies. The pace of rural out-migration has been so great in Bulgaria, once a classic example of rural overpopulation, that in one-half of its 28 administrative regions (okrugs) rural death rates are now higher than the corresponding rates of birth.[18] The indirect demographic losses in villages have not been counterbalanced as a rule by comparable population gains in the cities inasmuch as the young rural migrants to towns quickly acquire urban attitudes toward family size. Thus, the persistence of an intensive level of rural-urban migration despite a shrinking and rapidly aging village population base not only has affected the agricultural labor force but also contributed to a declining rate of reproduction of the national work force.

In recent years, birth incentives by the state have been increased and, also, migrational policies in many of the planned economies have been revised to encourage the retention in villages of technically qualified rural youth. Unfortunately, few large-scale successes in this respect have been recorded. The countries which have had the most stable level of rural population, such as Poland, Hungary, and Czechoslovakia, have been those in which a large percentage of the village population commutes to urban industries. In addition, the reversion in Poland and Yugoslavia, to a system of agricultural organization in which private farms are dominant also seems to have slowed the rate of rural out-migration. In the agrarian Albanian state, which is in an early phase of plannned development, the rural population is still increasing rapidly.

A conspicuous exception to the policies designed to keep young people on the farms is the continued planning goal to relieve the rural population pressure of lagging regions by spreading out industrial investment from the core regions to the growth centers and other settlements of peripheral areas. These measures have provided urban jobs for a large segment of the underemployed agricultural population in almost every centrally planned economy. However, as noted above, these job-creating measures have been unable to overcome certain deeply rooted ethnic and cultural barriers to rural mobility. In Soviet Central Asia, the rural population has been growing at a prodigious rate because birth rates are as high as those of lesser-developed nations and crude death rates are lower than those of advanced industrial societies.[19] Yet, a substantial expansion of jobs in local urban industries often has resulted in an intensification of the long-distance in-migration of Russian workers to fill these positions rather than increased intraregional rural-urban migration. In most of the centrally planned economies, the leveling out of the regional inequalities in development remains as only a partially attained objective.

Another migrational problem which has become increasingly apparent in recent years is the serious developmental impediment resulting from the inability of the U.S.S.R. to provide a permanent work force in Siberia large enough to meet the growing labor demands of this massive frontier region. The rapid pace of capital-goods industrialization and the continuing depletion of energy and ore supplies in European Russia have increased substantially the national reliance on the abundant resources of the East.[20] The settlement programs in the eastern frontier lands also might reflect the desire of the Soviet government to provide a large population in the areas of the U.S.S.R. claimed by China. Despite these impelling motives, Siberia has experienced a net loss of migrants recently because of an incredibly high rate of labor turnover derived from the return of migrants to European Russia after a brief stay in this cold environment.[21] Siberia and the Soviet Far East have a chronic labor deficit, estimated at two million to three million workers. Nonetheless, Siberia had a net migrational loss of 920 thousand people during the 1959-1970 intercensal period and in the Soviet Far East, which has much higher wages and bonuses for migrants than Siberia, the net migrational gains in this period were only 143 thousand.[22] To the dismay of Soviet planners, the migrants from labor-scarce Siberia often move to the labor surplus areas of the Soviet sunbelt. Particularly heavy inflows have occurred in the Northern Caucasus and the Ukraine, which have experience net migrational gains from 1954 to 1970 amounting to almost one-million people in the former region and nearly one-half million in the latter.[23] In a sense, Yugoslavia has experienced a similar kind of triangular migration inasmuch as a large percentage of the returnees among the Yugoslav citizens temporarily employed in Western Europe use their savings to build houses in the most developed areas of the country around Beograd and Zagreb or along the Dalmatian Coast rather than return to their town or village of origin.[24]

In the more advanced phase of industrialization in the centrally planned economies, the earlier legal restrictions against settlement in the largest cities designed to relieve overcrowding and temporary unemployment have been extended and reinforced. A wide range of economic measures has been introduced involving investment prohibitions in large cities and spatial reallocations of funds for new industrial development in small and medium-sized towns.[25] Moreover, a great deal of attention in planning literature has been devoted to defining the optimal size of cities, usually determined to be between 50 thousand and 300 thousand inhabitants, and new satellite cities have been created in urban agglomerations to reduce the dominance of central cities.[26] Despite these efforts, the difficulties associated with urban overcrowding and chronic shortages of housing and services have persisted and have been supplemented by newer problems, including the expansion of the magnitude and distance of commuting and the environmental impact of intensive urban industrialization.

By contrast to the earlier period, the phenomenon of temporary urban unemployment has been replaced by a pervasive shortage of labor in large cities. The combination of burgeoning demands for labor, wartime population losses, and, even more, the accelerated pace of the demographic transition has generated this increasingly severe urban labor deficit in the most industrialized planned economies. In Moscow, for example, the crude birth rate in 1973 was only 12.4 and the death rate was a high 10.0, reflecting the growing proportion of old people in the population.[27] The extremely low rate of natural increase of 2.4 makes it virtually impossible for the city to supply its own labor force in the future, despite planning directives to that effect. Because of unfilled labor needs, the largest cities and their suburbs have had to expand their migrational intake to meet planned economic assignments. Another approach used to mitigate the labor shortage in large cities has been the diversion to socialized-sector employment of a large share of the female population previously engaged in household or other personal work as well as a sizable part of the pensioneer age group.

The same motive has led to a substantial expansion of industrial construction in selected medium-size and small towns and, on a more limited scale, in some central villages in order to utilize both their reserve labor supplies and existing housing stock.[28] As a rule, only a relatively modest percentage of these smaller settlements received large-scale industrial investments under these programs. For example, the Soviet Union embarked upon an ambitious project to expand industrialization in small and medium-size towns in the 1966-1970 period. However, only 265 settlements out of a total of several thousand in these size categories were involved in the final planning for this program.[29] Many of the small towns in the planned economies are in the backwash of large cities and local administrative centers. Even the role of small towns as transitional learning centers in rural-small town-large city stage migration appears to be diminishing as the higher level of rural education and technical training has increased the radius of initial migration by encouraging rural youth to move directly to large cities. The small towns which have had the most persistent growth and have the brightest prospects for development are those in the agglomerations around major cities.

The complex problems which have emerged during the most recent migrational phase are, to a considerable extent, a result of the inability of the earlier planning approaches to migration to be transformed into viable policies capable of resolving the dilemma of individual preferences for migrational paths in directions contrary to ostensible national objectives. An examination of the methods used in attempting to resolve these problems in migrational planning should provide some insights into the difficulties of this task.

METHODS OF MIGRATION AND SETTLEMENT CONTROL IN CENTRALLY PLANNED ECONOMIES

The basic purpose of planning intervention in the migrational process is to guide the movement of people into channels which are consistent with national interests as perceived by planners. The attainment of this objective is an essential component of planning efforts to control the growth of the hierarchy of settlement and to resolve the problem of regional development.

Positive results have been achieved with respect to some of the national goals in the last migrational phase. For the most part, however, the migrational objectives of the centrally planned industrialized societies have not been met, in contrast to the impressive attainments resulting from the transfer of a large segment of the underemployed rural labor force to urban industries during the preceding phase of planned development. A closer look at the measures adopted by planners to control migrational flows will be followed by an examination of some of the reasons underlying the limitations of these approaches.

Some of the measures designed to control migration have been directed toward types of settlements and have been used either to generate or discourage movement. One of the approaches in the former category has involved the creation of new settlement forms or the alteration of older types of occupance. In the rural areas, many of the central villages of collective and state farms have been transformed into agricultural cities providing urban-type services and nonagricultural employment opportunities. This program also conforms to the tenet of Marxist ideology concerning the elimination of differences between the towns and the countryside.

Simultaneously, a large percentage of the numerous, small subsidiary villages are being allowed to wither away. This policy has been linked to the persistent consolidation of collective farms which has been carried out to such a degree in Bulgaria that the entire countryside now has fewer than 600 collective and state farms.[30] The central villages of these enormous farms are designed to serve as polarized rural growth centers draining off the population of the small villages.[31] The urbanization of the largest villages and the creation of rural jobs in nonagricultural sectors will be even more evident when a program grouping these farms into new administrative units, known as agricultural-industrial complexes, diffuses throughout the Communist Bloc of nations. This reform started in Bulgaria and has spread in more subdued form to the U.S.S.R. and other centrally planned economies. The anticipated migrational effect is that the urbanization of the central villages will deter the exodus of ambitious and technically qualified rural youth.

To some extent, the large number of specialized new towns created on the basis of a major new industry, administrative function, or transport role have attracted an influx of workers to small and medium-size settlements. A similar effect involving an elite urban population has been derived from the planned decentralization of scientific research in the U.S.S.R. The creation of many specialized science cities, including the famous Siberia Academy of Science city (Akademgorod), has induced a numerically small but important migrational flow. Most of the Soviet science cities, however, are located in the vicinity of Moscow and have contributed only to a small-scale, local redistribution of population. A more important contribution to the reduction of population pressure in the largest central cities has been the planned development of satellite cities, usually within a radius of 50 kilometers or less from the center.

Another planning approach to migrational control has been the preferential allocation of industrial investment to certain types of settlements in order to deter out-migration or to attract new migrants. The primary urban targets have been the small and medium-size cities designated as viable sites for major new industries. This has been a pervasive objective or regional planning. In the Eighth Five-Year Plan of the Soviet Union (1966-1970), 40 percent of the funds for new industrial investment were scheduled to be allocated to these smaller urban centers and 80 percent of all new industries were assigned to these small settlements.[32] A similar priority was established in Bulgaria. Accompanying these programs have been efforts to create new food-processing and building materials industries directly in central villages. In the case of Bulgaria, new village industries producing textiles, light chemicals, and diversified machinery also have been constructed.

The difficulties of establishing large industries in small settlements, which usually lack a developed infrastructure, have led to a recent shift in planning emphasis toward diverting sections of existing industries in large cities to nearby small towns and central villages. This program is designed to utilize the labor and housing supplies of smaller centers and, also, to derive the benefits of economies of scale without large-scale outlays for administrative overhead. These savings and the important social benefits, presumably, more than counter-balance the extra costs of coordination and labor training. To a considerable degree, the policy of establishing direct industrial linkages to small settlements offers an important way of increasing the spread of industrialization benefits from regional growth centers, which hopefully will counteract the polarization effects of investment in these centers.[33]

At least as important as the aggregative policcies oriented toward promoting migration or retarding out-migration have been the measures designed to discourage movement to certain categories of settlements and regions. Foremost among these approaches are the prohibitions imposed on new industrial construction in the largest cities. In the U.S.S.R., one of the earliest efforts to reduce the concentration of industries in the major urban areas was a little-heeded 1931 decree calling for all new industries to be located in the countryside. A more serious effort was made in 1939 to ban industrial construction in the major cities of European Russia. As discussed above, the list of cities with this type of developmental restriction has been expanded in the postwar era to include some of the largest cities in Siberia as well.[34] In some countries, some major industries have been evacuated from the largest cities, including Moscow and Sofia, although in many cases these enterprises only have been moved to suburbs in the same urban agglomeration.

All the migrationally related policies directed explicitly toward types of settlements have a strong indirect effect on individual migrational decisions. In addition, some of the most important planning devices used to control migration are incentives and disincentives designed to influence individual migrational choices directly. A wide range of measures has been used to guide migrational flows into directions viewed as desirable by planners. One of the most important of these is the use of material incentives. The most comprehensive set of these incentives has been applied by planners in the U.S.S.R. to encourage migration to Siberia, the Soviet Far East, and the regions north of the Arctic Circle. Among the specific measures are: regional pay supplements scaled by the priority of employment sectors; different types of bonuses; reduced taxes for both industrial and rural migrants; extra vacation time; and, state-paid transportation for vacations in European Russia.[35] In agricultural regions, a significant narrowing of the difference between the income of individual farmers and industrial workers in nearly all the planned economies, particularly when income from private plots is included, has been motivated in part by migrational considerations.[36]

Another approach used to encourage migration is the use of psychological incentives in which the mass media promote diverse forms of movement and, particularly, relocation to frontier regions. The Soviet Union has been the primary advocate of this device from the creation of the steel city of Komsomolsk in the Soviet Far East and the outpouring of youthful volunteers to work in virgin lands of Kazakhstan to the present-day "social calls" to youth to journey to Siberia in order to work in new industries, construction or resource exploitation. Accompanying efforts to meet the labor needs of the frontier regions with volunteers, a substantial percentage of university graduates in the U.S.S.R. have to work in a region designated by the state for a period from two to three years and, thus, provide an important cadre of badly needed specialists in Siberia and the Far East.

In general, the policies directed toward impeding certain types of individual migrational choice have been far more pervasive in the centrally planned societies than those encouraging migration. The most universally restricted form of movement is migration to the largest cities and, most of all, to the primate cities. The combination of an internal passport system and required registration in areas of in-migration has facilitated the implementation of this type of migrational control. If a potential in-migrant were able to obtain both employment and housing in a city on the restricted list, however, he often could receive permission to reside in the city on either a permanent or temporary basis.

As mentioned above, the planning measures used to redirect migrational flows into socially desirable channels have had only modest success and, to a surprising extent, the effective control of migration has not yet been established.

REASONS FOR THE PERSISTENCE OF MIGRATIONAL PROBLEMS

The most important reasons why the migrational problems of the industrialized planned economies have not yielded to facile planning resolution can be traced to the nature of planning objectives and procedures, the inherent complexity of these problems and the increasing role of individual choice.

Conflict of Planning Priorities

One of the basic problems associated with the centralized planning of migration is the conflict of priorities. Industrialization has been the dominant concern of all the centrally planned economies and remarkable gains have been made in this respect. However, this goal is not always compatible with migrational objectives. The cumulative advantages of concentrating industrial investment in large cities are based on their developed infrastructure, skilled labor, and proximity to markets, industrial suppliers, and administrative agencies. Generally, these have been assigned greater weight than urban diseconomies of scale in the decisions of industrial ministries. In part, this reflects the substantial savings derived from the expansion of existing industries in large cities as opposed to new investment in smaller settlements. Longer construction times, labor-quality problems and higher expenditures on secondary investment seem to characterize industrial development in nonsuburban, small towns located away from the main channels of economic and administrative communication. These disadvantages often are greater than the quantitative labor and housing advantages in the smaller settlements. Moreover, many of the extra costs resulting from excessive urban congestion are difficult to determine and, in any case, are not borne explicitly by the industrial ministries. This apparent ministerial preference for investment in large cities has resulted in the circumvention of many of the migrational control measures designed to reduce large-city dominance and increase the industrial role of smaller towns. For example, the use of the instrument of prohibitions on industrial construction in large cities as a means of population control usually has been ineffective because these restrictions have not been applied consistently to the expansion of existing industries, which is the major source of growth in the industrialized economies. In addition, numerous examples can be found of high-priority new enterprises receiving special permission to locate in the largest urban centers on the industrially restricted lists. Moscow, for example, has added hundreds of factories despite persistent prohibitions since 1931 on new industrial construction.[37] To meet the labor demands generated by industrial expansion, massive numbers of workers are allowed to migrate to the largest cities on a temporary basis and their residence often becomes permanent either formally or in a *de facto* manner. About one-seventh of the population of Sofia only has temporary residence permits, which have been extended for over a decade in many cases.[38] The temporary population of Moscow numbers several hundred thousand people. A similar population characteristic occurs in the other primate cities of the centrally planned economies as a response to legal restrictions on permanent migration.

The declining pool of potential additions to the labor force can be attributed in part to the accelerated demographic transition resulting from the large-scale population transfers from villages to housing-scarce cities. The labor-supply problem is intensified by the chronic inability to reach the ambitious, planned levels of labor productivity in industry and agriculture. The net result is a deeply rooted conflict between the labor needs of industrial enterprises and many migrational objectives. In addition to the difficulties of limiting large-city growth because of ministerial perceptions of comparative costs, the incessantly expanding demand for industrial labor, which enterprises must obtain in order to fulfill production goals, and the shrinking local supplies have made the administrative restraints on migration even more untenable from the point of view of enterprise managers. In rural areas, the growing shortages of industrial labor also have contributed to the continuation of the massive outflow of village youth to cities from areas now experiencing a critical scarcity of technically qualified agricultural labor. On a regional scale, the expanding labor needs of the cities in the old, industrialized regions have tended to discourage out-migration to the resource-abundant frontierlands characterized by a serious deficit of labor.[39]

From a different perspective, the increasing difficulties or procuring labor in the leading cities of the highly industrialized, planned economies have had a positive effect on the shift of industries to small towns with reserve labor supplies and in spurring industrial investment in the regional growth centers of areas with a substantial surplus of agricultural workers. In many cases, this labor incentive has outweighed the initial cost disadvantages and labor-quality problems characterizing the industrialization of smaller settlements. In the U.S.S.R., however, the priority assigned to industrial investment in medium and small towns recently has been reduced substantially because the earlier reserves of untapped labor in the industrially viable smaller settlements have been virtually depleted.[40] Both the Soviet Union and Bulgaria have made impressive efforts to divert investment to growth centers in the lagging, agrarian regions but these programs have been only partially successful in transferring surplus agricultural labor into more productive industrial employment because of the resistance of rural ethnic groups to the economic incentives encouraging them to migrate to culturally and ethnically distinct cities.

Another result of the preeminence of industrialization objectives is the frequent failure to divert a large enough portion of investment funds into activities designed to improve substantially living conditions in areas of desired in-migration. In the example of Siberia, the limited allocations for housing have resulted in shortages much worse than in European Russia and the assignment of most single migrants to dormitories rather than apartments. A similar neglect of consumers-goods production in Siberia not only has resulted in chronic local scarcities but also in a much higher cost of living than in European Russia.[41] On the urban level, the construction of new industrial towns has focused on the dominant industry and remarkably little investment in these towns has been allocated for the development of communal services which could persuade migrants to remain.[42]

Centralization Problem

The basic developmental plans are formulated in the center by national planning agencies and economic ministries to implement Party and governmental directives. Although this process involves close consultation with regional branches of these ministries, subordinate enterprises, and units of territorial political administration, a limited role is given to urban and regional planning organizations. The inadequate development of urban planning is reflected in the absence of national ministries concerned explicitly with cities, the frequent lack of effective zoning or even city plans, and the limited authority of city planners, usually urban architects, in urban decision-making. The urban planners rarely can resist requests for expansion or other planning variances emanating from high-priority industrial sectors. The major migrational results of the inadequate level of urban planning are that local efforts to control in-migration or city size have had little success and city administrations, in general, participate in only a peripheral manner in the formulation of important urban policies.

Regional planning is linked to the existing system of territorial political administration. The planning organs of these administrative units concentrate their efforts on integrating regional economies into the national economic plan. Earlier efforts to decentralize spatial economic administration have been replaced by a ministerial recentralization in most of the centrally planned economies and the creation of new organizational forms of industrial administration, some of which involve intermediate spatial associations of industries under the jurisdiction of the same ministry.[43] One of the serious shortcomings of the centrally directed and hierarchically structured administrative network is the commonly encountered absence of intermediate macro-regions between the center and the units of spatial political administration.[44] Even in the U.S.S.R., which has large union republics performing the role of macro-regions, the immense size of the Russian Republic and the lack of large administrative regions acting as mediaries between Moscow and the local districts (oblasts or their equivalent) has had a debilitating effect on the development of regional planning. An approach used to create regional economic plans which are not confined to the existing political-administrative network has involved the delimitation of a large number of special-purpose planning regions encompassing new industrial complexes and other major projects.[45] Although this has been an effective device for short-term industrial planning, it is not an adequate substitute for a macro-region with comprehensive planning and administrative functions nor is the existing network of large statistical-reporting regions. This regional gap in the hierarchy of spatial planning has impeded the creation of effective policies of migration on a broad regional basis and has magnified the disparity between centrally conceived migrational objectives and locally realized attainments.

Coordination Difficulties

The short-term and long-range economic plans normally contain general directives related to the direct and indirect coordination of economic objectives, population policies, and urban or regional developmental strategies. However, these aspirations usually are not embodied in operational plans is an integrated manner which would help attain the level of coordination necessary for effective migrational planning. For example, the efforts to improve the economic, social, and cultural quality of life in regions seeking to attract migrants are hampered by the absence of comprehensive plans in this area.[46] As cited above, massive investments in new industries very often are not accompanied by a level of outlays for housing and amenities necessary to reduce the incredibly high rate of labor turnover in these industries. Even the regional wage supplements, which are scaled by sectoral employment priorities, are not coordinated adequately with regional variations in the costs of living. In some cases, real income actually is lower in the frontier regions than in the area of origin for migrants working in the relatively low priority industries, such as those producing consumer goods.[47]

In cities with migrational prohibitions, the industrial enterprises have created large-scale demands for urban labor which can be satisfied only by in-migration and the legal measures designed to control this movement bear little relationship to economic realities. In part, this problem reflects the extremely difficult task city councils have in coordinating the activities of local industries which are subordinated to higher ministerial authorities. Another migrational effect associated with the structure of planning is the failure of most investment plans to consider population variables. The heavy concentration of capital-goods investment in Siberia has led to a preponderance of jobs for men and only modest employment opportunities for women. This has reduced family income substantially and has encouraged a high return-rate for families and single males. An employment imbalance in the opposite direction in the textile cities of the Central Industrial Region has resulted in the mass exodus of males.

Perhaps, the most conspicuous example of a coordination problem is the frequently encountered failure to incorporate commuting explicitly into the planning process despite the extraordinary importance throughout Eastern Europe and the U.S.S.R. of journeys to work from suburbs and villages to central cities.[48] Commuting possibilities have a direct influence on the formulation of settlement policies, the migrational flows to metropolitan areas, and the deterrence of the exodus from villages and small towns. Migrational planning can ill-afford to neglect the commuting variable.

Inadequacies of Existing Planning Mechanisms

The centrally planned economies do not have comprehensive migrational plans which embrace a wide range of flow characteristics and objectives. The basic instrument for planning population movement is the method of labor balances, which is similar in form to the approaches used to plan commodity movement. The basic balances measure the supply and demand of labor in general. Planners also employ special labor balances which are differentiated by age, education, technical skills, and sectors of employment for the nation as a whole and for certain political-administrative units (but rarely for cities).[49] The resulting surpluses and deficits are balanced and then implemented by the organized recruitment of labor carried on by ministries and individual enterprises.

Many problems are inherent in this approach. These include: the difficulties of projecting accurately labor shortages and, even more, surpluses; the inadequate degree of sectoral and areal disaggregation; and, the complexity of translating the labor-balance data into directionally specified migrational flows. Although labor input-output tables depicting sectoral and intersectoral labor coefficients are available in most centrally planned economies, they play a minimal role in the labor-planning process apparently because of the numerous unresolved problems associated with the use of this tool in planning projections.

One of the most important shortcomings of the use of labor balances in migrational planning is the implicit assumption that the jobs available in labor-deficit regions will be filled almost automatically by migrants from regions which have an ostensible surplus of labor. The modest role of organized labor recruitment in migration, accounting roughly for 10% to 20% of total flows reflects both the inherent limitations of the labor-balance approach and the growing dominance of individual migrational choice.[50] Recent migrational experience has done little to instill confidence in the use of labor balances as the primary instrument of migrational planning. Many East European and Soviet scholars are calling for new approaches and more comprehensive planning of the spatial labor-redistribution process.

The Inherent Complexity of the Migrational Problems

The task of formulating migrational policies which are in the national interest or meet broad societal objectives is at least as difficult as the problem of implementing these goals. Virtually every migrational program related to the structure of settlement has evoked controversy. This is most evident with respect to efforts to check the growth of large cities.[51] The opponents of these policies in Eastern Europe and the Soviet Union have cited the preeminence of the economic, social, and cultural advantages of large cities and metropolitan areas compared to smaller settlements. These critics maintain that the legal restrictions on the growth of the most productive, technologically advanced, and innovative centers not only are unrealistic but, also, impede the attainment of national economic objectives. The economic benefits obtained by industries located in large cities, which were discussed in the preceding section, are reinforced by advantages associated with the concentration of educational and labor-training facilities, scientific research, and cultural amenities in the major metropolitan areas. The increasing reliance of the centrally planned economies on the expansion of existing industries and gains in labor productivity also tend to strengthen large-city dominance. From a different perspective, the advocates of large-city growth point out that continued movement to major metropolitan areas on a major scale clearly conforms to the wishes of an imposing segment of the mobile population. Many Western scholars in recent years, also, have stressed the economic desirability of continued industrial investments in major metropolitan areas.[52]

The critics of urban decentralization in the planned economies have noted that optimality in city size is an elusive concept which scarcely provides a reliable basis for the formulation of viable policies of settlement planning. In his study on the economics of city size, Richardson has cited the futility of Western efforts to determine optimal population sizes for cities.[53] He emphasized that urban population magnitude is not a meaningful variable to use in an investigation of the problems of city size. The crucial factor is urban density and the basic unit of analysis should be the urban agglomeration or metropolitan area.

The proponents of urban decentralization, including the overwhelming majority of planners, support the continuation of efforts to curb large-city growth, although, as a rule, they would prefer to accomplish this goal by major improvements in smaller cities or rural areas rather than by unenforceable, legal constraints on migration. Very often, the housing shortages and the strains on urban services are more acute in the large cities than in medium and small towns. The problems of the increasing distance and volume of commuting and related problems of traffic congestion also are more pronounced in the larger settlements. The environmental problems induced by the heavy concentration of industries in the major metropolitan areas are intensified in some large cities, such as Sofia, by physical constraints on expansion because of local topographic conditions. The advocates or urban decentralization also have pointed out that large urban complexes not only reflect pervasive societal problems but often magnify them. Moreover, Gilbert has noted that even the frequently cited superiority of the largest cities in economic productivity could be reduced substantially if medium-size towns had a comparable degree of development of their infrastructure and educational or training facilities.[54]

Many regional growth theorists agree that per capita municipal expenditures are particularly high in large cities. The Bulgarian economist, Zakhariev, has stressed that if these and other indirect costs, including those linked to environmental disruption, associated with diseconomies of urban scale were considered explicitly in the cost calculations of industrial ministries, scarcely any new industries would be created in the largest cities.[55]

Despite the enormous amount of attention devoted to the problems of city size, the question of whether a city is too large appears to be indeterminate at our present state of knowledge and, perhaps, always will be a source of contention. The same types of methodological and practical difficulties occur in designing other policies with important migrational implications. The corollary of the debates concerning large cities is the desirable economic role of smaller towns. On a regional level, the determination of investment priorities for developed regions, the low-income agrarian areas, and the resource-rich frontierlands often has been associated with conflicts among divergent interest groups and the task of uncovering consistent criteria for these decisions is not an easy one. For example, the migrational problems of Siberia might be less severe if there had not been strong competition for capital investment funds and labor from enterprises in European Russia. Western literature on the theory and practice of regional growth has been more effective at pointing out the complexity of this process than in offering solutions in general and, even less, in suggesting remedies which could be transferred to centrally planned economies.

The imposing problems of determining societal migrational preferences are intensified by the inherent contradictions in some of the measures used to attain policy objectives. In frontier regions, for example, the higher wages used to attract the most needed workers also provide the in-migrants with an opportunity to accumulate substantial savings in a short time which they use to return to their region of origin, or frequently to a more desirable location. To entice skilled workers to migrate from large cities, the organized labor contracts often specify that the migrants can sublease their apartments in the metropolitan area of origin. The right-to-return is viewed as an indispensable incentive by migrational planners. Yet, an obvious effect of this policy is to aggravate the serious problem of a high rate of return migration. To make matters worse, the surprising extension of large-city migrational prohibitions to the major metropolitan centers of Siberia has retarded movement to the most attractive regional sites of migration, which also have the best prospects for retaining migrants. Presumably, these large Siberian cities were attracting too many migrants from local villages and small towns.

Perhaps the most serious dilemma is the increased propensity to migrate created by the enhancement of job-training and educational opportunities in areas or settlements hoping to retard the exodus of their young population by these improvements. Many of the migrants in frontier regions acquire labor skills which enable them to procure employment in more desirable areas. The departure of the youthful segment of small towns, also, has increased because the creation of new local industries has given workers skills which are sought in the labor market of large cities. In the villages of the planned economies, the striking growth of general and technical education has increased the incentives to move to cities. A frequent rural complaint is that the overwhelming majority of graduates from tractor and other farm machinery schools use their mechanical expertise in urban industry rather than in lower-paid and less prestigious farm work. Moreover, the transformation of some central villages into agricultural cities seems to have given rural youth a taste of urban life which ultimately is satisfied by relocation to large cities. Although the impressive attainments in urban education have not resulted in out-migration, they have contributed to a persistent influx of students and, also, of low-skilled labor into cities to fill jobs which the educated urban population regards as undesirable.[56]

Individual Migrational Preferences and National Objectives

The system of direct governmental incentives for migration is aimed primarily at participants in the organized recruitment of labor. As noted above, less than one-fifth of total migration in most of the planned economies falls into this category or within the scope of the national economic plan. Moreover, the relative share of organized labor flow is declining precipitously.[57] Most of the migrational movement reflects individual choices which do not necessarily conform to national goals. In fact, the individual perceptions of migrational desire lines often are diametrically opposed to those of planners. The large-scale abandonment of small towns, villages in labor-deficit agricultural areas, and frontier lands by their highly mobile young population in favor of large cities and regions with mild climates contravenes planning directives. These unplanned flows bear witness to the difficulty of reconciling individual and state preferences. Still another type of individual travel behavior, which is usually not encompassed in national plans, is the massive expansion of commuting from small towns and villages.

East European and Soviet scholars are cognizant of the policy implications of this migrational dilemma. Most of them have rejected an earlier view that individual migration represents nothing more than movement which is planned indirectly by the portion of the economic plan treating the distribution of productive forces. Obviously, the location of new investment will have a major effect on individual migrational choices. However, the widespread availability of jobs has given migrants a broad range of directional possibilities and has allowed other variables to play leading roles in these decisions.

The growing recognition in the planned economies of the importance of individual spatial preferences and the impossibility of formulating effective policies of migration without consideration of this variable have led to the implementation of alternative developmental strategies. One of these alternatives has been to change economic plans to conform to the realities of individual migrational desires. For example, the investment plans for Siberia of necessity have had to concentrate on types of capital-intensive industrialization which generate limited demands for labor. With respect to the large-city problem, the frequent encouragement in practice of industrial growth in the suburbs of large metropolitan centers represents an implicit recognition of the migrational attractiveness of the large urban agglomerations in the established regions.[58] In a rural setting, the massive outpouring of Bulgarian village youth to cities and the general level of rural depopulation have contributed to a drastic transformation of agricultural settlement and administration which has made Bulgaria a model of the Socialist agricultural society of the future.

Some basic changes, however, would have to be made in the planning process if viable migrational policies, reconciling individual, and societal goals, were to be attained. A starting point should be the formulation of consistent migrational objectives, which are interwoven with economic, settlement and demographic goals and which deserve to be implemented. The unrealistic administrative restrictions on migration should be eliminated. These migrational objectives should be embodied in comprehensive plans of regional, urban, and rural development. Among other things, these plans should implement in a coordinated manner the changes which are necessary in these areal units to draw in the desired level of permanent migrants from regions in which out-migration is a planning objective or, in other areas, to make local conditions sufficiently attractive to dissuade the youthful component of their population from leaving.

The planning of the migrational process will offer only partial remedies to a complex set of interrelated problems until the current planning approaches are expanded and improved substantially. In this context, the increased concern in Eastern Europe and the Soviet Union for more effective migrational planning provides evidence that major strides will be made in the future to resolve the problems induced by migration.

CONCLUSIONS

Both the centrally planned and Western societies have recorded impressive successes in effecting important and relatively straightforward migrational changes associated with urbanization and industrialization, the mechanization of agriculture, and the basic settlement of frontier regions. Conversely, neither type of society has yet been able to resolve adequately the complex set of migrational problems associated with developed industrial states and the preeminence of individual migrational preferences over societal goals. Just as the centrally planned economies are attempting to incorporate individual migrational factors into their plans and projections, the strengths and shortcomings of their experience in the planning of the migrational process should be considered in the efforts of Western economies to formulate national programs to aid in the possible resolution of the developmental problems linked to migration.

NOTES

1. An informative survey of this literature can be found in the study by R. Paul Shaw, *Migration Theory and Fact* (Philadelphia: Regional Science Research Institute, Bibliography Series, Number 5, 1975).
2. T.W. Schultz, "Reflections on Investment in Man," *Journal of Political Economy* 40 (1962), pp. 51-58; L. Sjaastad, "The Costs and Returns of Human Migration," *Journal of Political Economy* 40 (1962), pp. S80-S93 and "The Relationship between Migration and Income in the United States, "*Papers, Regional Science Association* 6 (1960), pp. 37-64.
3. Julian Wolpert, "Behavioral Aspects of the Decision to Migrate," *Papers, Regional Science Association* 15 (1965), pp. 59-169.
4. Some of the best general examples of this work are by Niles Hansen, ed. *Growth Centers in Regional Economic Development* (New York: The Free Press, 1972) and A. Kuklinski, ed. *Growth Poles and Growth Centers in Regional Planning* (The Hague: Mouton, 1972).
5. Shaw, *op. cit.*
6. The planning implications of demographic changes and migration in the GDR, Czechoslovakia, Poland, and Hungary are examined in an excellent paper by George Demko and Roland Fuchs, "Demography and Urban and Regional Planning in Northeastern Europe," presented at the Conference on Demography and Urbanization in Eastern Europe, U.C.L.A., February 6-9, 1976. (Boulder, Westview Press, forthcoming).
7. The magnitude of this problem in Eastern Europe prior to World War II is documented by Wilbert Moore, *Economic Demography of Eastern and Southern Europe* (Geneva: League of Nations, 1945).
8. L. Kosinski, "Migrations of Populations in East-Central Europe from 1939-1955." *Geographica Polonica* 2 (1964), pp. 123-137 and "Les Problemes Demographiques dans les Territories Occidentaux de la Pologne et les Regions frontieres de la Tchecoslovaquie," *Annales de Geographie* 71 (1962), pp. 79-98.
9. Minko Minkov, *Migratsiya na Naselenieto* (Sofia: Partizdat, 1972), pp. 110-116.
10. B. Khorev, "Osnovnyye Sdvigi v Rasselenii Naseleniya v SSSR i Nekotoryye Aktualnyye Problemy Migratsionnoy Politiki," in *Narodonaselenie* (Moscow: Statistika, 1973).
11. Kosinski, "Les Problemes Demographiques."
12. B. Khorev, *Migratsionnaya Podvizhnost' Naseleniya v SSSR* (Moscow: Statistika, 1974), p. 52.
13. Mincho Semov, *Potentsialnite Preselnitsi* (Sofa: Narodna Mladezh, 1973); V. Staroverov, *Gorod ili Derevyna* (Moscow: Partizdat, 1972); T. Zaslavskiy, *Migratisiya Sel'skogo Naseleniya* (Moscow: Mysl', 1970); Yu. Arutunyan, *Sotsial'naya Struktura Sel'skogo Naseleniya SSSR* (Moscow: Mysl', 1971); and, Elzbieta Iwanicka-Lyra, "Changes in the Character of Migration Movements from Rural to Urban Areas in Poland," *Geographica Polonica* 24 (1972), pp. 571-580, provide important examples of this type of survey.
14. Zaslavskiy, *op. cit.* pp. 208-214.
15. Semov, *op. cit.*, p. 41.

16. Robert Taaffe, "The Impact of Rural-Urban Migration on the Development of Communist Bulgaria." Paper presented at the Conference on Demography and Urbanization in Eastern Europe, U.C.L.A., February 6-9, 1976. (Boulder: Westview Press, forthcoming).
17. *Demografiya na Bulgariya* (Sofia: Nauka i Izkustvo, 1974) and *Naselenie* (Sofia: Tsentralno Statistichesko Upravlenie, 1973 and 1974).
18. *Naselenie: 1974.*
19. *Naselenie SSSR: 1973* (Moscow: Statistika, 1975), pp. 73-83.
20. *Problemy Razvitiya Vostochnykh Raionov SSSR* (Moscow: Nauka, 1971).
21. For detailed discussions of the migrational problems of Siberia, the Far East, and the Arctic areas see: V. I. Perevedentsev, *Migratsiya Naseleniya i Trudovyye Problemi Sibiri* (Novosibirsk: Nauka, 1966); A. Maikov, ed. *Narodonaselenie i Ekonomika* (Moscow: Ekonomika, 1967); A. Maikov, ed. *Migratsiya Naseleniya RSFSR* (Moscow: Statistika, 1973); L. Rybakovskiy, *Regionalnyy Analiz Migratsii* (Moscow: Statistika, 1973); and, B. Khorev, *op. cit.*
22. B. Urlanis, ed. *Narodonaselenie Stran Mira* (Moscow: Statistika, 1974), p. 411.
23. Ibid.
24. Ivo Baucic, "Regional Differences in Yugoslav External Migration," paper presented at the Conference on Demography and Urbanization in Eastern Europe, U.C.L.A., February 6-9, 1976 (Boulder: Westview Press, forthcoming).
25. In his article "Urbanization in East-Central Europe after World War II, "*East European Quarterly* 8 (1974), pp. 129-154, Leszek Kosinski estimates that medium-size cities in Eastern Europe will grow at the most rapid rate during the 1970s.
26. A review of the debate in Eastern Europe and the U.S.S.R. concerning optimal city size is provided by B. Khorev, *Problemy Gorodov* (Moscow: Mysl', 1975), pp. 167-178.
27. *Naselenie SSSR: 1973,* p. 97.
28. A wide range of views exists concerning the population magnitude of small and medium-size cities which are discussed in *Problemy Gorodov.* p. 147. The state planning and construction committees (Gosplan and Gosstroy) in the Soviet Union, however, group all urban settlements with fewer than 100,000 inhabitants into the category of small and medium-size cities without any differentiation between these urban classes. *Naselenie, Trudovyye Resursy SSSR* (Moscow: Mysl', 1971), p. 120.
29. *Naselenie, Trudovyye Resursy SSSR,* p. 124.
30. *Statisticheskiy Yezhegodnik: Stran-Chlenov CEMA* (Moscow: Statistika, 1975), p. 218.
31. Taaffe, "The Impact of Rural-Urban Migration on the Development of Communist Bulgaria," p. 25.
32. *Naselenie, Trudovyye Resursy SSSR,* p. 124.
33. Robert Taaffe, "Urbanization in Bulgaria," *Etudes Balkaniques,* III (1974), pp. 50-63.
34. A list of Soviet cities on the restricted list can be found in the official Gosplan handbook for preparing national economic plans: Gosplan, *Metodicheskie Ukazaniya k Razrabotke Gosudarstvennykh Planov Razvitiya Narodnogo Khozyaistva SSSR* (Moscow: Ekonomika, 1974), p. 293.
35 Zh. Zaionchkovskaya and D. Zakharin. "Problemy Obespecheniya Sibiri Rabochey Sili," pp. 43-61 in *Problemy Razvitiya Vostochnykh Raionov SSSR.*
36. *Statisticheskiy Yezhegodnik,* pp. 405-408.
37. Iu. Saushkin, *Moskva* (Moscow: Mysl', 1964), pp. 110-132.
38. Minko, *op. cit.,* pp. 110-116.
39. The migrational changes in the Ukraine are discussed by V. Onikienko and V. Popovka, *Kompleksnoe Issledovanie Migratsionnykh Protsessov* (Moscow: Statistika, 1973).
40. *Problemy Razmeshcheniya Trudovykh Resursov,* pp. 6 and 261.
41. G.V. Mil'ner, "Voprosy Sovershenstvovaniya Territorial'nogo Planirovaniya Urovney Zhizni," pp. 98-152 in *Problemy Ratsional'nogo Ispolzovaniya Trudovykh Resursov,* edited by A. Maikov (Moscow: Ekonomika, 1973).
42. *Problemy Razmeshcheniya Trudovykh Resursov,* p. 75.
43. A discussion of some of these organizational changes in the U.S.S.R. can be found in the study by Brenton Barr, "The Changimg Impact of Industrial Management and Decision-Making on the Locational Behavior of the Soviet Firm," pp. 411-446 in *Spatial Perspectives on Industrial Organization and Decision-Making,* edited by Ian Hamilton (New York: John Wiley, 1974).
44. V. Pavlenko, *Territorial'noe i Otraslevoe Planirovanie* (Moscow: Ekonomika, 1971) examines the Soviet approach to regional planning which has spread to many of the Communist nations of Europe. Discussions of regional planning in Eastern Europe can be found in studies by George Hoffman, *Regional Development Strategy in Southeastern Europe* (New York: Praeger Publishers, 1972) and Kosta Mihailovic, *Regional Development Experiences and Prospects in Eastern Europe* (The Hague: Mouton, 1972).
45. *Problemy Razmeshcheniya Trudovykh Resursov,* pp. 254-268.
46. Mil'ner, *op. cit.,* p. 105.
47. Ibid., pp. 98-104.

48. Among the numerous studies on this theme are: Yu. Pivovarov, "Commuting as an Aspect of Population Geography in Socialist Countries," pp. 73-76 in *Recent Population Movements in the East European Countries,* edited by Bela Sarfalvi (Budapest: Akademiai Kiado, 1970); G. Hadhazi "The Role of Commuting in The Economy of Budapest and the Method of Statistical Survey," pp. 111-118 and M. Klimczyk, "Methodological Problems of Commuting Statistics in Poland," pp. 140-149 in *Statistics in Urban Planning,* edited by Ken Williams (London: Charles Knight, 1973); and *Migratsionnaya Podvizhnost' Naseleniya SSSR,* pp. 109-122.
49. Some of the major studies of the labor-balance approach include: B. Breev and V. Kryukov, *Mezhotraslevoy Balans Dvizheniya Naseleniya i Trudovykh Resursov* (Moscow: Nauka, 1974); V. Kostakov and P. Litvyakov, *Balans Truda,* 2d ed., (Moscow: Ekonomika, 1970); E. Ruzayeva, *territorialnyye Balansy Trudovykh Resursov* (Moscow: Moscow University, 1967); and, G. Tsvetkov and A. Bairyamov, *Problemi na Balansoviya Metod v Selskoto Stopanstvo* (Sofia: Zemizdat, 1965).
50. Yu. Mateev, "Organizovannyy Nabor kak Odna iz Osnovnykh Form Planovo Razpredeleniya Rabochey Sili," pp. 62-78 in *Migratsiya Naseleniya RSFSR.*
51. Discussions of this controversy are found in: *Problemi Gorodov,* pp. 167-178 and 388-395; G. Konrad and I. Szelenyi, *The Future of Rural Communities* (Budapest: Akademiai Kaido, 1972); and N. Ulianova and V. Ushkalov, "Opit Issledovaniya i Regulirovaniya Rosta Gorodov v Polskoy Narodnoy Respubliki, pp. 111-120 in *Narodonaselenie: Prikladnaya Demografiya.*
52. Examples of this are: William Alonso, "The Economies of Urban Size," *Papers, Regional Science Association* 26 (1971), pp. 67-83; I. Houk, "Income and City Size," *Urban Studies* 9 (1972), pp. 299-328; K Mera, "On the Urban Agglomeration and Economic Efficiency," *Economic Development and Cultural Change* 21 (1973), pp. 309-324; Harry Richardson, *The Economics of Urban Size* (Lexington: Lexington Books, 1973); and, Lowden Wingo, Issues in a National Urban Strategy for the United States," *Urban Studies* 9 (1972), pp. 3-38.
53. Richardson, *op. cit.*
54. Alan Gilbert, "The Arguments for Very Large Cities Reconsidered," *Urban Studies* 13 (1976), pp. 27-34.
55. I. Zakhariev and I. Kozhukorov, *Naselenie i Trudovi Resursi v Teritorialnite Edinitsi na NR Bulgariya* (Sofia: BAN, 1973), pp. 164-172.
56. *Problemy Razmeshcheniya Trudovykh Resursov,* p. 58.
57. In the Soviet Union, 22 million workers were recruited by organized methods from 1930 to 1950 as compared to only 5.6 million between 1951 and 1970. Moreover, the decline is continuing. In the 1951-1955 period, 2.8 million migrants were in the organized category whereas the number of participants in the organized-recruitment programs dropped to just 571,000 in the recent five-year interval from 1966 to 1970. Mateev, *op. cit.,* p. 65.
58. *Problemy Gorodov,* pp. 307-323 and Ulyanova and Ushkalov, *op. cit.,* pp. 113-115.

Behavioral Geography

Introduction

The development of behavioral geography may have gained much of its early vigor from dissatisfaction with the unrealistic specifications of decision making incorporated in traditional models of spatial structure and from the failure of those models to provide satisfactory explanations for the behavior of individuals. Once under way, however, behavioral geography quickly shed its reactionary character and began to broaden the scope of geography by identifying new areas of geographic research as well as providing new approaches to traditional problems. Reginald Golledge points out, in the first article in this section, that behavioral geography is distinguished by its concern for process-oriented explanations of spatial behavior as opposed to structurally oriented explanations. The more traditional structurally oriented approaches relate overt acts of behavior to the characteristics of individuals and the characteristics of their environments. The emphasis in behavioral geography shifts toward the processes responsible for acts of behavior, including processes such as learning and attitude formation which are not readily observed in conjunction with overt behavior. These process-oriented approaches can provide more satisfactory explanations of spatial behavior but they also expand the range of geographic research into areas as cognitive mapping, attitude formation, and models of decision making.

The importance of attitudes and attitude formation in policy-oriented research is demonstrated in the paper by John Jakubs. The success of ghetto-dispersal policies and other controversial urban policies depends heavily on individual and community attitudes since the anticipation of negative impacts may be self-fulfilling. For example, fears of a decline in housing values may bring on panic selling which does lead to a decline in housing values. Knowledge of relevant attitudes is necessary in order to anticipate these impacts and to design strategies which may minimize the impacts. For the low-income surburban housing projects analyzed by Jakubs it is especially important to identify the specific negative impacts which residents anticipate, such as impacts on school quality, property values, or visual appearance, but reliable information on these attitudes is notoriously difficult to obtain. Surveys are difficult to conduct once a neighborhood is involved in controversy over a project but residents of unaffected neighborhoods are likely to display attitudes which differ from the attitudes they would hold if confronted with an actual project. Jakubs compares the results of field surveys in which residents responded to an actual project in their own neighborhood with surveys in which the residents responded only to hypothetical scenarios. His results indicate that carefully designed hypothetical scenarios can elicit useful information on attitudes and responses to controversial projects.

Behavioral geographers have used a limited array of models to describe process of individual decision making in spatial contexts and it is fair to say that they have been more concerned with constructing models which are useful in spatial frameworks than with including realistic descriptions of cognitive processes. David MacKay reviews multiattribute process models and shows how members of this class of models can provide more elaborate and satisfactory descriptions of cognitive processes without losing their utility as models of spatial behavior. Multiattribute process models treat individual decision making as dependent on how information is processed by the decision maker as well as the content of that information. This class of models includes composition and additive weighting models, which are familiar in behavioral geography, as well as lexicographic ordering models, discrimination net models, and generate-and-test models which are not. Each of these models contains a distinctive statement about how information is processed by a decision maker and MacKay shows how different models may be appropriate for different types of spatial behavior.

Behavioral Approaches in Geography Content and Prospects

Reginald G. Golledge
The Ohio State University

The main purpose of this paper is to discuss some characteristics of what is known as the behavioral approach in geography. Since reviews such as those by Golledge, Brown, and Williamson (1971), Golledge and Briggs (1973), and Downs and Stea (1973), adequately cover the development of a variety of behavioral approaches and itemize their historical content, I will not discuss the development aspect in this paper. Rather, I will first focus on some principal distinguishing characteristics and then discuss some of the critical problems and prospects facing those who wish to adopt this approach.

An ever increasing quantity of work in geography has shown us how human spatial behavior can be analyzed in terms of its physical properties. These properties are generally taken to be the directions, distances, frequencies, orientations, and forces associated with the actions of people. We have seen abundant evidence of how volumes of flow of people between origins of destinations can be quickly and easily summarized by using various types of gravity models (Olsson 1965, Wilson 1970, Shaw 1975). We have been shown that, "en masse," spatial behaviors can be grouped, clustered into heirarchial systems, allocated to specific nodes, and correlated with a vast array of environmental variables. In all these cases, researchers analyzed overt behavioral activity as it was recorded a posteriori. To this extent we can describe such analysis as being structurally oriented: that is, human behavior is related to the physical structure of the environments in which behaviors take place. The search for explanation of spatial regularity in the aggregate behavior of humans was assumed to be possible because man is a reasoning being who, given sufficient opportunity, can learn to differentiate rationally between alternative situations and over time can develop behaviors which are repetitive, relatively invariant, difficult to extinguish, and consequently can be explained and predicted.

As opposed to the "structurally oriented" approach, I plan to discuss here the "process-oriented" approach sometimes called the "behavioral approach." For the most part, it tries to explain certain types of human spatial behavior rather than to describe it. These explanations rely heavily on sociological and psychological constructs rather than on physical properties of the systems in which behaviors take place. This approach, for example, argues that in order to understand the nature of man's built environment we need to know about the decisions and behaviors which influence the location and arrangement of the phenomena in such environments. In developing this approach, it is assumed that the spatial structure of human environments is the logical outcome of sets of behavioral actions, and in order to understand fully the structure of these environments, we need to understand the decision-making process that produce them. Consequently, an argument is developed that the analyses of both environments and people are equally important in understanding man's actions in space.

Many of the major theories of geography have a behavioral base. Sometimes this base is explicitly recognized, as in the assumption of least effort, and sometimes it is unearthed only when detailed analysis of some empirical problem is undertaken. For example, central place theory explicitly incorporates the least effort assumption, or spatially rational man assumption, into its behavioral postulates. The Clark-Berry density decay function for the distribution of urban populations on the other hand has its behavioral base made explicit only when we try to answer the question "what makes people behave in this manner?"

CONCEPTS OF HUMAN SPATIAL BEHAVIOR

The concept of behavior is not an easy one to define because there are so many interpretations, definitions, and arguments that it is difficult to sift through them all. As a beginning, I shall adopt a very broad definition: that behavior is a sequence of life processes which can occupy time. Such a definition does not limit our interpretation to humans, but it does eliminate the actions of inanimate objects such as rocks falling, glaciers advancing, and winds blowing. This definition allows us to infer that behavior is caused, has directedness, motivation, action, and achievement (Golledge 1970). Associated with this definition of behavior is the concept of a "behaving organism." If we adopt Lewin's point of view (Lewin 1936), this organism can be considered as a geometrical point moving about in some sort of life cycle which is affected by various pushes and pulls. In the course of its locomotion, it will encounter and try to circumvent barriers of one sort or another as it moves between origins and destinations. The location and behavior of this organism at any particular point in time is thus partly determined by the physical forces which impinge upon it and partly determined by its own desires. We can imagine each behaving organism having some mobility characteristics and it can be assumed that each organism moves through its life space towards vectors which are adient (or which have a positive valence) and moving away from vectors which are abient (or which have a negative valence). The adient and abient vectors may then be represented as attractive and unattractive goals existing within a space-time system. Part of the behavioral approach is to identify the processes which influence the behavior of any particular organism and to identify those elements of the physical structure which produce the pushes and pulls in various life spaces.

At this stage differences between overt and covert behavior can also be explored. Overt or explicit behavior is that which is easily observed by another person and which can be measured. To be observed, of course, it must have an external existence. Overt behavior takes the form of actions which can be observed at some scale of operation. Covert behavior, on the other hand, is not easily observable by another person without the aid of special sensitive measuring instruments. Such behavior is internal and is often inferred or implicitly recognized to have taken place. Covert behavior can exist both independently, and as a part of overt behavior. In the past, geographers have concentrated almost exclusively on overt behavior; the behavioral approach attempts, at least in part, to bring out some characteristics of the interface between covert and overt activities.

These introductory comments have not as yet brought out some of the specific characteristics of the behavioral approach which make it different from other approaches that have been used in geography. Let us turn now to a discussion of some of these characteristics and follow this by providing a weak rationale for the development of a behavioral approach by examining the structure of human decision making and pointing to the contribution that the behavioral approach may make in terms of understanding the decision-making process.

MAJOR CHARACTERISTICS OF THE BEHAVIORAL APPROACH

Perhaps the most fundamental characteristic of the behavioral approach in geography is that it is process oriented. In particular, it focuses on *human behavioral processes* such as learning, perception, attitude formation, and all the cognitive processes. Thus, the approach encompasses the human actors in systems rather than the systems themselves or their physical structure. As such, it is comparatively easy to distinguish between the behavioral approach and a purely geometrical approach to the analysis of geographical problems. However, it is more difficult to decide where the differences lie between the behavioral approach and the approaches used by other human geographers in their studies relating to human spatial activity. Perhaps an attempt could be made to distinguish between these by arguing that the focal point of interest in many geographical studies is the *act* of behaving or moving from point to point in space, while the behavioral approach focuses more on the *processes responsible* for such behavior acts and on behavior changes over time. Of course, the behavioral approach does not neglect the overt acts themselves, but it does seek an explanation of overt acts not entirely in terms of the mechanics of the spatial structure in which the acts take place or factors external to individual decision making, but also in terms of the cognitive processes which help to produce the behaviors.

A second major characteristic of the behavioral approach is that in many cases it works with the lowest level of aggregation, that is, the *individual human actors* themselves. As with other research approaches in the field, there is a desire to search for generalizations. In order to make generalizations, there has to be some means of aggregating behavior. Those concerned with the behavioral approach tend not to use already aggregated data such as might be found in censuses, but to aggregate individuals only up to a level that is meaningful in behavioral terms.

Another difference between the behavioral approach and other approaches in geography relates to the *types of data* and the *analytical methods* that are used. In the previous paragraph, we mentioned the emphasis placed on individually collected data. These data frequently consist of items such as preferences, perceptions, habits, cognitions, and attitudes. The massive effort involved in collecting these data has resulted in a lack of large data banks for the behaviorally oriented geographer to use. For the most part, individually collected survey data is the building block on which analysis is performed.

Just as the type of data collected varies so does the method of collecting data vary. The behaviorally oriented geographer uses a variety of sociometric, psychological, and psychometric methods of data collection. Typical survey methods include the collections of paired comparison judgments, rankings of items, group cohesiveness indices, single stimulus data sets, and so on (Golledge, Rivizzigno, Spector 1975; Wish 1972). These data are invariably collected using carefully constructed experimental designs (see Spence 1976; Arabie 1976).

Along with the somewhat different data and the methods of its collection, researchers using a behavioral approach in geography have introduced and/or emphasized a number of different analytical methods. These include both unidimensional and multidimensional scaling methods, a concern for noneuclidian geometries, an ever-increasing use of stochastic models (particularly those relating to choice and search), and widespread use of nonparametric statistics. The overall effect of the use of these analytic methods has been to widen the scope of spatial analysis and to help operationalize many of the poorly defined subjective concepts that have been used for years in the discipline.

MOTIVES FOR THE DEVELOPMENT OF THE BEHAVIORAL APPROACH

The behavioral approach has often been dismissed as "only an attempt to introduce more realistic assumptions into existing geographic models." This in fact *was* one of the prime reasons for the initial development of the approach, and it must be applauded as being a worthy aim. It is certainly not the *sole* reason or purpose for the approach nor does this motive dominate current behavioral work. The particular assumption which triggered early behavioral work was the assumption relating to economic man and the spatial rationality of behavior. Increasing concern with the limited nature of rational man assumptions led to the investigation of the worth of using "satisficing" and "boundedly rational" assumptions in geographic models. A search was undertaken for assumptions which allowed behavioral variability rather than assumptions which denied it and assumed constant (habitual) behavior across all members of a population.

A second motive that influenced the development of the behavioral approach was that of introducing new variables into the traditional geographic explanatory schema. These variables included things such as attitude measures, learning parameters, psychological distance measurements, preference rankings, and a range of objective variables related to cognitions of large scale external environments.

Another motive influencing behavioral research can be found in the argument that because the structure of man's built environment is an outcome of human decision making, then some concern for the decision-making process itself appears valid. Researchers espousing this idea examined human decision-making processes in detail in an attempt to understand the types of decisions that may have produced the environments that we currently observe. This also spawned another research concern which concentrated just on examining rules of choice and decision making. For example, attention was paid to determining sets of rules by which choices of locations might be made or by which the mechanics of migration flow might be understood. In a sense, the geographer adopting this latter point of view loses part of his geographical identity and becomes much more of an interdisciplinarian. This happening was part of a movement that was not unique to geography. In fact, the early 1960s saw a general interdisciplinarian movement not only in the social and behavioral sciences but also in the physical and engineering sciences. Geography has traditionally relied on sister sciences such as economics and anthropology (and to some extent sociology) for a selection of variables and as a source for many of its theories. One immediate outcome of the behavioral approach was to forge a strong link between geography and psychology and to introduce psychological theory and models to the geographer.

I should also point out at this stage that one of the many points raised by critics of the behavioral approach is that it was a "reaction against the statistically and mathematically biased approach common in geography in the 1960s." I would argue that this is certainly *not* the case and that the the behavioral geographer by and large is by no means less rigorous in a mathematical and statistical sense than his geometrically minded counterpart in other areas of the discipline. This tendency becomes obvious when one considers the complex experimental designs that are needed in order to undertake good-quality survey research. For example, fractional experimental designs, factorial designs, complete and incomplete cyclic designs, and other efficient data collecting methods have had to be examined in great detail for incorporation into the surveying practices of geographers. Associated with this concern for rigorous experimental design have been discussions of noneuclidian geometry, and unidimensional and multidimensional scaling methods, a variety of stochastic learning and decision-making procedures, game theoretic models, and of course the widespread use of distribution free statistics. These analytical devices have been made necessary by a change in the type of data collecting from the ratio and interval scaled data commonly collected in large data banks such as the census, to nominal and ordinal scale data which reflect people's preferences, choices, attitudes, and so on. I will, therefore, state quite categorically that behavioral geography is not necessarily nonquantitative, nonstatistical, or nontheoretical. It is oriented towards all these things and if practiced properly is a viable and rigorous area of research in the discipline.

I have at this stage provided a few critical characteristics of the behavioral approach and defined some of the motives for the development of the approach. Let us now turn to describing a weak rationale for the existence of the behavioral approach and in doing so a range of problems that are particularly suited to this type of approach will be uncovered.

A CONCEPTUALIZATION OF A DECISION-MAKING PROCESS

A brief outline of several of the major states in decision making is as follows. Assume that the basic behavioral unit consisted of the members of a quiescent population. Fundamentally, behavior can be considered to be either motivated or unmotivated. Initially I shall assume that this population is unmotivated (i.e., quiescent) such that the only behaviors that are allowed to exist consist of neural or synaptic responses—purely physiological actions. At present, these are of little interest to us and we shall not pursue them further. We shall want to motivate our population, and in order to do this we will confront it with a stimulus which will start the chain of decision making. However, prior to the presentation of the stimulus, the population still has critical sets of characteristics *which can be used to describe it in an inventory sense.* Thus, antecedent to the creation of the desire or antecedent to the presentation of the stimulus, we can assume that the population could be described by the following sets of abstract characteristics:

a. Personal structural characteristics: such as the age, sex, height, of the members of the population.
b. Personal functional characteristics: for example, personalities, mental ability, personal habits, attitudes, and so on.
c. Existence variables: for example, the location of each population member and the spatial structure of the environment or system in which the individuals are placed.
d. Socio-psychological and cultural variables: including things such as education, occupation, income, role status, personality traits, and so on, and
e. Prior learning: including any prior learning associated with the activities of existing in various environments which might provide a reservoir of experiences which can be transferred and generalized to help cope with a particular stimulus situation.

These antecedent conditions exist in a premotivated organism. Such a population could be analyzed in a wide variety of ways to give a very accurate description of its nature and content.

One of the more common ways of analyzing such data sets is to use grouping techniques such as factor analysis, discriminate analysis, and other multivariate clustering devices.

Let us assume now that we allow our population to become motivated by presenting it with a stimulus. A stimulus condition may be either physiological or nonphysiological and may take the form of either a drive or a cue. Both drives and cues may be conscious or unconscious. For example, a nonphysiological conscious drive may be the stimulation of a sense of competition; a physiological conscious cue may be the presence of food. Although little attention has been paid in our discipline to the role of different types of stimulus conditions, we accept the fact that they exist and produce the variety of overt behaviors that are taken to be critical indicators of spatial activity. Now given that the population becomes active in some way, we assume that something which shall be called the *first motivated response* occurs.

The initial or first motivated response to the existence of a stimulus is the search for and acquisition of information concerning the nature of the stimulus and the potential goals that accompany its presentation. Information concerning both the stimulus and potential goals can be obtained from: latent recall of previously learned information; activation of responses due to some sort of conditioning in the prelearning stage; and perhaps of most interest to geographers, the tapping of external information flows and the collection of relevant information about those stimuli and goal conditions. This latter action involves getting access to the types of information that are dispensed by various agencies concerning stimulus and goal. In a practical sense, it may involve locating a potential supply point or a potential goal object in an environment and finding out the characteristics of the stimulus in order to make assessments of whether or not a goal reaching behavior is feasible. A tremendous volume of work has taken place in the discipline of geography in the last fifteen years relating to the tapping of information flows and the role of various information dispensing agencies in all manner of decision-making processes (Brown [1968], Gould [1969], Hudson [1972]). This area perhaps more than any other has become a dominant area of research in geography.

As the basic information about the stimulus and a potential goal is being collected, the motivated individual begins to fit both stimulus and goal into its behavior space. This is the first filtering undertaken to see whether or not a particular act designed to achieve a goal in response to the presentation of a stimulus can be feasibly accomplished. This involves relating the stimulus and goal characteristics to the psychological and physiological capabilities of the organism and towards short and long-term goals consistent with its life space and its place in a social system. In other words, an individual decision maker looks inward at his own life space. Within his life space there exists a behavior space which encompasses those behaviors which lie within the feasible limits of performance by the individual. If the connection of a particular stimulus and an achievement of a goal involves a behavior that appears to be outside the limits of physical or socioeconomic performance of an individual, goal gratification might be achieved by reverting to a covert action such as daydreaming. Given, however, that a particular overt act may be undertaken in order to achieve a goal, then the individual is able to reconcile a stimulus situation to a possible behavior space. In the course of this reconciliation, the decision maker constructs a mental image of the processes involved in goal gratification. The critical part of this is the movement imagery, which involves relating both stimulus and goal to cognitions of the external environment in which the action is to take place (Huff 1960). In the course of this relating, acts such as selection of a goal destination, formalization of reward expectancy, and recognition of barriers to goal gratification occur. At another level, the individual decides on things such as the mode of spatial activity (or the mode of transportation) which may be necessary in order to achieve goal gratification, and all the societal and spatial problems of formalizing the cognitive structure of the environment in which those acts take place (see Downs and Stea 1973, and Moore and Golledge 1976).

The first motivated response, therefore, involves reconciliation of stimulus object and goal seeking behavior in order to determine if an individual is capable of achieving certain goals. At this point critical assumptions relating to the motives of the decision maker are introduced. These assumptions include such things as least effort, economic maximization, bounded rationality, and spatial irrationality. Once this first motivated response has been completed, however, the decision maker starts the process of goal achievement and consequent goal gratification. This begins the *second motivated response.*

The second motivated response can be divided into three basic components. The first of these is designated as the overt spatial act. This consists of spatial movement of one sort or another and may be labeled as the beginning of search or "provisional try" behavior. At this level a path of behavior that has been mapped out at the previous stage of decision making is actuated and followed and some type of overt act such as a movement between an origin of destination occurs. Of course, it is this movement which identifies specific origins or destinations and gives a spatial component of distance or direction to a particular act. Once a provisional try has been undertaken, the decision-making individual moves into an assessment phase. In other words, there is a comparison of actual and expected rewards resulting from the act. The expected reward depends both on attitudes and the level of aspiration that the decision maker sets. The actual reward can be used to give a positive or negative valence both to the particular goal that is selected and to the executed behavior pattern. This second motivated act is generally considered to be the start of learning activity. Following pioneering work by Gould (1965) and Golledge and Brown (1968), a considerable amount of work has been undertaken in the discipline on the nature of spatial search (e.g., Adams 1969; Brown and Holmes 1970). Of course, the provisional try is but a first step in the whole search process and undertaking this first overt act together with the achievement of an outcome for the act constitutes the basic unit of learning.

Following the presentation of the stimulus condition the entire learning process began. With the completion of the overt act and goal gratification the decision process has been carried to the end of its first stage. Drawing on this experience an individual may consequently restructure the entire decision process that produced a particular overt act, or he may proceed with repetitious behavior whenever the same or similar stimulus conditions arise again. For example it is well known that only a limited amount of searching takes place prior to the development of a repetitious pattern in the journey to work and for some shopping activities (Golledge and Zannaras 1973, Marble and Bowlby 1968). It is of particular importance to the geographer to identify in what stage of the learning process a given act can be located. If it is early in the learning process then it might have a low probability of repetition. If it is late in the learning phase, then it may well represent a repetitive act which can be easily explained. Of course, the identification of the probability of repetition of an act is a very critical part of all geography. The achievement of a first unit or learning, starts a feedback process which may change some of the personal functional, existence, or sociocultural variables of a population.

CONCLUDING REMARKS

One of the things that must be immediately obvious to the nongeographer reading geographical literature is that for the greatest part of our existence we have attempted to explain overt or spatial behavior only in terms of the functional and structural characteristics of individuals and sets of existence variables, such as the location of the individual and the spatial structure of the system in which he finds himself. When one considers this carefully, one is inevitably forced to the conclusion that we have probably accepted many coincidental relations as explanations. In other words, we have found groups of individuals with similar functional, structural, or existence characteristics and have searched in their external environment for some set of physical environment factors that correlates highly with the sets of individual characteristics. The next step in many cases has been to assume some type of causal relationship between the two and to explain one in terms of the other.

It is quite significant that geographers in general have little knowledge about the cognitive structures by which individuals operate, or if they do have the knowledge, they have consistently ignored it. Realistically, we should be very unsure about the types of data that we collect by observing spatial activities or by asking people what types of spatial activities they undertake. For the most part, the data that the geographer has used in the past have been cross sectional in nature, nonsequential, and have had little emphasis on the cognitive process associated with the behavior. If, for example, we were to take a cross section of behavior acts in a randomly selected population, it would be very difficult to decide as to the uniqueness of the acts which people in the population were performing at that particular time. In other words, there would be doubt about whether an observed act represents a regular behavior pattern that can be described and predicted with some degree of reliability; a part of a recently initiated search procedure instituted by an individual in an attempt to find some satisfactory response; a repeat of a previously tried situation which was on its way to becoming an established habit; or, indeed, a particular observed behavior that would never be repeated. Traditionally the geographer has overcome these uncertainties by working with aggregate data and assuming that if he takes a large enough sample of a population then there would be a high probability that he would get general evidence as to the nature of the repetitive acts in the population. In effect, it would be saying that if we take a large enough population there should be a high *probability* of finding acts which are repetitive, relatively invariant and predictable. The degree of success that has been enjoyed with some of the more simple models developed in geography in the past has given us some proof that this type of assumption is not entirely invalid.

However, our degree of confidence in our ability to explain spatial behaviors is probably much higher than it should be. This is simply because we really do not know what it is that we are trying to explain. By examining cognitive information, we should begin to comprehend the nature of the data with which we work, and we may begin to understand some of the actions which are described in those data. Accordingly, although much behavioral work is very much piecemeal (and some very badly formulated), the hope that lies behind the continuation of this approach is that by expanding the relevant set of variables commonly used in geography, our explanatory power will be increased, our predictions will gain more validity, and our ability to produce theories also will increase.

At this time, therefore, it is pertinent to speculate as to some of the problems and prospects of work in the behavioral area of geography.

Perhaps the most significant deficiency at the moment (and the area that would give the greatest repayment for effort extended) is work of a fundamental nature attempting to structure theory related to man's behavior in large scale external environments. In particular, the theory that should be examined would relate to the cognition of these environments and to the importance of cognitive processes in relating man's acts to an underlying environmental system. In speculating about the development of such theory there is also a need for developing appropriate methods for the testing of hypotheses generated from theory and designing experiments so as to apply whatever elements of theory appear relevant and useful.

Of the many and varied areas into which the behavioral approach has extended, others that are likely to give significant immediate rewards for invested research time are the following:

Attitudinal research—such as that found in the literature on hazard perception, choice of recreation activities and sites, and both producer and consumer attitudes in location theory; Discovering the meaning of environments—for example, the discovery of meaning through repetory grid analysis and personal construct theory;

Information search—particularly the continuation of work in both the diffusion and adoption stages of innovation diffusion, the development of marketing plans, and a continued interest in the *actions* of different information dispensing agencies;

Cognitive cartography—this includes a rigorous analysis of the geometry of cognitive representations (or mental maps) and exploration of the most appropriate ways to represent cognitive information;

Epistomology—including a detailed analysis of the role of process in knowledge generation and the importance of understanding which "realities" are being dealt with; and

Representations of behavior and environments in art, prose, and poetry.

I believe that the prospects for the continuation of the behavioral approach in all the different areas of geography are good. That is why geographers have been experimenting with a diversity of approaches over the last few years in many subareas of the discipline. All the experiments undertaken have been united in a common concern for the building of geographic knowledge which is logically sound, meaningful, and has considerable scope.

REFERENCES

Adams, John A. "Directional Biases in Intra-Urban Migration." *Economic Geography,* 1969, pp. 302-323.

Arabie, Phipps. "Clustering Representations of Related Data." In R.G. Golledge and J.N. Rayner (eds.) *Multidimensional Scaling and Large Data Sets.* Ohio State University Press, Columbus, Ohio, 1976.

Brown, L.A. *Diffusion Process and Location.* Philadelphia: Regional Science Research Institute, 1968.

Brown, L.A., and Holmes, J. "Search Behavior in an Intra-Urban Context: A Spatial Perspective." Paper presented to the 1970 annual meeting of the Population Association of America.

Downs, Roger M., and Stea, David (eds.) *Image and Environment: Cognitive Mapping and Spatial Behavior.* Chicago: Aldine, 1973.

Golledge, R.G. "Process Approaches to the Analysis of Human Spatial Behavior." Discussion Paper no. 17, Department of Geography, Ohio State University, 1970.

Golledge R.G., and Briggs, Ronald. "Decision Processes and Locational Behaviors." *Highspeed Ground Transportation Journal* 7 (1973):81-99.

Golledge, R.G.; Brown, L.A.; and Williamson, F. "Behavioral Approaches in Geography: An Overview." *The Australian Geographer* 12 (1972):59-79.

Golledge, R.G., and Brown, L.A. "Search, Learning, and the Market Decision Process." *Geografiska Annaler* 49B (1967):116-124.

Golledge, R.G.; Rivizzigno V.; and Spector, A. "Learning about a City: Analysis by Multidimensional Scaling." In R.G. Golledge (ed.) "Cognitive Configurations of a City: The Case of Columbus, Ohio." OSU Research Foundation, Department of Geography, Columbus, Ohio, 1975.

Golledge, R.G., and Zannaras, G. "Cognitive Approaches to the Analysis of Human Spatial Behavior." In W. Ittelson (ed.) *Environment and Cognition.* New York: Seminar Press, 1975, pp. 59-94.

Gould, Peter. *Spatial Diffusion.* Association of American Geographers Commission on College Geography, Resource Paper no. 4, 1969.

———. "Wheat on Kilimanjaro: The Perception of Choice Within Game and Learning Model Frameworks." *General Systems Yearbook* 10 (1965):157-166.

Hudson, John C. *Geographical Diffusion Theory.* Northwestern University Studies in Geography, no. 19, 1972.

Huff, David L. "A Topographical Model of Consumer Space Preferences." *Papers and Proceedings of Regional Science Association* 6 (1960):159-173.

Marble, Duane F., and Bowbly, S.R. "Shopping Alternatives and Recurrent Travel Patterns." In F.E. Horton (ed.) *Geographic Studies of Urban Transportation and Network Analysis.* Northwestern Studies in Geography. Evanston, Illinois: Northwestern University Press, no. 16, 1968, pp. 42-75.

Moore, Gary T., and Golledge, R.G. *Environmental Knowing: Theories, Perspectives and Methods.* New York: Dowden, Hutchinson and Ross, 1976.

Olsson, Gunnar. *Distance and Human Interaction: A Review and Bibliography.* Philadelphia: Regional Science Research Institute, 1965.

Shaw, R. Paul. *Migration Theory and Fact.* Philadelphia: Regional Science Institute, 1975.

Spence, Ian. "Incomplete Experimental Designs for Multidimensional Scaling." In R.G. Golledge and J.N. Rayner (eds.) *Multidimensional Scaling and Large Data Sets.* Ohio State University Press, Columbus, Ohio, 1976.

Wish, M. "Notes on the Variety, Appropriateness, and Choice of Proximity Measures." Murray Hill, New Jersey: Bell Labs Workshop on Multidimensional Scaling, 1972.

Controversy and Low-cost Housing Strategies: The Use of Hypothetical Scenarios in Policy-directed Research

John F. Jakubs
Indiana University

Urban housing policies for the economically underprivileged in the United States are of two broad types: "ghetto dispersal" approaches and "ghetto enrichment" policies. The former attempt to diminish residential segregation by economic status by offering low-cost housing in peripheral areas. Secondarily, programs may be mounted to develop centralized housing for the more affluent. Ghetto enrichment, on the other hand, concentrates efforts within current poverty regions. No attempt is made to alter the present high levels of residential segregation by economic status.

Difficulties have arisen with respect to ghetto dispersal. One has been public outcry, real or anticipated, against dispersal by present residents of peripheral areas (Downs, 1973). Another problem, especially important for private market ventures, is the decreasing availability of land for any program other than ghetto enrichment. This results from suburban zoning regulations and building codes which indirectly stipulate minimum levels of wealth or income necessary for residence in a particular area by increasing the overall cost of the residential package of land and dwelling unit. Both reasons may be viewed as indicative of opposition by suburbanites or by residents of peripheral areas within central cities to an influx of disadvantaged people from the urban core.

This paper examines the potential contribution of a type of survey research to this problem. Specific emphasis is on approaches to identifying policy levers, in terms of housing program features which can influence the response of inhabitants of so-called "receptor" communities. Such knowledge may aid in shaping public policy which minimizes anticipated negative impact on suburbanites while performing its main function of providing the economically disadvantaged with an alternative to residence in more centralized low-income ghettos.

However, prior to discussing such issues, the need for peripheral low-cost housing opportunities must be established, if only briefly.

THE NEED FOR GHETTO DISPERSAL

Socioeconomic mixing, either stated directly as a goal or seen as an indirect byproduct of ghetto dispersal, has been assessed both positively and negatively. Dispersal to achieve this end is questionable because increased interaction across social classes may not occur in the first place, and were it to happen, this might result in increased friction rather than in mutual understanding and communication. Further, the promotion of such interaction by governmental action is attacked as a form of cultural imperialism (Marrett, 1973) by many who believe that the eradication of widely different life-styles, personal tastes and habits should not be a goal of policy.

It is incomplete, however, to assess dispersal on the basis of social interaction considerations alone; economic aspects also are relevant. The geographic location of the poor within metropolitan areas increasingly is becoming a disadvantaged one. It has been shown elsewhere that in low-income ghettos the cost of providing public public goods is higher than average and ghetto residents, who form a small part of a large central city electorate, have less discretion than do suburbanites as to the specific mix of public goods provided (Cox, 1973; Kee, 1968). Prices of privately produced and sold goods and services in inner city poverty areas are greater than the norm (Caplovitz, 1967), and accessibility to employment opportunities is increasingly a problem as these opportunities decentralize (Davies and Albaum, 1972; Rubinowitz, 1974).

For these reasons, soon after the 1968 Civil Rights Act, the Department of Housing and Urban Development proposed a two-pronged attack on inner city problems, including both ghetto dispersal and ghetto enrichment efforts. Implementation of ghetto dispersal programs has proven difficult in many instances and has caused considerable controversy (Glazer, 1972, Goodman, 1972; Haar and Iatridis, 1974). This controversy results in large part from the responses of residents of "receptor" communities, which are usually white and middle-class locales in peripheral areas where low-cost housing would be constructed. Specifically, these residents may anticipate unwanted changes in characteristics of their communities, such as school quality, personal safety, property values and the like.

These changes may or may not actually occur; however, the *anticipation* is as important as the actual change. In the first place, for certain community characteristics, anticipation of changes may cause them to occur in fact. This aspect of self-fulfillment can be illustrated for the case of property values. Fears of property value decline may cause a decreased demand for local housing, which would result in the original fear being actualized. Secondly, opposition to ghetto dispersal programs often occurs before implementation and thus, before residents can assess actual community change. It is the set of *anticipated* changes upon which such opposition typically rests. To counter local opposition, three possibilities present themselves. First, peripheral low-cost housing programs might be designed to minimize anticipated changes in receptor communities. Second, it may be possible to convince residents of receptor communities that at least some anticipated changes are unrealistic and will not occur in fact. This educational campaign might be accompanied by formal governmental guarantees. Third, public policy might be formulated which compensates residents of receptor communities for those anticipated undesirable changes which become actualities. Clearly the first step should be to investigate the extent to which housing programs can be designed to deflate the problem at the outset. Equally evident is that the first of these three approaches should be taken—to the extent that it can.

DECENTRALIZED LOW-COST HOUSING PROGRAMS: POLICY LEVERS

Residential satisfaction is functionally dependent upon a set of attributes pertaining to the dwelling unit and the neighborhood such as cost, size, and architectural style. Each household conceives its residential "bundle" to contain specific amounts of such attributes; a certain amount of school quality, a certain distance from the workplace or workplaces, and the like. Further, each attribute possesses a specific level of importance to the household. Thus, the residential package can be viewed as having a *conceptual* component, the level of quality of local schools, for instance, and a *preference* component, the extent to which school quality is important to the household.

Neighborhood changes can effect changes in the levels of these attributes. Anticipated changes often are believed to cause future changes in attributes. Decentralized low-cost housing, for example, might be believed to cause school quality to diminish. Attempting to alter households' preference patterns for attributes likely would be a fruitless task in addition to being morally questionable; preferences are intrinsic to one's basic values and philosophies. Clearly more hope lies in the investigation of household conceptions of attribute change, with preference patterns viewed as constraints.

This implies that efforts to design housing programs to minimize receptor community resistance must begin by determining preference structures for the community in question in order to identify the most important residential attributes. Following this should come efforts to identify the extent to which various housing programs are believed to cause negative changes in these critical attributes. If these anticipated changes were heavily dependent upon housing program parameters, then such parameters could act significantly as policy levers.

Research along these lines has been undertaken and completed in a number of communities. Two examples will be discussed briefly. The first took place in the greater Dayton, Ohio, region (Gruen and Gruen, 1972). A "fair share" housing program was adopted by the Miami Valley Regional Planning Commission whereby each municipality within the metropolitan Dayton region was to provide some low-cost housing locally. A preliminary study was carried out to identify ways by which this might be accomplished with a minimum of controversy and with a minimum of community instability. The study consisted of a survey research approach in selected affluent suburbs and inner city poverty areas. Points of agreement were found between residents of receptor communities and potential occupants of low-cost decentralized housing, especially in terms of housing style. The general method employed was to confront suburbanites with alternative and hypothetical housing programs and to solicit their reactions to these. Characteristics of potential occupants were examined as were architectural considerations. Respondents indicated that with certain guarantees in the areas of property value maintenance, free additional social services, school quality maintenance, and others, they would not be adverse to some of the programs.

A problem with this research is of course that there may well be differences between respondent reactions to hypothetical programs when interviewed and reactions to actual policy when proposed and implemented. A similar effort was undertaken in a suburb of Columbus, Ohio (Jakubs, 1974). Findings indicated a low tolerance level for low-income households, independent of both architectural style and the degree of scattering versus clustering of low-income households. Participants were offered hypothetical programs and asked to respond and again the question is raised: "Are responses meaningful and realistic?"

There have not been sufficient efforts to examine the actual impact of decentralized low-cost housing to allow us to draw summary conclusions from actual experience. In addition, it was noted above that community preference patterns are critical in determining anticipated impacts. Because communities are likely to vary considerably in preferences of residents for attributes, one must be careful to attempt to draw conclusions on the basis of cross-community comparisons. Thus, it seems that if research is to provide us with policy guidelines, the question of assessing "hypothetical program" type research must be addressed. If community residents react differently to hypothetical constructs than to actual housing programs, such preliminary studies are meaningless. The remainder of this paper presents a study designed to analyze this question, though on a small scale.

THE INDIANAPOLIS STUDY

Preliminaries

The general approach taken in the research which follows was to identify two peripherally located and relatively affluent communities which were similar in all respects but one: the presence in one community of publicly-supported low-cost housing. Responses then were obtained from residents of both communities. In one case participants reacted to a development which actually existed; in the other case questions were structured to insure that the participants knew they were responding to a hypothetical program. Features of this scenario corresponded as closely as possible to those of the existing program.

The area selected for study was a community in the northwestern sector of Indianapolis in which three apartment complexes recently had been constructed, including one under the old section 221d3 designation and two as part of the Department of Housing and Urban Development's Section 236 program. These were meant primarily for "moderate" income occupancy, although rent supplement programs existed in two of the three to provide for a mixing of low and moderate-income households. Immediately to the North and East of these complexes were three other large scale apartment developments and beyond these were single-family detached houses averaging $31,000 in value as of 1970. Immediately to the West of the subsidized apartments was a commercial area and to the South was another single-family detached residential district exhibiting substantial heterogeneity in value but generally consisting of homes slightly less costly than those in the district mentioned above.

This entire area was canvassed as part of a larger research effort which was an attempt to measure social interaction among residents of various subareas. A sample of approximately 50 respondents in single-family detached units and 60 respondents in privately developed apartment complexes was selected. These subjects were asked whether they knew of any apartment complexes located nearby which were built in part or in full by government funds. Those who responded affirmatively were asked to identify such developments and were confronted with a series of questions investigating community impacts which they believed resulted from this housing. Specifically they were asked the following questions:

1. Do you think the construction of (this complex/these complexes) you named above has affected your neighborhood in a good way, a bad way, or not much at all?
2. Do you think this housing has caused any of your neighbors to move away or not?
3. Are you thinking of moving away because of this housing or not?
4. Has this housing caused your property value to increase, to decrease, or has it had no effect on your property value?
5. Before this housing was completed, did you think it would cause your property value to increase, to decrease, or did you think it would not affect your property value?
6. Do you think this housing has caused the crime rate around here to increase, to decrease, or hasn't it affected the crime rate?
7. Do you think this housing has made the community more attractive to look at, less attractive to look at, or has it had no effect?
8. Do you think this housing has caused schools around here to be better in quality, worse in quality, or hasn't it affected the schools?
9. Are there any other effects, good or bad, which this housing has had?

Only one particular complex was identified frequently enough for further analysis. Interestingly, this was the only one of the three which did not participate in a rent supplement program. Thus, it completely excluded low-income households and was populated by those whose incomes were classified as "moderate."

Those who had moved from the area during the past six years were identified by means of Indianapolis City and Suburban Polk Directories. Of the movers, those who had remained within the Greater Indianapolis area were contacted by mail and were presented with the same questions as noted above.

The 1970 Census block statistics were employed to identify matched residential areas. Housing values and racial distributions were matched to the single family areas described above. In addition, visual field checks were made to estimate age of housing and to investigate changes which might have occurred since 1970. From blocks selected, a sample of 200 households was drawn at random but stratified by block-type for matching purposes. These people were mailed a brief questionnaire, a cover letter, verbal and pictoral descriptions of the subsidized apartment complex in question, and a prestamped postcard on which they were to place their answers to the series of questions.

The Analysis

The above procedures generated a set of comparative data (Table 1) involving perceived community impacts in six areas: general impacts, overall negative impacts, impacts related to migration, and negative impacts involving school quality, property values, crime rates, and visual appearance. Results were arranged in 2 × 2 contingency tables which were subjected to chi-square tests to discover whether the two cases, actual and hypothetical, exhibited differences large enough to allow one to conclude that something other than chance variation was operating. The results appear in Table 2.

It can be seen that for school quality, crime rate, visual appearance of the community, and migration behavior, differences were slight and statistically not significant. Respondents to the hypothetical construct did have a tendency to anticipate general negative impacts more often than did those in the second group. The basis for this quite likely is the large difference noted for the case of property values. Thus, the social types of impacts demonstrated stability from group to group, while the economic factor was associated with greater levels of anxiety on the parts of respondents to the hypothetical construct.

Clearly it would have been preferable to have found no such difference. Yet if we view the differences as exaggerations on the part of the hypothetical construct group, it is fortunate that such exaggerations occurred for an economic type of impact rather than for one of the social ones. This is the case because governmental efforts to deflate such exaggerations, perhaps by a program of property value guarantees in this case, probably would be easier to implement for economic than for social factors. While a property value guarantee program would involve a dollar and cents compensation, it is difficult to imagine, for instance, what a school quality guarantee would entail or to assume that such a program would satisfy residents. It would depend on their definitions of school quality.

CONCLUSION

This study is small in scale and should be taken as a prototype rather than a definitive statement. Nonresponse rates were high (Table 1) and alternative modes of response-gathering were employed. If further studies were to exhibit similar findings, however, this would bode well for hypothetical construct types of research in this area at least. Generally, a common research strategy is to take an interview approach with subjects responding to idealized or simplfied

Table 1
Background Statistics to the Study

	Hypothetical-Construct	Actual
I. Comparative Descriptive Measures		
Age (mean)	42.1	39.8
No. of Children at Home (mean)	1.48	2.00
Proportion with at Least One Child Living at Home	.759	.784
II. Sample Representation		
A. Personal Interviews		
Attempted	52	
Refusals	8	
Interviews	44	
Percent Refusals	15	
B. Movers		
Numbers of Movers in Past Six Years	120	
Total Number of Dwelling Units	389	
Movers Remaining in Indianapolis and Contacted	35	
Returns	7	
C. Hypothetical Construct Sample		
Questionnaires Mailed	198	
Return to Sender	22	
Net Mailed	176	
Returned	43	
Percent Returned	24	

situations. While these situations may bear a close correspondence to the real world in the mind of the researcher, it is not safe to assume that this is the case for the respondent. In the area of controversial housing programs this is particularly critical. In this paper we have attempted to provide the beginnings of a verification process which must be continued and expanded before policy development can rely on such research.

ACKNOWLEDGMENTS

Field research reported in this paper was carried out in the Fall of 1975 by students in G480, a course in spatial analysis in the Department of Geography at Indiana University. The following people participated in questionnaire design, study area selection, sampling design and interviewing: Joe Blair, Joachim Burdack, Gaston Bouquett, Ron Dexter, Debra Embry, Jim Fall, Norma Kuypers, Jerr Irving, Scott Lycan, Jeff McReynolds, Gus Pachovas, Phil Reeves, Pamela Scanlon, Susan Shannon, and Nanda Shrestha.

Table 2
Results of Comparative Analysis: Single-family Detached Areas

	Hypothetical		Actual		Chi-square Value
	Yes	No	Yes	No	
General Impacts	38	5	31	15	4.48**
Effects on Migration (of respondent or neighbors)	28	15	19	18	1.04
Effects on Property Values	37	4	14	20	18.37**
Effects on Crime Rate	32	8	28	7	0.08
Effects on Neighborhood Appearance	32	10	26	9	0.01
Effects on School Quality	31	10	20	11	0.58

**Significant at the .95 confidence level.

REFERENCES

Advisory Commission on Intergovernmental Relations. *Fiscal Balance in the American Federal System,* vol. 2. *Metropolitan Fiscal Disparities,* Washington, D.C.: USGPO, 1967.

Caplovitz, David. *The Poor Pay More: Consumer Practices of Low-Income Families.* New York: Free Press, 1967.

Cox, Kevin R. *Conflict, Power and Politics in the City: A Geographic View.* New York: McGraw- Hill, 1973.

Davidoff, Linda; Davidoff, Paul; and Gold, Neil N. "The Suburbs Have to Open Their Gates." *New York Times Magazine,* November 7, 1971, pp. 40-60.

Davies, Shane, and Albaum, Melvin. "The Mobility Problems of the Poor in Indianapolis." In Richard Peet (ed.) *Geographical Perspectives on American Poverty.* Worchester, Mass.: Antipode, 1972.

Downs, Anthony. *Opening Up the Suburbs.* New Haven: Yale University Press, 1973.

Glazer, Nathan "When the Melting Pot Doesn't Melt." *New York Times Magazine,* January 2, 1972.

Goodman, Walter. "The Battle of Forest Hills—Who's Ahead?" *New York Times Magazine,* February 20, 1972.

Gruen, Nina, and Gruen, Claude. *Low and Moderate Income Housing in the Suburbs: An Analysis for the Dayton, Ohio Region.* New York: Praeger, 1972.

Haar, Charles M., and Iatridis, Demetrius S. *Housing the Poor in Suburbia: Public Policy at the Grass Roots.* Cambridge, Massachusetts: Ballinger Publishing Company, 1974.

Jakubs, John F. *Ghetto Dispersal and Suburban Reaction.* Columbus, Ohio: Department of Geography, The Ohio State University, unpublished Ph.D. dissertation, 1974.

Kalachek, Edward. "Ghetto Dwellers, Transportation and Employment." In John F. Kain and John Meyer (eds.) *Conference on Poverty and Transportation.* Springfield, Virginia: National Technical Information Service, U.S. Department of Commerce, 1968.

Kee, Woo Sik. "City-Suburban Differentials in Local Government Fiscal Effort." *National Tax Journal* 21 (1968):183-189.

Marrett, Cora B. "Social Stratification in Urban Areas." In Amos H. Hawley and Vincent P. Rock (eds.) *Segregation in Residential Areas.* Washington, D.C.: National Academy of Sciences, 1973, pp. 172-188.

Meyer, John R. "Urban Transportation." In James Q. Wilson (ed.) *The Metropolitan Enigma.* Cambridge, Massachusetts: Harvard University Press, 1968.

National Advisory Commission on Civil Disorders. *Report of the National Advisory Commission on Civil Disorders.* Washington: USGPO, 1968.

Rubinowitz, Leonard S. *Low Income Housing: Suburban Strategies.* Cambridge, Massachusetts: Ballinger Publishing Company, 1974.

Cognitive Mapping and Processing of Spatial Information

David B. MacKay
Indiana University

The "intendedly rational" man postulated by much of the cognitive mapping literature is conceptually attractive to empirically based geographers. Most of the development so far, though, has concentrated on means and methods for measuring cognitive content and has essentially ignored the related issue of cognitive process. This may be due to the attractiveness of statistical models which provide numerically precise procedures for data description and analysis but assume away the question of process or it may be due just to the difficulty and bother associated with analyzing data in the detail that is needed to define truly how individuals process spatial information. The purpose of this paper is to review a class of models, termed *multiattribute process* models, that can be used to investigate both the content and processing of spatial information and to present the results of some recent studies that illustrate their application to the common geographical problem of store selection by consumers.

INTRODUCTION

The rationale of behavioral geography lies in the belief that the psychological processes of an individual are important determinants of spatial behavior. A glance at any introductory psychology book tells us that there are many psychological processes and it would seem that most of them can be related to spatial behavior. Thus, it is possible to discuss spatial motivation, learning, perception, cognition, and attitude. In addition, we can investigate how these concepts are mediated by communication, personality, and culture. Most if not all of these areas have been touched on by geographers, though none has been explored in depth.

Recently, a great deal of attention has been focused on cognitive mapping or cognitive structure. Broadly defined, "cognitive mapping is a process composed of a series of psychological transformations by which an individual acquires, codes, stores, recalls, and decodes information about the relative locations and attributes of phenomena in his everyday spatial environment" (Downs and Stea, 1973). In psychology, the cognitive movement grew out of the works of Burnswik, Lewin, Tolman, and others. Their contributions emphasize the goal-directed actions of man. Concern is placed on the role of expectations and values in behavior determination, differing sharply from the dominance of learning and stimulus-response concepts in the behavioralist school. Along with the cognitive psychologists' formulation of an expectancy-value model of motivation has been the development of a host of techniques for measuring and describing the mentalistic states and processes that are fundamental to the cognitivist viewpoint.

Despite the interest that has been expressed in exploring cognitive-behavioral models, the difficulties in operationally defining models which have a sufficient degree of predictive or explanatory power and which are plausible and consistent with a cognitive outlook are considerable (Harvey, 1969). There are, though, a number of possible avenues, several of which will be discussed under the rubric of *multiattribute process* models. Multiattribute process models are concerned with both information content and process. They are empirical, decision-oriented models that operate at an idiographic level yet try to capture the processes and mechanisms that are common to all persons. Criteria, whether behavioral or attitudinal (defined to include the components of cognition, affect and conation [Fishbein, 1967], are modeled as being dependent on how stimulus information is processed. Stimulus information in most geographic situations is of a multiattribute nature. In keeping with their idiographic orientation, multiattribute process models are non-stochastic, though they may use model phenomena that are commonly described in a stochastic framework.

Multiattribute process models consist of both composition and decomposition models. As their names suggest, composition models build a model of behavior or attitude out of attributes that are determined from outside the model while decomposition models start with data that define a mentalistic state and then break the data down into two or more subsets, each with its own identity. Decomposition models are usually associated with conjoint measurement techniques, particularly multidimensional sealing (MDS), while composition models span a broad range of methodologies.

Decomposition models have seen a lot of attention in geography during the last few years, (Burnett, 1973; Golledge, 1975, Louviere, 1975; Mackay, 1976). In contrast, composition models have received scant notice. Therefore, this paper will begin with a review of the types of composition models that have been used in social science research. The models will be seen to vary substantially in their ease of implementation and in their ability to capture the dynamics of information processing. They also differ significantly from the majority of the search and learning models in geography (Golledge, 1969; Schneider, 1975) which have their theoretical roots planted in behavioralist traditions and are usually stochastic in nature. A brief discussion of the future prospects for using multiattribute process models in spatial research will follow.

COMPOSITION MODELS

A number of multiattribute composition models have been proposed in behavioral geography. For example, Appleyard (1970) has proposed that cognitive structure is a function of education, familiarity, travel mode, and gender. Downs (1970) suggests that the cognitive structure of an urban downtown shopping center is based upon nine cognitive components. Hudson (1974) has used repertory grid methods to describe grocery store images on the basis of choice attributes and also used regression techniques to explore consumers' spatial search patterns. Walmsley (1974) proposes that spatial behavior can be explained by a model that represents needs as a function of a need set and aspiration level and Jackson and Johnston (1972) have tried to recover the components of urban images from data collected with semantic differential scales. Other examples could be cited.

All of these authors used models which compose an image from a set of predefined components yet these multiattribute models are not actually process models because they are not defined and analyzed at the level of an individual subject. In fact, all of the studies, except for Appleyard's which relies upon summary statistics from frequency tables, depend upon correlation matrices for their analysis, thus assuming that individual variations can be represented by a stochastic departure from the model defined by the population parameters.

ADDITIVE WEIGHTING MODELS

In a more recent article, Cadwallader (1975) has proposed a model of consumers' spatial decision making that incorporates a measure of spatial attractiveness. Spatial attractiveness is defined by each subject with respect to four attributes of five supermarkets. This four by five matrix is then multiplied by a vector of standardized weights that reflect the importance of each attribute in selecting a supermarket. The result is an attractiveness score for each supermarket. Unfortunately, the individual matrices are aggregated before the multiplication operation thus losing a lot of the information content. A similar process was used in an earlier (MacKay, 1972) study of shopping behavior in which the individual character of the data were preserved but results from the specific operation are not reported. Both of these articles make use of attitude models which are essentially of the form

$$A_{im} = \sum_{k=1}^{n} I_{km} \cdot B_{ikm}$$

where: A_{im} = the attitude of subject m for store i
I_{km} = the importance subject m places on the kth attribute of the store
B_{ikm} = subject m's belief as to the extent which attribute k is offered by store i.

Additive weighting models of this sort may be said to define an elementary multiattribute process model. Their strength lies in their methodological simplicity and their robustness. They have been shown to be reasonably good predictors of attitudes and they can boast of an extensive empirical evaluation in marketing that has followed their original development in social psychology (Fishbein, 1967; Rao, 1974). In fact, the volume of discussion surrounding these models in marketing is fast approaching that of gravity models in geography. Despite its apparent simplicity, numerous issues concerning the operationalization of the model and its precise theoretical lineage have developed. Reviews of the literature may be found in Mazis et al. (1975) and Wilkie and Pessmier (1973). Parenthetically, it should be noted that several authors (Hughes and Naert, 1970; Pessmier and Wilkie, 1974) include models where the importance weights are estimated by multiple regression or discriminant analysis in the same category as models where importance weights are estimated directly by the subject. This in effect adds a third component that is derived from group level data. Such models are, therefore, not multiattribute process models. For this reason, the clinical judgment literature in applied differential psychology (Dawes and Corrigan, 1974; Einhorn, 1971; Goldberg, 1971) is largely ignored here.

While additive weighting models are relatively good predictors and robust, they have serious problems of plausibility. Because they make use of multiplication and addition, they require all attribute values to be comparable (intervally scaled) and importance weights to be ratio scaled. It is indeed questionable if people can really employ ratio scale weights in comparing diverse attributes. These models also imply that irrelevant (nonsalient) attributes can be offset by high belief valences and that a store which is weak with respect to a consumer's evaluation on one attribute can be compensated for by a high evaluation on a second attribute. Thus, a store that has low quality merchandise might be compensated for by being below average in price. The plausibility of this requirement is dubious as is the assumption that the required judgments can be made in the first place. An interesting attempt to evaluate the construct validity of additive weighting models using Anderson's (1974) integration theory has been reported by Bettman et al. (1975) but their evidence largely is speculative (Shanteau and Troutman, 1975).

Another way of looking at the plausibility problems of additive weighting models is to consider the general structure of an information processing system. A structural diagram of such a system (adapted from Newell and Simon) is shown in Figure 1. Central to the information processing system (IPS) is the short-term memory which is thought to have a limited capacity of five to nine symbols (Miller, 1956; Sandusky, 1974). It is doubtful that this "working memory" can simultaneously do all of the things that additive weighting models demand. Similar implications may be derived from the brief review of some memory models provided by Betak (1975).

LEXICOGRAPHIC MODELS

A class of models which portray decision makers in a more plausible fashion are lexicographic models. Lexicographic models portray decision makers as evaluating one attribute at a time. The attribute chosen first is the one to which they give the most importance. If there were one stimulus (e.g. supermarket) with the highest attribute value for the most important stimuli, then that stimulus would be the one chosen. If there were multiple stimuli with tied maximal values on the most important attribute, then the decision maker would go to the next most important attribute and evaluate the set of stimuli that was tied on the most salient at-

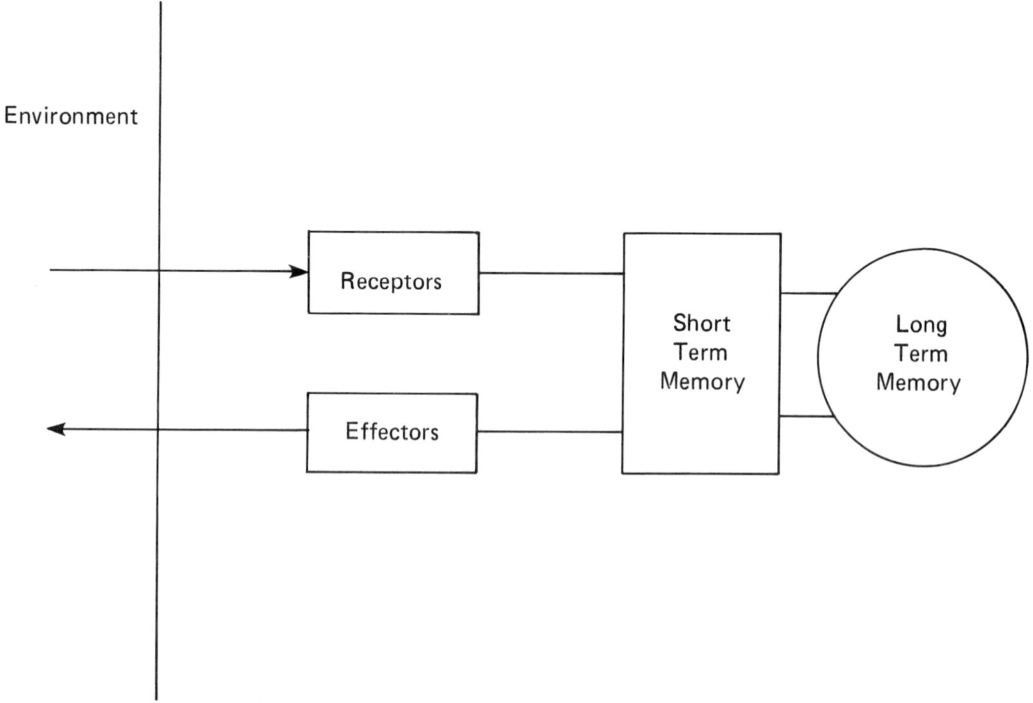

Figure 1. General Structure of an Information Processing System.

tribute. In this manner, the set of stores under consideration diminishes at each step. The term "lexicographic" comes from the correspondence between the mechanics of this procedure and the way in which one would go about looking up a word in a dictionary.

To state this model a bit more formally, suppose that a_{ij}: i = 1, . . . , m; j = 1, . . . , n is the value assigned to the ith attribute of the jth stimulus and that the index set i = 1, . . . , m is ordered so that i = 1 is the most important attribute and i = m is the least important attribute. Then if A_j represents the original set of stimuli, let

$$A_k = \underset{A_j}{\text{max'er}}\ a_{ij}$$

where max'er is a maximizer expression that designates a member (s) of the index set rather than a value in the domain of maximization. If A_k contains more than one element, then consider

$$A_\ell = \underset{A_k}{\text{max'er}}\ a_{ij}.$$

If A_ℓ has multiple stimuli, continue in the same manner.

Lexicographic models can also incorporate numerous variations. One variation, referred to by Russ (1971) as a lexicographic semi-order model, gains its value by recognizing that the important attributes may not be the discriminable attributes. Thus in the case of supermarkets, a consumer may believe that price is the most important attribute in making a decision about which supermarket to patronize but all supermarkets may be perceived as having the same price structure. The consumer may thus discriminate between supermarkets on the basis of other attributes: ones with less importance but more discriminability. To illustrate this distinction, consider the data for four attributes of four supermarkets which are defined for a hypothetical subject in Table 1.

Table 1
Attribute Data for Four Supermarkets

	Attributes		Supermarkets			
Definition	**Importance**	**Discriminability**	**A**	**B**	**C**	**D**
Price	4	L	5	5	2	4
Distance	2	H	2	3	5	1
Quality	3	L	4	3	4	2
Variety	1	H	4	5	2	3

Each supermarket in Table 1 is described by the attributes of price, distance, quality, and variety. These attributes are presumably cognitive appraisals as opposed to being physical measures. The values are magnitude estimates which, for simplicity, are made on a five point scale. In each case, a five refers to the most preferred attribute value (low price, close distance, high quality, and large variety) and one refers to the least preferred attribute value. (It may of course be true that the correspondence between these psychological scale values and the corresponding physical attribute values are nonlinear or perhaps even nonmonotonic.) Attribute importance is designated by a rank order with high values again designating the most important attributes while discriminability is simply designated as H (high) or L (low).

With the simple lexicographic model the subject would first find the most important attribute, price, and then reduce the set of relevant stimuli from four to two (A and B) which are tied with the highest valence on price. The subject is then modeled as proceeding to the second most important attribute, quality, and then selecting supermarket which was the higher valence on this attribute. In contrast, the lexicographic semi-order model would first model the subject as dividing the set of four supermarkets into two subsets, the first containing the attributes that are considered to vary significantly between stores (distance and variety) and the second containing those attributes that do not have high discriminability (price and quality). Each subset is then reordered on the basis of the importance rankings and the sets are merged. The new ranking of attributes would then be distance, variety, price, and quality. Our hypothetical subject would now proceed as in the lexicographic model. Store C which is closest would be selected on the first pass since no other stores are tied with it.

Before contrasting the lexicographic models to the additive weighting model a few points about measurement issues should be made. First, it should be noted that the measurement of discriminability is very flexible. One can, as in this example, use a direct ordinal classification or one could estimate discriminability on the basis of the spread of values given for each attribute across stimuli. The latter approach of course implies that comparisons across stimuli can be made but this need not be the case. If one thought that it would not be possible to compare a magnitude estimate for distance with a magnitude estimate for price, then a separate measure of discriminability would be preferred. Second, it is possible with lexicographic models to use only rank order data for the evaluation of the a_{ij} on each stimulus. The same comment applies to the perference data which are represented this way in our example. On some occasions the importance rankings may be tied. In this case, the experimenter may either break the ties randomly or combine the attributes by averaging the a_{ij} values across the tied attributes i. The choice again depends on the measurement properties one attributes to the data.

If an additive weighting model were to be applied to the data in Table 1, steps would have to be taken to ensure that the importance measures were ratio scaled. Suppose this were done and the values were proportionally transformed so they summed to 1.0, then the additive weighting model would predict that the subject would go to store B. All three models discussed so far thus give different answers. (In actual application, we have found that the lexicographic semi-order model gives more importance to spatial stimuli in the decision-making evaluation than the other models do. This is because differences in distance are apparently much more recognizable than are differences in other variables.)

A number of operational difficulties also attend these multiattribute models. Two of the more severe problems are the tendency to observe halo effects when the favorite store is rated high on attributes or stores with which the subject is marginally familiar are rated low or average on all attributes and the tendency for subjects to respond in a manner that is considered socially acceptable. Quality is thus said to be of most importance in the great majority of cases even though quality does not seem to be as big a determinant of actual behavior as one would expect.

DISCRIMINATION NET MODELS

The lexicographic models discussed in the previous section can also be described in terms of a discrimination net. As an example, the data for the subject using the lexicographic model in

Table 1 are portrayed as a discrimination net in Figure 2. While there is substantial variety in the ways discrimination nets may be constructed, most models contain a branching structure where nodes are cues (in this case attributes) relating to a specific stimulus. The branches from each cue indicate whether the store will be patronized or rejected, or if new information will be sought. Discrimination net and lexicographic models posit a strict sequential order of analysis that is defined individually for each subject.

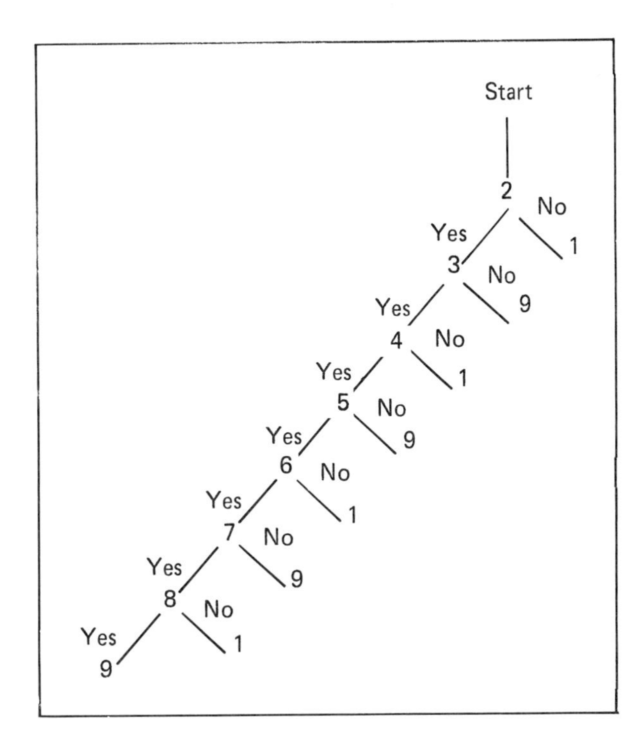

1—Reject store.
2—Does this store have the lowest prices in town?
3—Are any other store prices equally low?
4—Of the stores with equal prices, does this store have the highest quality?
5—Do any other stores with equal prices have equal quality?
6—Of the stores with equal quality and prices, is this store closest to home?
7—Are any other stores with equal quality and price equally close to home?
8—Of the stores with equal quality, price and distance, does this store have the best variety?
9—Go to store.

Figure 2. Store Choice Discrimination Net.

There is of course no reason why a discrimination net must adhere to the limitations of lexicographic models. Lexicographic models assume that a subject makes a choice on the basis of knowledge they possess about all stimuli. This may be too severe an assumption. In fact, the rules a decision maker uses may be much more elementary. Instead of asking, as in Figure 2, whether a particular store has the best prices in town, a subject may just ask whether it has satisfactory prices. Discrimination net models also allow for a much higher degree of specificity than do lexicographic models. As a result many more trade-offs are allowed. This is illustrated in Figure 3 which shows part of a discrimination net for a subject who is evaluating the location attribute of a particular supermarket. The information recorded in discrimination nets can also be expanded to consider things such as the amount of time a subject spends at each node.

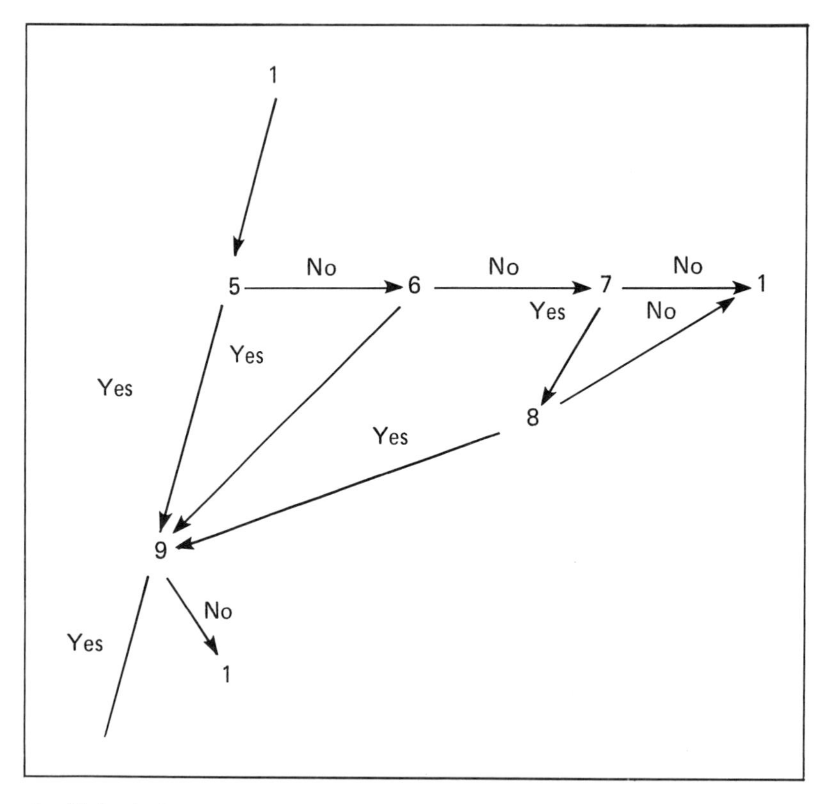

1—Reject store.
5—Is the store within two miles of home?
6—Is the store between two other places that I must visit?
7—Is the store near other stores that I would like to shop?
8—Do these other stores make the extra distance worthwhile?
9—Does the store have adequate parking?

Figure 3. A Location Evaluation Discrimination Net.

The flexibility of discrimination net models does not come without a price. Numerous problems beset discrimination net procedures but some of these problems may be overcome as acquaintance with the procedure increases. One of the biggest problems is the cost of gathering the data. While data may be obtained from questionnaires or from retrospective interviews, the best data are thought to come from protocols that are obtained in the field while subjects are actually engaged in the act of shopping. As a result, discrimination net studies are commonly characterized by very small samples. Another factor which adds to the cost of the data is the need to use experienced interviewers. Even with experienced interviewers, the opportunities for bias to creep in due to interaction between the subject and inteviewer are considerable. A related issue has to do with unravelling the protocols. Several researchers have suspected that discrimination nets may reveal more about the experimenter than about the subject (Haines, 1974; Mitroff, 1969). In a pedagogical example a few years ago, the differences between students who were constructing discrimination nets from identical protocols were as great as the difference between subjects (Haines, 1974).

Most geographers will probably feel that discrimination nets carry the quest for individual level analysis a bit too far. The high level of idiosyncracy that pervades discrimination nets appears to eliminate any ability to generalize or discuss substantive issues. While this is indeed a severe problem, the difficulty has some possibility of being solved or at least ameliorated. Most efforts in this direction have concentrated on attempting to describe the structure of discrimination nets and then tried to determine the extent to which certain structures are prevalent among different subjects or population segments. Nets can be characterized in different ways. Intuitive measures include things such as "branchiness," "depth," and time per node (Wagner and Scurrah, 1971). Graph theoretic measures can also be used to assess the overall degree of similarity between discrimination nets (Bettman, 1971). Despite these advances, generality is still a major problem with discrimination net procedures.

On a more theoretical level, some researchers have questioned whether discrimination nets provide an adequate theory of memory structure. This critique comes from the dependence of discrimination nets on a highly structured decision process that may contain as many as fifty cues. It is not very plausible that even one subject can be expected to duplicate such a sequence in a retest situation. Unfortunately, the ad hoc procedures that have been used to unravel protocols and describe cues to date have prevented a systematic analysis of discrimination net reliability. Nakanishi (1974) has suggested that the strict sequential model posited by discrimination nets is due more to the data file structure of the digitial computer that is used to simulate decision processes than to human information processing. He suggests that humans have a "cluster" data file structure. Instructions are not executed in a strict sequence, but are unpredictably executed from the center to the periphery of the data file. He also suggests that "when there is more than one contiguous instruction to which control can be passed, the stimuli that are dominating the perceptual field of the individual may influence the sequence of retrieval." This is simply another way of stressing that the decision process is probably very dependent upon environmental factors. The possibly dominant role of the environment is one reason why geographers have the opportunity for a significant contribution in this area since environmental or situational factors are largely ignored in most disciplines. This point will be commented on at greater length later.

GENERATE AND TEST MODELS

As defined by Newall and Simon (1972), the generate and test procedure consists of a generation process and a testing process that are completely independent of each other. The generator may be random but in other cases there may be a solution path which takes advantage of information in the task environment to produce a more efficient search procedure.

In the store selection problem we have been looking at so far, there are a number of heuristic search procedures which could be put forward. A recent study (Olshavsky and MacKay, 1975) at Indiana University proposed a model in which the generate process was defined on the basis of the cognitive distance separating supermarkets from the consumers' home locations. Each supermarket, starting with the one nearest home, was then tested. The first supermarket to pass the test was designated as the shopper's primary supermarket. Results using the generate and test model (G-T) were compared to models employing additive weighting (ADD), lexicographic (LEX) and lexicographic semi-order (LSO) procedures. As is shown in Table 2, the generate and test model did substantially better in predicting store choice. Other comparative tests using more sophisticated measuring devices substantiated the superiority of the generate and test model in this situation.

With the generate and test model, the question of how the generating mechanism is to be formed is an empirical question. The same holds true for the testing mechanism. In the supermarket evaluation, for example, stores were judged to be acceptable unless they were below average on critical attributes. Critical attributes were those that the subject evaluated as having greater than average salience and having high discriminability. Going back to the data in Table 1 to illustrate this process, the subject would first generate an itinerary of stores which in this situation would be stores C, B, A, and D (high cognitive distance stores are associated with places that are close to home). Since none of the four attributes is both greater than average in salience (average salience = 2.5) and highly discriminable, the first store, store C, is patronized. If some attributes had been salient and discriminable, the first store would have had to be better than average on all such attributes. If the store were not acceptable, the subject would then have gone on to evaluate store B. While it might seem that such a process could end in a situation where all stores failed and the customer went hungry, no such situations occurred in this study of approximately eighty subjects. The possibility of such an event occurring, though, indicates a logical problem with the model.

Table 2
Number of Correct Predictions by Model and Information Set

Information Set[1]	Model			
	ADD	G–T	LEX	LSO
Internal	40	49	32	38
Objective	21	38	22	30

1. Internal information sets used cognitive attribute measures and objective information sets used primarily physical attribute measures.

Besides predicting store patronage well, the generate and test model has other desirable attributes. It postulates a sequential mode of decision making and does not run into the plausibility problems that the additive weighting model does. On the other hand, it does not demand a rigorous sequence of events as does the discrimination net or even lexicographic models. The generation procedure is the only ordered sequence in the model. The testing process is sequential but no specific order is mandated and it may even be random. The generate and test model seemed to distill the essence of the bulk of the interview data yet maintained a structure that was generic enough to apply to different consumers in different shopping situations. By basing the generation process on cognitive distance the importance of spatial environmental variables is again stressed.

Before turning to decomposition models, it should be noted that the generate and test as well as the other models mentioned in this section use cognitive or internal data as opposed to physical data. This reflects central belief of behavioral geography, namely that the way a person thinks about phenomena is more important in explaining his behavior than the physical properties of the phenomena. A quick illustration of the importance of this view is found in Table 2. All of the four models reported in this table incorporated data on four attributes of a set of supermarkets. Three of these four attributes were then replaced with a set of objective information—cognitive distance was replaced with physical distance, Likert scale values of price and variety were replaced with a price index reported by a local consumer action group and a square footage measure of the scales area of each store. No appropriate objective measure of quality could be found so it was left as an internal measure. Values were rescaled across stores by attribute to be of a comparable magnitude and then input into each of the four models. In each situation, the number of correct predictions made with objective data was far worse than those made with internal data.

OTHER COMPOSITION MODELS

The four classes of composition models mentioned so far are not the only multiattribute process models in this category but they represent the main thrusts of research in this area. Other models are possible. A number of multiplicative and configural decision models have, for example, been suggested in the applied psychology literature on clinical judgment analysis. While most of the tests that have been used employ aggregative data, these models could in many cases be reformed as multiattribute process models (Einhorn, 1971; Goldberg, 1971). MacCrimmon (1968) has also shown how basic decision theory rules, such as Maximin and Maximax strategies may be reformulated to a multiattribute situation. However, these procedures do not present a very plausible picture of the information processing procedure. A more promising field of inquiry may prove to be in the artificial intelligence literature. Some of the models developed in this area also fit under the title of multiattribute process models, many being quite similar to the discrimination net models mentioned earlier. At present, most of these efforts have been spent in problem areas that are at best indirectly related to spatial information processing. An exception to this statement, though, may be found in a provocative article by Moran (1973).

DECOMPOSITION MODELS

Instead of building up toward an estimate of behavior or attitude, decomposition models start with observations obtained under what may be thought of as experimental conditions and then seek to find values of components which, when combined, provide good estimates of the original observation. These estimates also provide a scale for the original set of observations. Since both scale values and values of components are obtained, the underlying operation is referred to as conjoint scaling. To be more precise, let ${}_hO_{ij\ldots m}$ represent an observation obtained under the hth replication of experimental conditions i, j . . . , m. Following Young's development (1973), let i be a particular level of experimental condition I, j be a particular level of experimental condition J, etc. Let components A, B, . . . parallel the experimental conditions I, J, . . . and denote the scale values of particular component levels by a_i, b_j. . . , allowing each a_i or b_j to be single or multidimensional. Conjoint measurement then relates a composite estimate of $_hO_{ij...m}$, designated $c_{ij...m}$, to the scale values by some functional equation or combination rule f. Thus,

$$c_{ij...m} = f(a_i, b_j, \ldots, d_m).$$

It is common to speak of conjoint measurement and multidimensional scaling as two distinct forms of decomposition models, yet this is incorrect because multidimensional scaling is a form of conjoint measurement. To see how this relates to our definition of conjoint measurement consider the following combination rules:

$$f(X) = [\sum_{a=1}^{n} |X_{ia} - X_{ja}|^c]^{1/c}; \qquad i, j = 1, 2, \ldots, p$$

$$g(X) = [\sum_{a=1}^{n} |X_{ia} - X_{ja}|^c]^{1/c}; \qquad i = 1, 2, \ldots, p_1; \quad j = 1, 2, \ldots, p_2$$

where: r = number of dimensions
c = the Minkowski constant ($c \geqslant 1$)
p_1 = number of stimuli
p_2 = number of subjects or ideal points

Function f(X) represents the combination rule used in the multidimensional scaling of similarities judgments and g(X) represents the combination rule used in the multidimensional unfolding of preference judgments. With respect to our earlier terminology, the original observations on similarities or preferences correspond to the $O_{ij...m}$, the X_{ia} and X_{ja} are the components, and the distances derived from the components through f(X) or g(X) are either the $c_{ij...m}$, or a monotonic function of $c_{ij...m}$, depending upon the particular algorithm one chooses to use. It is apparent that it is possible to have many forms of conjoint measurement. One form, with which we shall be concerned later, is known as monotonic analysis of variance and its combination rule may be expressed as

$$f(X) = x_{il} + X_{jl} + \ldots + X_{ml}$$

This says that an observation may be thought of as being composed of its main effects. Unlike multidimensional scaling models, each component level is limited to a single dimension. Other monotonic analysis of variance models are also available. Some include interaction effects (Green and Wind, 1973) and others are multiplicative in design (Johnson, 1974).

Multidimensional scaling models have been used in a variety of ways in behavioral geography. Reviews of the literature may be found in Brummell and Harman (1974) and Golledge and Rushton (1972). At the risk of oversimplification, MDS studies in geography may be divided into spatial applications where the distances reflect cognitive estimates of the physical or temporal distance between points and aspatial applications where the distances are psychological distances, of a similarities of preference nature, with no necessary relation to spatial separation. In the former case the dimensions commonly correspond to the standard North-South and East-West axes (at least after a suitable transformation) while in the latter case the dimensions reflect the aspatial attributes of the stimuli that are being studied.

It is easy in the aspatial case, where the attributes may be defined by dimensions such as price, quality, distance, etc., to see how the output may be interpreted as reflecting cognitive content. Since all MDS models can be defined at an individual level, they also meet the microanalytic requirement for multiattribute process models. It is less easy to see how these models may be used to portray the cognitive processes involved in information handling. By itself, MDS says little about the processing of spatial information. It may, though, be profitably used in conjunction with other procedures to assist in defining parameters of the information handling process. Rao (1974) has outlined a basic information processing model for applying MDS. Rao's model illustrates a number of research areas which may be of interest in exploring the information handling process. These include the relation of different information sources to attribute determination and stimuli configurations, the role of individual and situational characteristics in explaining perceptual and preferential differences, and the relation of perception to preference.

In the spatial case, where the resulting cognitive distances correspond to a measure of separation between points in space, Rao's model may also be used to illustrate how perceptions of separation vary. In addition, the cognitive distance measures may be used in other models to measure cognitive content. An example of this would be in the case of the information processing models reported in Table 2. All these models employed a measure of cognitive distance which, in this case, happened to be derived from a multidimensional scaling of subjects' similarity judgments on spatial separation. The same process could have been used for the other dimensions of price, quality, and variety as well but the tediousness of such a task precluded this. While not included in the study from which these data came (Olshavsky and MacKay, 1975) an interesting test would be to see what, if any, advantage is obtained when using MDS procedures to determine the distance of stores from home as opposed to using a Likert scale as was done with the rest of the attributes.

Other than MDS applications, conjoint measurement methods have seen almost no discussion in the geographic literature so far, but their potential is as great as MDS procedures. In contrast to MDS, conjoint methods in the social sciences often derive their inputs from experiments in a behavioral laboratory or from an evaluation of hypothetical stimuli. While this may seem to run counter to the strong empirical directions in geography, it may be that an evaluation of hypothetical stimuli would contribute substantially to an understanding of how individuals go about making choices in real life environments. To give an example from a study that was cited before (Olshavsky and MacKay, 1975) subjects in this project were asked to make rank order preferences between hypothetical stores that were defined on the basis of possessing different amounts of the four attributes cited earlier. This was done in three stages. The first stage had stores defined on just two dimensions (price and distance), the second had stores defined on three dimensions (price, distance, and variety) and the third had stores defined on four dimensions (price, distance, variety, and quality). Five values of each attribute were used and Latin square and Greco-Latin square designs were employed in the last two stages respectively to keep the number of observations at twenty-five per stage.

Results from this study cannot be reported since the analyses are still incomplete, but it is valuable to look at some of the questions that can be asked with these data. For each attribute, one can determine if it is valued in a manner that is linear or even monotonic to the change in the physical values of the treatment levels. By looking at the success of the scaling procedure, an idea of the appropriateness of a linear main effects model under different conditions of informational complexity is obtained. This might not be apparent with a traditional analysis of variance procedure since a resulting interaction could be a function an ordinal data set being used with metric procedures and not indicate a true interaction. It is also possible to take the difference between the extreme values reported for any attribute and use the absolute value of this difference as an estimate of the significance of the corresponding attribute in the decision process. The degree to which the salience of an attribute changes as it is embedded in increasing accounts of information is an unexplored area in behavioral geography. Finally, one can, as in MDS methods, use the scaled value of the ordinal input data for metric analysis in other situations.

There are a number of ways in which monotonic analysis of variance methods can be used to assist other models. One example would be in the information processing models cited earlier. All these models made use of a measure of preference for each attribute. These were assumed to be derived by a magnitude estimation method but the lexicographic and generate and test procedures could simply have used rank ordered data. Another way of estimating these values would have been to use the spread of the treatment values for each attribute. The applicability of such a procedure would, of course, depend upon the ability to relate the data from the behavioral laboratory to the real world judgments on which the earlier models were based. Such a comparison whether positive or negative, would be of interest for future studies.

While conjoint measurement procedures are conceptually attractive, a number of caveats should be kept in mind when using them. One caveat is that many of plausibility problems inherent in additive weighting models are also present in MDS procedures. MDS methods may be more sophisticated in terms of their abilities to derive metric scale data from ordinal inputs but they still assume that preferences or similarities are determined by additively weighting the distances along each derived attribute. The assumptions which underlie nonmetric MDS models also have some other disturbing implications. Such implications are noted by Zinnes and Wolff (1975) who show that in extending the unidimensional Thurstonian equations (Torgerson, 1958) on which most MDS processes are predicated to the multidimensional setting a subject must make comparative judgments between distance pairs which may theoretically be negative. As a result, subjects can report that one stimulus pair is more similar than another (because their distance is smaller) when in fact the absolute distance separating the stimulus pair is larger. These conclusions may suggest the need to develop metric and interactive forms of MDS that are based on a more plausible set of information processing assumptions. On a more empirical note, the reliability and validity of MDS solutions at the microanalytic level is a topic that has received very little attention in any discipline. Most studies have dealt with either artificial data or aggregate empirical data. A recent study in marketing has suggested that there may be an overwhelming amount of noise in the direct similarities data which serve as input to most MDS algorithms (Summers and MacKay).

Many pages could be written about the real and potential problems of conjoint measurement methods. Some of these problems are inherent in the technique itself while others are due to the fact that the stimuli geographers deal with are more complex (except possibly for distance) than the stimuli used by the psychometricians who developed MDS procedures. In all applications, the predictability of our results and the plausibility of our assumptions should be carefully investigated.

SUMMARY AND PROSPECTS

A class of models—multiattribute process models—has been described as possessing the potential for simultaneously investigating the cognitive mapping and processing of spatial information. The models differ substantially in terms of their morphology and plausibility. Even though many of the models use the same inputs and derive similar outputs, there have not been enough theoretical or empirical comparisons of these models to determine under what set of conditions alternative models are most appropriate. One empirical problem in evaluating these models is that they will often give identical predictions for a large number of subjects. The similarity of results combined with data that commonly exhibit a high amount of noise makes the common practice of using statistics such as chi-square or the Kolmogorov-Smirnov statistic to find the probability that deviant observations could have arisen by chance rather dubious. As a result, a number of mathematical statisticians have suggested that significance tests should be dispensed with in evaluating these models. If this were done, the burden of evaluation would fall on finding "critical" experiments (Gregg and Simon, 1967) that separate alternative hypotheses radically. This is not always easy to do.

Despite the problems inherent in their application, multiattribute process models would seem to be a fruitful area of inquiry for behavioral geographers. Behavioral geographers are usually concerned with situations that are related by their common occurrence in space yet distinctive in the mix of psychological processes that are employed in each situation. Thus, the processes that are employed in a migration decision are probably quite distinct from those employed in store choice or journey to work decisions. Multiattribute process models have the ability to adapt to different situations. They are very flexible and can incorporate a wide range of situational characteristics. Further, there is reason to suspect that multiattribute process models may be used quite successfully by geographers. This is due to geographers' familiarity with methods for logically and quantitatively describing the environment, an area which has been too often neglected in other fields. Application of these techniques should extend our insights into spatial behavior processes.

REFERENCES

Amadeo, Douglas, and Golledge, Reginald G. *An Introduction to Scientific Reasoning in Geography.* New York: Wiley, 1975.

Anderson, Norman H. "Algebraic Models in Perception." In Edward C. Carterette and Morton P. Friedman (eds.) *Handbook of Perception, vol. 2.* New York: Academic Press, 1974, pp. 215-298.

Appleyard, Donald. "Styles and Methods of Structuring a City." *Environment and Behavior* 2 (1970):100-117.

Betak, John F. "Syntactic and Subjective Complexity: A Simple Test of a Two-dimensional Language Model." *Geographical Analysis* 7 (1975):285-293.

Bettman, James R. "A Graph Theory Approach to Comparing Consumer Information Processing Models." *Management Science* 18 (1971):114-128.

Bettman, James R.; Capon, Noel; and Lutz, Richard J. "Multiattribute Measurement Models and Multiattribute Theory: A Test of Construct Validity." *The Journal of Consumer Research* 1 (1975):1-15.

Burnett, Pat. "The Dimensions of Alternatives in Spatial Choice Processes." *Geographical Analysis* 5 (1973):181-204.

Brummell, A.C., and Harman, E.J. "Behavioural Geography and Multidimensional Scaling," McMaster University, Department of Geography, Discussion Paper no. 1, 1974.

Cadwallader, Martin. "A Behavioral Model of Consumer Spatial Decision Making." *Economic Geography* 51 (1975):339-349.

Dawes, Robyn M., and Corrigan, Bernard. "Linear Models in Decision Making." *Psychological Bulletin* 81 (1974):95-106.

Downs, Roger M. "The Cognitive Structure of an Urban Shopping Center." *Environment and Behavior* 2 (1970):13-37.

Downs, Roger M., and Stea, David. *Image and Environment.* Chicago: Aldine, 1973.

Einhorn, Hillel J. "Use of Nonlinear Noncompensatory Models as a Function of Task and Amount of Information." *Organizational Behavior and Human Performance* 6 (1971):1-27.

Fishbein, Martin. "A Behavior Theory Approach to the Relations Between Beliefs About an Object and the Attitude Toward the Object." In M. Fishbein (ed.) *Readings in Attitude Theory and Measurement.* New York: Wiley, 1967, pp. 389-400.

———. "Attitude and the Prediction of Behavior." In M. Fishbein (ed.) *Readings in Attitude Theory and Measurement.* New York: Wiley, 1967, pp. 477-492.

Goldberg, Lewis R. "Five Models of Clinical Judgment: An Empirical Comparison Between Linear and Nonlinear Representations of the Human Inference Process." *Organizational Behavior and Human Performance* 6 (1971):458-479.

Golledge, R.G. (ed.). "On Determining Cognitive Configurations of a City, vol. 1." Unpublished manuscript completed for the National Science Foundation, 1975.

———. "The Geographical Relevance of Some Learning Theories." In K.R. Cox and R.G. Golledge (eds.) *Behavioral Problems in Geography: A Syposium.* Evanston, Ill.: Northwestern University, Department of Geography Studies in Geography, no. 17, 1969, pp. 101-145.

Golledge, R.G.; Briggs, R.; and Demko, D. "The Configuration of Distances in Intra-Urban Space." *Proceedings: Association of American Geographers*, 1969, pp. 60-65.

Golledge, R.G., and Rushton, G. "Multidimensional Scaling: Review and Geographical Applications." *Association of American Geographers,* Technical Paper no. 10, Washington, D.C., 1972.

Grant, David A. "Testing the Null Hypothesis and the Strategy and Tactics of Investigating Theoretical Models." *Psychological Review* 69 (1962):54-61.

Green, Paul E., and Wind, Yoram. *Multiattribute Decisions in Marketing: A Measurement Approach.* Hinsdale, Ill.: The Dryden Press, 1973.

Gregg, L.W., and Simon, H.A. "Process Models and Stochastic Theories of Simple Concept Formation." *Journal of Mathematical Psychology* 4 (1967):246-276.

Haines, George H., Jr. "Process Models of Consumer Decision Making." In G. David Hughes and Michael L. Ray (eds.) *Buyer/Consumer Information Processing.* Chapel Hill: The University of North Carolina Press, 1974, pp. 89-107.

Harvey, D.W. "Conceptual and Measurement Problems in the Cognitive-Behavioral Approach to Location Theory." In K.R. Cox and R.G. Golledge (eds.) *Behavioral Problems in Geography: A Symposium.* Evanston, Ill.: Northwestern University, Department of Geography, Studies in Geography, no. 17, 1969, pp. 35-58.

Hudson, Ray. "Images of the Retailing Environment: An Example of the Use of Repertory Grid Methodology." *Environment and Behavior* 6 (1974):470-494.

Hughes, G.D., and Naert, P.A. "A Computer-Controlled Experiment in Consumer Behavior." *Journal of Business* 43 (1970):354-372.

Ittelson, William H. "Environment Perception and Contemporary Perceptual Theory." In W.H. Ittelson (ed.) *Environment and Cognition.* New York: Seminar Press, 1973, pp. 1-19.

Jackson, L.E., and Johnson, R.J. "Structuring the Image: An Investigation of the Elements of Mental Maps." *Environment and Planning* 4 (1972):415-427.

Johnson, Richard J. "Trade-Off Analysis of Consumer Values." *Journal of Marketing Research* 11 (1974):121-127.

Louviere, Jordan J. "The Dimensions of Alternatives in Spatial Choice Processes: A Comment." *Geographical Analysis* 7 (1975):315-326.

MacCrimmon, K.R. "Decision Making Among Multiple-Attribute Alternatives: A Survey and Consolidated Approach." Rand Corporation Memorandum RM-4823-ARPA, 1968.

MacKay, David B. "The Effect of Spatial Stimuli on the Estimation of Cognitive Maps." *Geographical Analysis* 8 (1976):439-452.

———. "A Microanalytic Approach to Store Location Analysis." *Journal of Marketing Research* 9 (1972):134-140.

Mazis, Michael B.; Ahtola, Olli T.; and Klippel, Eugene R. "A Comparison of Four Multi-Attribute Models in the Prediction of Consumer Attitudes." *The Journal of Consumer Research* 2 (1975):38-52.

Miller, G.A. "The Magic Number Seven, Plus or Minus Two: Some Limits on our Capacity for Processing Information." *Psychological Review* 63 (1956):81-97.

Mitroff, Ian. "Fundamental Issues in the Simulation of Human Behavior." *Management Science* 15 (1969):B650-B664.

Moran, Thomas P. "The Cognitive Structure of Spatial Knowledge." In Wolfgang F.E. Preiser (ed.) *Environmental Design Research, vol. 2.* Stroudsburg, Pa., 1973, pp. 336-373.

Nakaniski, Masao. "Decision-Net Models and Human Information Processing." In G. David Hughes and Michael L. Ray (eds.) *Buyer/Consumer Information Processing.* Chapel Hill: The University of North Carolina Press, 1974, pp. 75-88.

Newell, Allen, and Simon, Herbert A. *Human Problem Solving.* Englewood Cliffs, N.J.: Prentice-Hall, 1972.

Olshavsky, Richard W., and MacKay, David B. "Towards an Information Processing Theory of Consumer Behavior." Unpublished manuscript, Indiana University, Department of Marketing, 1975.

Pessemier, Edgar A, and Wilkie, William L. "Multi-Attribute Choice Theory: A Review and Analysis." In G. David Hughes and Michael L. Ray (eds.) *Buyer/Consumer Information Processing.* Chapel Hill: The University of North Carolinia Press, 1974, pp. 288-330.

Rao, Vithala, R. "Multidimensional Scaling Models for Research on Consumer/Buyer Information Processing." In G. David Hughes and Michael L. Ray (eds.) *Buyer/Consumer Information Processing.* Chapel Hill: The University of North Carolinia Press, 1974, pp. 373-383.

Rosenberg, Milton J. "Cognitive Structure and Attitudinal Affect." *Journal of Abnormal and Social Psychology* 53 (1956):367-372.

Russ, Frederick A. "Consumer Evaluation of Alternative Product Models." Unpublished Ph.D. dissertation, Carnegie-Mellon University, 1971.

Sandusky, Arthur. "Memory Processes and Judgment." In Edward C. Carterette and Morton P. Friedman (eds.) *Handbook of Perception,* vol. 2. New York: Academic Press, 1974, pp. 61-83.

Schneider, Clarke H.P. "Models of Space Searching in Urban Areas." *Geographical Analysis* 7 (1975):173-185.

Shanteau, James, and Troutman, Michael C. "Commentaries on Bettman, Capon and Lutz." *The Journal of Consumer Research* 1 (1975):16-19.

Summers, John O., and MacKay, David B. "On the Validity and Reliability of Direct Similarity Judgments." *Journal of Marketing Research*, in press.

Torgerson, W.S. *Theory and Methods of Scaling.* New York: Wiley, 1958.

Wagner, Daniel A., and Scurrah, Martin J. "Some Characteristics of Human Problem-Solving in Chess." *Cognitive Psychology* 2 (1971):454-478.

Walmsley, D.J. *The Simple Consumer Behaviour System.* Armidale, Austrialia: University of New England, Department of Geography, Research Series in Applied Geography, #38, 1974.

Wilkie, William L., and Pessemier, Edgar A. "Issues in Marketing's Use of Multi-Attribute Attitude Models." *Journal of Marketing Research* 10 (1973):428-441.

Young, Forrest W. "Conjoint Scaling." Chapel Hill: University of North Carolina, the L.L. Thurstone Psychometric Laboratory, Report no. 118, 1973.

Zinnes, Joseph L., and Wolff, Ronald P. "Single and Multidimensional Same-Different Judgments." Indiana University, Department of Psychology, Report 75-8, 1975.

Applied Social Geography

Introduction

The 1960s will be marked in the history of geography as a period of major change and growth in response to the set of innovations which have been labelled the "quantitative revolution." The 1970s may eventually become equally noted as the decade when the interests of geographers expanded dramatically and began to include a far broader array of research problems than at any time in the past. This dramatic expansion is especially evident in the area of social geography where many of the new research problems have particular social or political relevance.

The possibility of a disastrous accident at a nuclear power plant is one of the most alarming prospects facing American society and the proliferation of nuclear reactors makes the analysis of their impacts, including the possibility of disaster, a crucial societal problem. Julian Wolpert reports on a research project on the problem of evacuating cities following a nearby nuclear accident. The major objective of the evacuation study is to identify the size and nature of buffer zones which should separate nuclear reactor sites from major settlements but Wolpert's results raise disturbing questions about the location of existing nuclear power plants, about the candor of the nuclear power industry, and about the decision processes involved in siting nuclear facilities.

The geography of social problems is only beginning to develop but Harold Rose makes a major contribution to it in his analysis of the geography of homicide. Rose shows how important insights into social problems can be gained by careful spatial analysis at several scales. He examines behavioral as well as environmental approaches to violent behavior in order to develop a framework for analyzing the occurrence of homicides and provides an ecological analysis of homicides in St. Louis and Detroit. He recognizes that personal as well as environmental information is necessary for a complete understanding of the homicide problem but his ecological analysis shows how the act of homicide is strongly associated with social disorganization and stressful environments.

The geography of poverty has attracted some interest as a topic of social relevance but little serious research has been done in this area. Ernest Wohlenberg reviews the problems and possibilities for research in the geography of poverty. The problems begin with the task of defining poverty in an operationally useful way and include the analysis of alternative policies for relieving poverty. These policies include rural industrialization, relocation of the population of poverty areas, growth center strategies, and direct income transfers. Geographers should find a role in policy analysis because these alternative policies have distinct spatial aspects.

The geography of education is another topical area which has been relatively neglected and Alan Backler has provided an exploratory analysis using mainly data for states. He investigates the relationships between educational attainment and a set of variables including race, family income, and state commitment to education where commitment is measured by state expenditures on education. Backler also analyzes the effect of migration in spatially redistributing the benefits of educational expenditures.

Ethnic geography is a more firmly established field and one which has attracted considerable research interest. In the final paper of this section Don Bennett provides an extensive review of recent work in ethnic geography including the topics of migration, intergroup interaction, acculturation, and the formation of ethnic neighborhoods. Bennett notes that, despite the extensive geographic literature on ethnicity, studies have mainly been confined to a few ethnic groups and that very few studies have been set within wider theoretical frameworks.

Evacuation from the Nuclear Accident

Julian Wolpert
Princeton University

I recall as a child being frightened by the rumblings of what turned out to be a minor earthquake and wondering why it had occurred. My image was that the quake was a necessary by-product of some useful activity. I naturally had assumed that a rumbling under the ground had its origins under the ground and that one could put an end to the rumbling by digging a tunnel to the control room and telling those in charge that *they* had been discovered and were doing more damage than good. (My control room for the hurricane, as you might imagine, was a balloon just above the clouds.) How different it is to visit a nuclear power facility. It looks more like a war gaming room, only staffed by engineers and technicians. The facility is not unattractive structurally. It can be hidden rather charmingly far from the main roads and it gives off no yellow or black smoke. Its discharge is colorless, odorless, and tasteless, even when most lethal. It is increasingly common for the public to be invited to visit a nuclear power plant just prior to the start of operation. How could it be harmful?

I discovered later that my naive image of the harmful origin of earthquakes and hurricanes required another explanation than the intervention of man. Their consequences, perhaps cannot be as easily foretold or prevented as is the case with nuclear plants. Is the threat of nuclear accident comparable to that of being struck by lightning or a meteorite as has been claimed? If the threat were significant and preventable, could the controls be safeguarded from human errors? Engineering solutions to the problems of hazard, after all, are designed by man.

What are the risks of a major accident, in which containment fails and radioactive material is spread across the countryside? Can engineering solutions be found to insure that the level of risk is held to a minute point, e.g., say the chances for any one place to be hit by a large meteorite? Are the risks primarily institutional rather than controllable through fail-safe engineering solutions? I do not know the answers to these questions but I do know that I will not become better informed by reading the official reports of pseudo-rigorous research written by the power companies.

I remained quite neutral on the issue of nuclear reactors during our studies in the late 1960s of noxious facilities. The problems of thermal pollution seemed at the time not to be beyond solution and the nuclear share of power generation was still modest. Nuclear energy received another spurt after the 1973 Mid-East War but the trends of the past year have become more alarming. I felt that more study was necessary after reading about the doubts and defections of some of the nuclear cadre. Our research is only in its preliminary stage but I would like to share with you some of our plans and speculations with the hope that you would also examine the evidence and form your own judgments.

The proliferation of nuclear reactors and the shortcomings of the impact analysis format should be considered with some alarm. I am more distressed, however, about the lack of candor by nuclear power officials and by the overly vehement denials of danger. The process of hazard analysis involves the stringing together of events which themselves have low probabilities of occurrence. The overall probabilities of cooling system failure, core melt down, and containment rupture appear, therefore, to be remote. Allowances cannot be made, however, for unlikely human errors. The defensiveness of officials makes me fear that precious warning time would be wasted underestimating the scope of hazardous procedures and minor or serious accidents.

THE EVACUATION MODEL

I am working with a group of researchers to examine how cities threatened by nuclear power plant accident might be evacuated.[1] What is the probability that a large or small city might have to be evacuated? I do not know. The chances are small now. Only 59 reactors are operational now, but 68 more are under construction and we may have a total of 1,000 reactors by the year 2000. Approximately, 500 nuclear reactors are already operating within 45 countries of the world. To be licensed, each new reactor site requires an impact statement, some analysis of the risk likelihood, as well as an evacuation plan. The impact statement for each reactor, however, is a document quite independent from impact statements of other plants. The risk of an incident occurring, however, is additive and perhaps multiplicative because of the transfer of fuel. The Price-Anderson Act limits liability claims due to accidents to the artificial amount of $560 M. It was designed to protect the utilities and deny full compensation to victims of a catastrophic nuclear accident. The true actuarial rate is probably not affordable. The Class 9 type of accident according to the Nuclear Regulatory Commission could result in 3,300 immediate deaths, 45,000 cancer fatalities during the succeeding 30 years, $14 billion in property damages, and the long term abandonment of 290 square miles of land. According to Von Hippel, 10 immediate deaths implies 7,000 cancer deaths, 4000 genetic defects, 60,000 thyroid tumor cases, 3,000 square miles of land contamination. The true damage could be infinite. The larger reactors could in the most serious accidents yield lethal effects up to 100 miles away for a long period of time.

I want to confine my attention here purely to the evacuation issue and leave aside other dangers such as the impacts of:

a. low-level radioactivity emitted into the air and water around nuclear stations;
b. the problem of sabotage;
c. the movement of nuclear fuel by road and railroad; and
d. the problem of waste disposal.

These issues merit independent examination as well. I will also not discuss the minor accidents and engineering shortcomings which have occurred at many of the existing plants because they are already documented.

Why examine the evacuation issue at all? Perhaps the need for evacuation will be so slight or the success of evacuation so questionable that we may safely exclude that prospect from consideration. An evacuation plan could not even be tested in a drill. Research on evacuation planning is justified primarily by the need to provide *peripheral zones around settlements for the exclusion of nuclear reactor sites.* How large a barrier zone should be provided around a nuclear site? What restrictions on land use should be made in this buffer?[2] What alternative types of major accidents can be anticipated and with what degree of forewarning? Will the escaping radioactive material diffuse through the atmosphere, or through ground water or water bodies? Will the discharge diffuse rapidly or slowly and in only one direction or radially? These factors are all important in the evacuation problem.

We elected to use a simulation approach to analyzing the issues of evacuation that would allow for a controlled set of parameter testing. The hardware is an interactive computer graphic system with a CRT, plotter and digitizer which has been used hereto primarily for traffic flow analysis and the siting of public facilities. The remaining software package evolved from earlier work I did on coal mine disasters. We start with a "rationally" controlled evacuation plan so as to be able, later in the analysis, to incorporate varying degrees of breakdown and panic response for measuring the plan's performance under less than ideal conditions. The diffusion of the radioactive cloud for airborne leakage has its origin point at the reactor and its movement if governed by atmospheric conditions. The cloud's evolution can be displayed as an overlay on a large-scale map of the affected area.

The simulation has 5 linked submodels governing the:

a. explosion and radioactivity diffusion;
b. information diffusion;
c. population reaction;
d. evacuation; and
e. casualty elements.

The explosion model generates the simulated physical effects of the accident, including the spread of fallout. The information model withholds, spreads, or distorts the news about the accident to officials and the public via radio, siren and other means. The evacuation plan controls the flow of information in a preset or interactive mode. The population reaction model is dependent upon population distribution which will vary with time of day, day of the week, and season. A network model specifies a graph-theoretical representation of urban arterial streets, highways and transit routes. Households are assigned probabilities of selecting given modes of response. Evacuation routes and probable congestion effects are determined in this way. The evacuation model employs the population reactions to calculate congestion flows under alternate traffic control schemes and their effects in producing blockage, delay, and accidents. The casualty model calculates probabilities of injuries and death resulting from the explosion, fallout, and evacuation. Other evacuation plans can be tested to determine their "yield" of casualties. Other types of accidents with a variety of differing atmospheric conditions will be examined as well. The key in the sensitivity analysis, of course, is to withdraw systematically the nuclear site from the populated area to assess the importance of the buffer.

The simulation model, as described, is not so difficult to implement. Each major population center threatened by a nearby reactor in operation or planned for a local site should have its own contigency plans. I would predict that many communities would be distressed by the outcome of their study and planning effort.

Our preliminary simulation designs point to the severe problems of evacuation. An airborne radiation cloud cannot be monitored remotely with present technology. Held in suspension, carried by winds, subject to rapid changes in direction or deposited on the ground surface, the variability in the course of the radioactive material makes statistically programmed evacuation plans unsuitable.

If the source of danger cannot be seen and its course predicted, how can guidance be provided to the population? How effectively can a transportation network designed for the journey to work move a large population in one direction. Will mass transit work? What is the best way to inform the population? Our early indications point to the extreme hazards of evacuation, the high likelihood that many will choose not to leave or will be immovable.

Larger cities create an exponential growth of problems. The early experimentation points to remote sites, just how remote we cannot tell yet, but we suspect that some existing plants are already too close. It may be best also to restrict the development of industrial complexes at the reactor sites, despite the obvious advantages of being able to utilize the heated water from the reactor in industrial processes.

Should sparsely settled areas be victimized by reactor sites because accident hazard and the possible need for evacuation would affect fewer people? If insufficient numbers of remote sites which satisfy the other requirements for plants can be found in the more densely settled regions of the nation, what will be the effect of an interregional transfer of power capacity on the movement of industry, jobs, and labor force? Must the northeast quadrant accept nuclear reactors to maintain its comparative position? These issues pertain to the safe distance problem. Our goal for study here is to proceed from buffer rings around settlements, the exclusion zones, to examine the tradeoffs of a lesser number of sites each containing a cluster of reactors.

Should reactors be clustered or dispersed among many sites, given the danger of evacuation? What criteria should be used to order or not order an evacuation and how should discretion for an evacuation plan be allocated? Should sites be shifted to sparsely settled areas, even if they are in the close proximity to major recreational zones? The Bailly Nuclear Generating Station next to Indiana Dunes State Park has a daily peak of 87,000 visitors on a hot summer weekend. What special evacuation problems are posed in this case? Long Beach Island on the New Jersey shore attracts 250,000 visitors to a site only a few miles from the Oyster Creek Station. The single road to the mainland passes within a mile of the reactor—the wrong direction for evacuation! The Indian Point plant in New York has 66,000 people living within a 5 mile radius, 900,000 within 20 miles, and 16 million within 40 miles.

Other issues are more serious. Where should the evacuated population go—how far and in what direction? Should a plan for reception centers be developed? What about reentry into the contaminated area? When can reentry occur and under what conditions? What kinds of decontamination processes would have to precede reentry? What kinds of other safeguard processes would have to precede reentry? Premature reentry defeats the purpose of evacuation. As you know, only massive doses of radiation poisoning lead to immediate death. The dangers of smaller doses may take their toll over two decades or more, showing up in increased incidence of cancer and genetic aberrations.

All the 59 licensed nuclear power plants must have specific evacuation plans, as a requirement of the AEC/NRC approval process. Are these plans adequate and feasible, given what we know about evacuation difficulties? The evacuation plan for one of our local facilities covers only the immediate plant area. The plan excludes the nearby town of Salem City and the metropolitan areas of Wilmington and Philadelphia, which are respectively only 30 and 50 miles away and downwind.

Our simulation is being carried out for the Salem plant under conditions of the worst possible accident; requiring the evacuation of a 400 square-mile area. Among the possibilities to be examined are the increased hazards resulting from clustering more plants at the same site. Clustering has become a more prominent notion because of increased public opposition to nuclear plants. Shifting and licensing problems from 350 sites (the required number of new sites in the next 25 years) to 20 clusters of 20 plants each would reduce considerably the opposition problem.

My objective in this research on evacuation and in reporting to you the details of our research plan are prompted by my deep concern that independent studies of the dangers must be done. I do not believe, as has been maintained, that "risks from accidents have been generously overcompensated by cautious design, careful operational analysis, accident analysis, and incorporation of multiple, independent safety systems, to the point where these risks are too low to estimate." I do not trust the engineering solutions and, even less, the computation of the human factor. Too many very unlikely accidents have already occurred with respect to core cooling systems. The Rasmussen Report indicates that only one individual of the 15,000,000 people living in the vicinity of 100 reactors could expect to die from a reactor accident, i.e., one chance in 300,000,000. With the assumption of 100 plants in operation his data indicated one core-melt accident every 170 years on the average and only one in ten such accidents leading to serious danger of immediate fatalities.

I do not have the data to refute these estimates but neither was Rasmussen able to revise his studies after improbable accidents occurred. He further assumes successful and rational evacuation procedures. No acts of God can be permitted.

The issue of accidents is easy to distort and to sensationalize but I have little reason to believe analysis and data exclusively generated and evaluated by an industry which has little incentive to be candid. The evacuation plan must incorporate the knowledgeable participation of the local communities, with full liabilities negotiated.

CONCLUSION

Examining the problem of evacuation is disquieting. It recalls wartime bombing, the post-World War II nuclear scare and flood, earthquake, and volcanic eruptions. Urban evacuation related to accidents encapsulates within a brief time span many of the urban decay dynamics which have taken decades to unfold. Another form of vulnerability is revealed which affects most profoundly the older industrial cities of the northeast. Perhaps most significant, however, is the additional vigilance which can be imposed upon the nuclear sector as well as an enhanced incentive for greater candor in an open planning framework with involvement of local citizens. Extensive buffer zones, more remote sites, greater built-in redundancy in plant operations, and more rigorous safety checks impose greater costs for nuclear energy, but a single serious accident will jeopardize the continued operation of all reactors.

NOTES

1. The research team includes: David Morell, Jonathan Hazony, and Jerome Lutin of Princeton University.
2. The Environmental Commissioner in New Jersey used the following criteria:
   a. not more than 500 people per square mile within a 30 mile radius of the plant;
   b. unpopulated zone of 4 mile radius around single plant, 6 miles from twin plants;
   c. no settlement of 25,000 or more population with 40 mile radius.

Urban Violence: The Case of Homicide

Harold Rose
University of Wisconsin
at Milwaukee

It is frequently stated that the America of the last fifteen years represents an increasingly violent society. Since 1960, the incidence of violent crime has risen from approximately 160 per 100,000 to a level approaching 400 per 100,000 in 1972. A growing societal concern with rising levels of violence can be observed from the results of professional pollsters and the posturing of political candidates running for both local and national office. While numerous theories have been proferred to explain this sudden upsurge in national violence levels, those responsible for the provision of public safety have had little success in bringing the epidemic under control. This paper will focus attention on a single facet of the violence syndrome which has fostered a public outcry for improved safety in residential environments. The aspect of violence which has been selected for attention is homicide, usually considered the most serious form of violent behavior. A recent study by Rossi and others revealed a general consensus, among all subgroups in their single city sample, that murder received very high seriousness scores (1974, p. 227). The study did point out however, that the nature of the event and the victim-offender relationship did influence the seriousness score (rank) assigned by individual respondents.

HOMICIDE EXPLANATIONS: ENVIRONMENTAL AND BEHAVIORAL ORIENTATIONS

Homicide as an element of violence has received only scant attention in the geographical literature and this is understandable, given that the field of social geography is yet in a youthful stage of development. Even those contributors to the growing literature on the geography of crime have failed to give this topic more than limited treatment. The treatment which follows is at best a preliminary attempt to address the question of urban environmental behavior which is undergirded by violence. This does not purport to represent a contribution to the geography of crime, but instead it is intended to represent an excursion into the realm of behavioral geography.

Homicide, as other patterns of violence, climbed steadily from the mid-sixties through the early seventies. There is now some preliminary evidence that the peak period has been reached, and that we are now on the down side of the cycle. Though this may be the case, there is still little definitive evidence to explain the sudden upsurge after more than thirty years in which homicide showed general evidence of decline, although marked by minor periodic fluctuations. A recent investigation of homicide rates in the fifty largest cities of the nation attributed the general growth rate in homicide to a national trend, rather than to any unique characteristics which might be associated with individual places (Barnett, et. al., 1975, p. 93). The investigators suggest that causes which are national in scope, such as the proliferation in handguns, greater leniency in the criminal justice system, etc., are likely to play a more important role in the homicide growth rate than specific factors present within individual cities. Needless to say, while such explanations may aid in explaining changes in national trends they fail to assist the geographer who is concerned with the role of specific environments on the promotion of the homicide act.

Another series of explanations were recently offered on the individual behavioral level. Waldron and Eyer (1975), who were primarily interested in the increase in the homicide rate among 15-24 year olds during the sixties, attributed the rise to increased alcohol consumption, impulsive rage, and the general breakdown in respect for societal institutions. They admitted the tentative nature of these explanations and urged caution in accepting them as being definitive, because of the inadequacy of the data used to describe victims and perpetrators.

There has been no detailed investigation of homicide on a national basis during its recent resurgence as an important contributor to death rates. A number of studies detailing the pattern of homicide within individual urban places have been produced during the last twenty years, but most of these described the pattern prevailing before 1965, and most were designed to replicate the study of Wolfgang (1958) whose detailed analysis of homicide patterns of Philadelphia is now recognized as a classic. The work of Block and Zimring (1973) and Block (1976) are among the few studies that have addressed themselves to the topic during the recent period. None of these studies addresses itself specifically to the homicide environment, aside from that of Wolfgang.

Most homicide studies are conducted by researchers whose approach is either ecological or clinical, but seldom is there an attempt to integrate these two orientations as a means of attempting to ascertain the joint contribution of internal and external influences on the commission of the act. Bullock writing more than twenty years ago suggested the need for such a research direction. In making this case he stated:

> Using urban homicides as a form of criminal behavior, this paper seeks to suggest a direction toward the development of such a theory by tracing out the natural manner in which assailant, victim, and place become organized into a complex of ecological and interpersonal situations that result in homicide (1955, p. 566).

This need is still present, lest we be left with the macroscale explanations of Barnett and others and simply withdraw from attempts at microscale analysis.

THE HOMICIDE ENVIRONMENT

It is generally recognized that homicide occurs more frequently in some environments than in others. As a matter of fact, homicides, unlike suicides, tend to be concentrated in a limited number of environments. Block (1976) in describing the spatial pattern of homicide in Chicago in 1971 indicated that 2% of the blocks accounted for 22% of the homicides. Thus it appears that homicide environments can be easily delineated within urban areas. Segments of ghetto areas in large urban environments are the most frequent sites for homicide. Bullock (1955) attributed this to patterns of ecological segregation which concentrate people possessing common characteristics to a common niche. It is apparent that the Bullock type perspective is supportive to Wolfgang and Ferracuti's concept of a subculture of violence. The latter writers suggest the following: there is a potent theme of violence current in the cluster of values that make up the life-styles, the socialization process, the interpersonal relationships of individuals living in similar conditions. However, there is opposition to this notion. One writer recently stated that while this thesis has received a measure of acceptance, there is evidence to suggest that it is questionable (Erlanger, 1974). Regardless of the merits of the thesis, it is true that certain environments lend themselves to the commission of acts of violence and it seems we are not yet in a position to partition the role of the diverse factors on homicidal behavior. The integration of man and his environment remains poorly understood, but the work of the ecological and environmental psychologist is beginning to provide additional insights that could aid in the elimination of this veil of ignorance.

THE HOMICIDE ENVIRONMENT AND THE ROLE OF ETHNICITY

Just as there is a propensity for specific environments to have a higher probability for the occurrence of homicide, there is likewise a higher probability that the homicide victim and offender are likely to be members of ethnic and oppressed minorities. The principal victims, in terms of risk, during the recent period have been black males. The ratio of black to white homicides utilized by Barnett and others (1975) to describe the differential propensity of being a homicide victim is 8 to 1. They contend that this ratio shows signs of stability across all cities. In 1972, black male homicide rates were set at 83.1 per 100,000, while the corresponding rate for white males was 8.2 per 100,000 (Klebba, 1975, p. 197). Females of both races have relatively low homicide rates, although the rate for black females is almost twice as high as that for white males and six times that of white females.

It was recently revealed that non-white males in the state of Michigan experienced an increase in their mortality rate during the decade of the sixties and a corresponding decrease in life expectancy. All other race-sex groups experienced an improvement in life expectancy as mortality rates fell during the decade. The major contributors to increased black male mortality levels were the rising incidence of accidents and homicides among black males 15-44 years of age. It is reported that homicide was responsible for reducing black male life expectancy by 2.3 years during the period 1969-1971 (Gorwitz and Ruth, 1976, p. 144). Thus behavioral causes of death are beginning to exert an undue influence on the life expectancy of young black males. Members of the age cohort born between 1948-1952 who died in 1972 experienced homicide rates of 152.7 per 100,000. White males in the same cohort possessed a rate of 13.2 per 100,000 (Klebba, 1975, p. 197). 40% of the homicide victims in 1973 were males 15-29, while almost 60% of the offenders were found in this age category (Klebba, 1975, p. 197). Thus homicide has become an act that is disproportionately engaged in by black males in the first two-thirds of the life cycle, who reside in the most depressed urban environments in our largest central cities.

Victim and offender are most often members of the same race, but increasingly whites are being victimized by blacks. The latter trend reflects a change in the pattern of homicide and also implies a difference in homicide environments, based on the relationship between victim and offender. Details of the latter type are not easily available as there is no single source of public information which identifies the race of the victim and offender by urban place. While studies of a few individual places provide insight into the racial or ethnic character of victims and offenders, this usually represents information secured from individual police departments. Thus the race specific character of a growing cause of death is generally unavailable for individual cities. The unavailability of this kind of information leaves the public unaware of the contribution of this behavioral cause of death to the overall pattern of death among blacks and other low-income ethnics. Homicide currently ranks as the fifth ranking killer of black males within the United States and within some local environments it may rank as high as the third or fourth major killer. But more significantly it results in death at an early age and from a community impact perspective often results in the creation of two sets of dependents, those of the victim and those of the offender.

REGIONAL VARIATIONS IN HOMICIDE LEVELS

The risk of becoming a victim of homicide has long been associated with one's place of residence. Regional variations in risk, as well as size of place variations, are reasonably well known. At the regional scale, the South continues to hold a hegemony on homicide victimization rates. Likewise, higher homicide victimization rates are associated with the nation's larger urban centers. As a matter of record the nation's larger urban centers represent the model homicide locations at the present time. In 1973, the South continued to lead the nation in homicide levels. At the latter date, the homicide level for the South was 14.1 per 100,000, a level which approached almost twice that prevailing in other regions (see Table 1). Some would contend that the differences in homicide levels in the South vis-a-vis the rest of the nation result from the higher percentage of blacks in the population. Klebba seems to imply such a relationship by illustrating that the black percentage in the South, vis-a-vis other regions, is quite similar in magnitude to the differences in homicide rates between the South and other regions. This position has been stated more explicitly in the past, as Wolfgang's review of the literature illustrates. Wolfgang (1958, p. 41) cities one author who states "that it is the presence of the Negro in the South that makes for a high homicide record, and that southern whites compare favorably with other sections of the country."

Table 1
Homicide Rates by Major Geographic Division, 1973

Region	Rates per 100,000	Number of Homicides
Northeast	7.7	3,839
Northcentral	7.6	4,362
South	14.1	9,316
West	8.1	2,948

Source: Klebba, A. Joan "Homicide Trends in the United States, 1900-74," *Public Health Report,* Vol. 90, May-June 1975, p. 203.

IS THERE A SOUTHERN CULTURE OF VIOLENCE?

Needless to say, others have demonstrated that both blacks and whites of southern origin have historically exhibited a high propensity for homicide holding structural variables constant. Cash, in his treatment of the Old South, indicated that the closeness of the backcountry South to frontier conditions spawned a strong sense of individualism in the white non-plantation-owning segment of the population. This phenomenon, he asserted, led those who had grown up in the backwoods to manifest a strong tendency toward violence (Cash, 1941, p. 55). Likewise, Hackney (1969, p. 51) has demonstrated that southern whites, at least at one point in time, manifested higher homicide rates than whites from selected rural non-southern states. On the basis of his analysis, the latter writer attributed southern violence to cultural patterns which are not associated with current influences. Thus the question of a regional culture of violence has been raised which is thought to be practiced by both blacks and whites of southern origin.

The question of a southern subculture of violence based on the concept developed by Wolfgang and Ferracuti has been challenged by Erlanger (1976). The latter writer is of the opinion that there does not exist a normative culture in the South which sanctions violence, and thus while gun ownership is higher there than elsewhere this should not serve as evidence that violence is sanctioned by such individuals. Erlanger is critical of the concept and indicates that non-cultural variables should be investigated more thoroughly in attempting to explain higher levels of violence in the South.

The varieties of support or nonsupport for the subculture of violence thesis seem to be derived from both methodological and observational experiences. The issue is not likely to be resolved soon, but homicide levels continue to climb. However, between 1962 and 1972 the rates of increase in homicide levels were generally higher outside the South. For instance the Middle Atlantic Census Division experienced a 159% increase during the ten-year interval, while the South Atlantic Division recorded a 70% increase (Metropolitan Life Statistical Bulletin, 1974). Nevertheless, the higher growth rate region possessed a homicide level in 1972 that was essentially comparable to that prevailing in the slow growth region ten years earlier.

THE HOMICIDE PHENOMENON IN LARGE URBAN ENVIRONMENTS

The homicide level prevailing within the nation's larger central cities singles them out as environments in which the problem is most serious. In 1973 the criminal homicide rate stood at 25.6 per 100,000 in cities with populations greater than 250,00, but for black males residing in central cities within metropolitan counties the level was placed at 85.4 per 100,000. The level for black males was unduly high in all place categories, reaching its lowest level within suburban places. Abrahamson (1974) employed murder rates along the other violent crime rates in a factor analysis model in an attempt to derive the social structure of a sample of urban places. The factor which he identified as Disorganization-Deviance was strongly associated with the violent crime variables. He states that this factor might alternatively be called "ghetto development." Once again this brings us back to the question of culture. Is Abrahamson implying that the expansion of the black population within cities leads to expected outcomes? To the contrary, he is explicit in his reference to the role of social deprivation, but we are still uncertain regarding how such individuals devise styles to cope with life conditions which are themselves penalizing.

Curtis (1975) takes the position that in order to define the role of a black subculture on criminal violence we must acquire an in-depth knowledge of black life-styles which transcends the study of delinquency, juvenile gangs, and persons in poverty. While the latter writer's work is preliminary, it does offer a hypothetical construct that might result in shedding additional light on the behavioral dimensions of the problem.

VARIATIONS IN THE HOMICIDE EXPERIENCE IN SELECTED CITIES

Even within cities in the same size class there is evidence of large variations in the homicide level prevailing in 1973, as well as the homicide growth rate over a seven- to eight-year period. Among the fifty largest cities in the nation, more than one-third possessed homicide levels which exceeded the national level by a factor of two or more. At the same time, there were fourteen cities whose levels fell below that of the national average cities, with the highest levels occurring in places where blacks constituted approximately four-tenths or more of the population. These are places that this writer has previously identified as first and second generation ghetto centers. Cities with homicide rates at least three times the national average include Atlanta, Detroit, Cleveland, Newark, Baltimore, Washington, and St. Louis. At the same time, such large urban places as Milwaukee, San Diego, Minneapolis, Portland, and Omaha, all possessed homicide levels below the national level.

RACIAL COMPOSITION AND CENTRAL CITY HOMICIDE PATTERNS

What factors were at work other than racial composition in promoting this outcome cannot be fully explained at this time. Since we are without information which identifies the race of the victim at the level of individual urban places, we cannot even explain how much variation exists within race from place to place among these cities. It is clear however that black homicide rates in high homicide level cities exceed the national average for blacks, while those in low homicide level cities evidence signs of falling below the national level for the group. In 1970, Chicago experienced 834 homicides of which 74% of the victims were black. But had the national black homicide level prevailed in Chicago at this time, it would have resulted in approximately 160 fewer black victims. Thus the maximum homicide level cities were the places of residence of black populations that more frequently found themselves in situations which led to homicidal outcomes (see Table 2).

By 1975, however, there was evidence that homicide levels had begun a downturn in five of the high homicide level cities, with 1974 representing the peak year in most. Detroit and Cleveland, have since 1971-1972, surpassed Atlanta as the most homicide-prone city in America. Needless to say, the cities in the high homicide rate class during the last eight years have increasingly represented environments in which life was cheap and 70% or more of all homicide victims were black. What is it about these environments that has contributed to the sudden upsurge in the willingness of individuals to take the life of others? This question cannot be answered at this time because of the nature and complexity of the problem. Nevertheless, we find it necessary to suggest a research orientation which will possibly advance our understanding of a problem whose dimensions are often ignored or misunderstood.

Table 2
High and Low Level Homicide Rates in Large American Cities: 1971-72

High Homicide Rate Cities	Rate (per 100,000)	Low Homicide Rate Cities	Rate (per 100,000)
Atlanta	48.79	San Diego	4.88
Detroit	38.93	San Jose	4.83
Cleveland	38.47	Wichita	5.81
Newark	36.52	St. Paul	5.60
Washington	34.39	Tuscon	5.89
Baltimore	36.08	Portland	6.82
St. Louis	34.16	Seattle	7.92

Source: Barnett, Arnold and others, "On Urban Homicide: A Statistical Analysis," *Journal of Criminal Justice,* Vol. 3, 1975, p. 90.

THE ROLE OF PLACE IN HOMICIDE RESEARCH

At the present time, much of what we know about urban homicide has been gleaned from a few well-studied places. The major homicide research during the previous twenty years seems to have focused on Philadelphia and Chicago. There seems to have been little in the way of a systematic effort to compare a larger sample of places along a common dimension. The absence of comparative studies of this sort stems in part from a lack of resources, as well as the disciplinary and research orientations involving the subject. From a geographic perspective, one would wish to include in such a proposed study cities drawn from each of the major regions of the nation. Since homicide does not represent a random event, but tends to be highest in those places with large black concentrations, the places selected for sample inclusion should reflect this. Recently, a small sample of cities was purposely selected which might serve as an appropriate set for the study of changing patterns of homicide since 1960. Only twelve places were selected for inclusion in this group, including three from each major region (see Table 3). Eight of the selected cities were characterized by changes in homicide levels of greater than 300% during this 14-year period. In most cities, however, the period of maximum change took place between 1965 and 1970. Changes in Pittsburgh and Milwaukee were more uniform over the interval than those of other places.

Whatever forces were at work, each of these places registered higher levels of homicide at some interval during this period than was true during the base year, even though every city except Houston and Los Angeles lost population during the interval.

DEMOGRAPHIC SHIFTS AND HOMICIDAL PATTERNS

It is true that black population growth as a result of large-scale in-migration was underway in many of these places. However, St. Louis and Pittsburgh, which are found on opposite poles of the homicide growth curve, did not experience significant black in-migration during the interval. It is obvious though that large-scale white out-movement, especially of younger families, and large-scale black in-movement, concentrated in the younger age bracket, changed the

Table 3
Changes in the Number of Homicide Deaths in a National Sample of Cities: 1960-1973

The Northeast Region					The South				
	1960	1965	1970	1973		1960	1965	1970	1973
Boston	27	57	114	135	Atlanta	67	100	242	263
Newark	47	68	143	163	Houston	113	139	289	263
Pittsburgh	28	40	63	48	Memphis	39	41	91	153
The North Central Region					**The West**				
Detroit	150	188	495	672	Los Angeles	154	249	395	489
St. Louis	67	138	266	215	San Francisco	36	57	111	107
Milwaukee	15	27	50	66	Denver		28	37	96

Source: Selected National Vital Statistics Reports, Part B.

demographic base in favor of homicide in most places, although Barnett and other (1975) indicate that less than one-tenth of the national rise in homicide rates during the sixties can be attributed to changes in the demographic structure of the population. While a description of the sort undertaken here is a primary first step in an attempt to gain greater insight into the homicide problem, we are still left without much real understanding of the problem.

Because of a lack of information describing the race of victims in the above sample, it is difficult to ascertain how important the racial makeup of the population is in influencing the homicide rate in individual places. Barnett, cognizant of the differing probabilities for ethnic group members to become homicide victims, has developed what he terms a demographically adjusted index that would take into consideration the percentage of the population in both the high and low probability victimization groups as a means of specifying the impact of high victimization groups on homicide levels within individual places.

A simple regression model using percent black as the independent variable and homicide rates as the dependent variable produced a standardized regression coefficient of .925 when applied to data from the above sample, but this level of analysis is too crude to allow any meaningful assessment of the problem. In this particular case the predicted homicide levels for Newark and St. Louis using percent race as the independent variable produced large discrepancies in the residuals, with percent black serving as a good predictor in St. Louis, but not in Newark. Likewise, Boston and Milwaukee showed a similar pattern with the percent black overpredicting homicide levels in Milwaukee by a large margin, while underpredicting in Boston. This seems to suggest that while the percent black in the population can serve as an indicator of the direction of homicide levels, this variable is much too gross to assist in explaining much else. For instance, Savitz (1973) reports that in 1940 blacks were arrested for committing 40% of all homicides in the nation, but by 1969 they constituted 62% of those arrested. Do these changes simply represent class improvements in the white population, or life-style changes in the black population, or a shift in regional cultural practices associated with the large-scale migration of blacks from the South during the post-1940 era?

HOMICIDE PATTERNS AT THE URBAN MICROSCALE LEVEL

One approach to clarifying the problem is to shift from a macroscale evaluation to a microscale evaluation. This would lead us to attempt to identify homicide environments within the cities chosen for analysis. Given what we already know about the association between homicide levels and the racial composition of places, a logical search for environments which might lend themselves to heightened incidences would focus primarily on black residential communities. Yet, we find specific behavioral settings tending to serve as places in which one homicide motivated offense is likely to occur vis-a-vis another. For instance, are the locational characteristics of the behavioral settings similar or dissimilar when the parties to the act are known or unknown to each other? There is growing evidence that fewer homicides are being committed by friends and acquaintances of the victim, while more are being committed by persons previously unknown to the victim. In Chicago during the period 1965-1973, the trend was in the direction of a higher percentage of victimization by strangers. Approximately one-third of the offenders were unknown to Chicago victims in 1973, while one-fourth possessed this status in 1965 (Block, 1976, p. 498). This pattern generally reflects an increase in the incidence of homicide committed during the commission of a felony. Nationally 30% of all homicides fell into that category in 1972.

Homicides involving strangers have a greater tendency to be committed in public spaces than in private spaces. Thus, it is clear that the motive surrounding the act influences the behavioral setting in which the act occurs. Motive is often related to the nature of the personality of the individuals involved, while personality is partially conditioned by the cultural milieu in which the individual has been socialized. Given the changes which have occurred in the larger environment relative to the status of blacks in American society, it seems only logical to expect that black subcultural differences and their attendant personality modifying force would alter the mix of individuals in the society that would respond to external stimuli in highly differentiated ways. There is no dearth of studies detailing the personality characteristics of homicidal actors, but these studies seldom place the individual in an environmental context that would permit one to observe the interaction effect of persons with their social and cultural milieus.

AN ECOLOGICAL VIEW OF HOMICIDE PATTERNS

This study, like most undertaken by geographers at this scale of analysis, tends to be ecologically oriented. In attempting to define the homicide environment and determine the strength of individual variables on behavioral outcomes, a traditional approach has been employed. Because of the difficulty of securing homicide data at the microscale, it is only possible to search for homicide environments in the selected places which make up our national sample. Homicide environments have been identified in each city from our Midwest sample at the neighborhood scale. For other cities in the sample, the internal location of homicidal events varies in terms of the scale of the data collecting unit (e.g., Police Districts, Patrol Beats, Health Districts, etc.). In both Detroit and St. Louis, a series of neighborhoods have been evaluated within one or more health districts and/or police districts.

In the Detroit case, the health district possessing one of the highest homicide levels in the city in 1970 was selected for analysis (see Figure 1). In this year, it was the scene of approximately one-fifth of all of the homicides within the city, but even among the neighborhoods in this district there is much variation in prevailing homicide levels. This should not be surprising given the district was the place of residence of 184,000 people. In one neighborhood in this district the homicide rate was approximately 300 per 100,000. We have chosen to identify homicide environments as those in which the approximate homicide level exceeds 100 per 100,000. This level is more than double that of the national black homicide level and thus can be thought to constitute a high risk environment. Munford and others (1975) also employed this base level to identify high risk homicide environments in Atlanta. Eight of the thirteen neighborhoods included in our analysis can be labeled high risk environments.

A standard multiple regression model was employed to determine the contribution of seven independent variables on the number of observed homicides within each neighborhood during 1970. The independent variables employed included percent female-headed households, percent females divorced, median family income, percent residing elsewhere in 1965, percent population aged 15-34, percent population in poverty, and total population. A multiple correlation coefficient of .940 resulted, with an associated corrected coefficient of determination of .722. Females divorced, female-headed households, and family income were the variables most strongly associated with homicide. Thus, family instability and subsequent breakdown, and limited financial resources seem to set the stage for the kind of conflict that often leads to homicide. The ecological conditions found within these Detroit neighborhoods are not explanations of the existing homicide levels, but they do point up conditions in which the probability of the individual response to the situation in which one finds oneself could serve as the triggering mechanism.

THE HOMICIDE ENVIRONMENT AND THE PRESENCE OF STRESS

Harburg and others (1973) previously evaluated one of the neighborhoods among our set of high risk environments and described it as a high stress environment in which there was evidence that the resident population was more vulnerable to high blood pressure than white or black residents from low stress environments. If the pressures of one's life condition can be demonstrated to impact on aspects of physical health, then we can expect stressors of the type present in the Detroit neighborhood to influence behavior. James and Kleinbaum (1976) recently replicated the Harburg study in North Carolina as a means of ascertaining the impact of high stress environments on the differential black-white hypertensive mortality rates. They found the social disorganization component of the high stress environments of blacks, but not of whites, influenced hypertension-related deaths more significantly than did the income component of the environment. We assume that these stressors are at work in our Detroit high risk neighborhoods and one of the outcomes is a behavioral, rather than a physical cause of death. Between 1970-1973, homicide levels have continued to rise within this set of neighborhoods, indicating that the environment has become even more stress-filled than it was three years earlier.

THE ECOLOGY OF HOMICIDE: THE CASE OF ST. LOUIS

The St. Louis neighborhood analysis parallels that of Detroit. The same seven independent variables explained slightly less variance than in the Detroit case; 60% as opposed to 72%. The

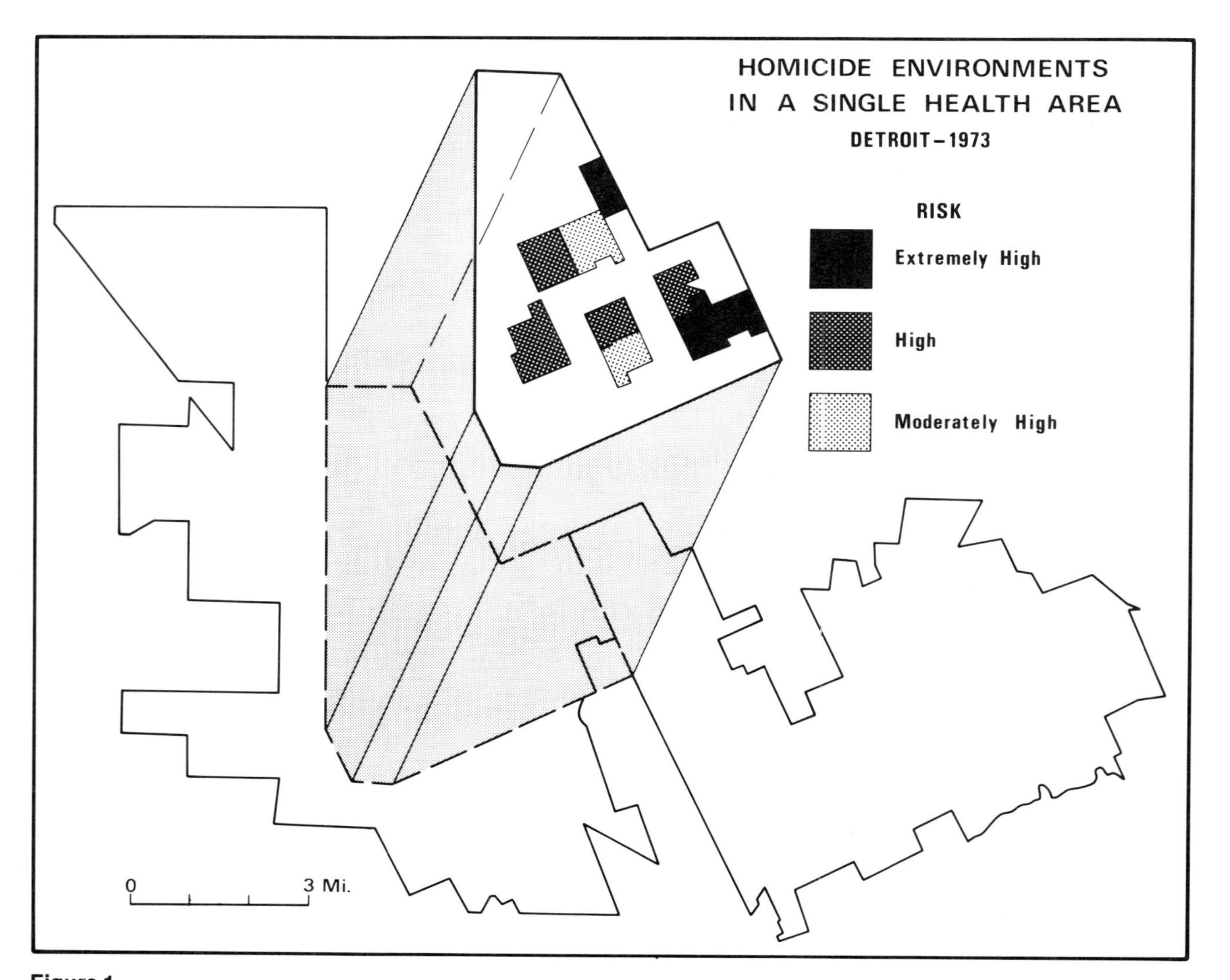

HOMICIDE ENVIRONMENTS
IN A SINGLE HEALTH AREA
DETROIT–1973
RISK
Extremely High
High
Moderately High
0
3 Mi.

Figure 1.

only variables showing significance at the 10% level were family income and percent residing elsewhere five years earlier. The high correlation between family income and the family stability variables no doubt minimized their impact on significance levels. But these variables (family stability) contributed most of the uncorrected coefficient of determination. This reflects a major difference in the role of ecological variables in the homicide environments of Detroit and St. Louis. The simple correlations between female-headed households and family income in St. Louis and Detroit are –.843 and –.305 respectively.

The highest homicide levels prevailed in the lower income environments and an unusually high number of incidences of homicide occurrred in the environment in which large-scale public housing projects were found (see Figure 2). Block (1976) recently observed a similar ecological relationship in homicide patterns in Chicago. Homicide tends to be become more commonplace in the western section of the St. Louis black community in the middle seventies. In 1974, 43% of all homicide related deaths were concentrated within four of St. Louis' 26 health districts. All but one of these were located in the western part of the city. There was a drop in the total number of homicides in St. Louis between 1970 and 1974, but this is probably offset, in terms of its effect on incidence, by the population decline of the city. What appears to have happened during this interval is the spread and intersification of the risk of homicide in environments which were previously classified as moderate risk environments.

An ecological analysis of the homicide environment provides some additional insight into the problem of homicide, but it also leaves much to be desired. We are still uncertain regarding the strength of the environment in promoting conditions which lead to homicidal outcomes. Moos, in a recent review of the role of the environment on human populations, pointed to studies which illustrate the influence of environmental press. Environmental press was defined as the directional tendency attributed to a specific environment (Moos, 1976, pp. 325-376). The personality of the environment is thought to be important in terms of its contributions to the promotion of individual goals and life-styles but this dimension of the role of the environment does not seem to be well understood at this point. There has been, however, a more fully developed understanding of the role of the individual personality in aggression.

PERSONALITY DEVELOPMENT AND THE HOMICIDE ISSUE

Southern-born blacks are (were) thought to inhibit feelings of anger more often than their northern-born counterparts (Crain and Weisman, 1972, p. 43). A more important problem in terms of our interest in violent behavior is how the individual learns to handle aggression that bears on the question of homicide. Even though a large segment of the black population has learned to inhibit aggressive tendencies, there is evidence that when anger is expressed it is displaced onto safe targets, usually other blacks. The constant frustration growing out of the presence of threatening life conditions often leads to a reduction in ability to control ones aggression and such situations tend to encourage the commission of acts of violence. Crain and Weisman take the position that persons who have been arrested or have been in a fight are less likely than others to have control over their aggressions. Likewise, they contend that because of poor control over aggression, such persons are more likely to become school dropouts. The latter position finds some support in the research results of Robins (1968), who has demonstrated that the best predictor of death among her sample of black males was truancy. Robins also indicated that we may have confused social class with behavioral problems in terms of its impact on adult death.

The extent to which one is willing to resort to aggression is also thought to be related to the notion of internal control. Those persons who exhibit signs of low internal control are those that most often feel that they are powerless to exert control over the external environment. Crain and Weisman, indirectly, deduced that such persons are more likely to have a record of

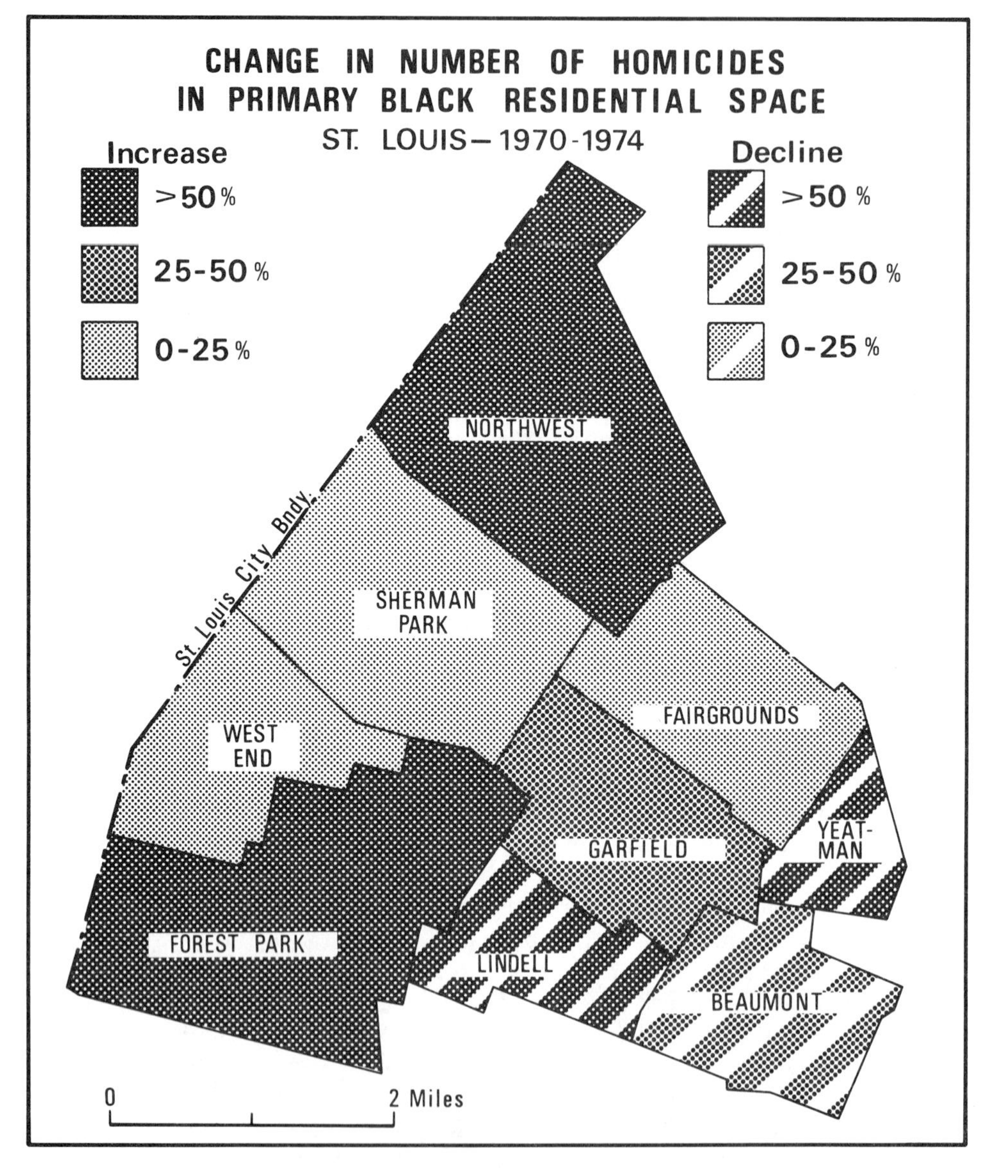

Figure 2.

engaging in escapist behavior via alcohol or drugs. Although they point out that they have no hard data to support this position they have provided empirical evidence that both internal control and self-esteem are associated with patterns of aggression. Because of the past prevalence of more blatant practices of discrimination in the South, Southern blacks are thought to have lower self-esteem than northern-born blacks. When low self-esteem is coupled with low inner control, as it sometimes is, this leads to the internalization of anger. Although such individuals tend to inhibit aggression, they are frequently provoked to engage in aggressive behavior (Crain and Weisman, 1972, pp. 80-83).

It seems evident that a disproportionate number of blacks socialized in the South during an era of strict racial segregation might have developed what Megargee describes as the over-controlled personality. Such persons develop strong inhibitions against aggression and must be severely antagonized before they are able to overcome these inhibitions (Megargee, 1966). Once such defenses have been overcome previously nonaggressive individuals display extremely assaultive behavior, often leading to brutal forms of homicide. The undercontrolled personality on the other hand, is essentially governed by external rather than internal controls of aggression and is thus expected to become involved in acts of aggression with only minimal provocation. It is unknown to what extent these two basic personality types are to be found in the black community, but the mix of such persons should be expected to influence the nature and pattern of overt aggressive behavior.

The predominance of individuals socialized in the North, in frustrating environments, should be expected to lead to increases in the frequency of violent outbursts which may or may not lead to death. It is suggested here that the overcontrolled personality type would exhibit the greater likelihood of being involved in an act of violence with intimates, whereas the under-controlled personality type would be more prone to become involved in instrumental acts of violence with unknown persons. This general description of the aggressive personality is noted simply to highlight the role of personality on behavior, and to illustrate the shortcomings of any attempt to evaluate homicidal behavior on the basis of an ecological analysis alone. The ecological approach would not likely prove to be highly productive if the goal of the research were to assist in the abatement of the described behavior.

THE ROLE OF PERSONALITY AND ENVIRONMENT

Each environment provides the individual with resources required for sustenance. One writer has identified the basic attachments, which must be secured from the environment in order to insure survival, as attachments to: food; a clear concept of self-identity; persons, groups, and roles; money; and a comprehensive system of meaning (Hansell, 1976, p. 33). When any of these attachments are severed, the individual must engage in attempts to adapt to the inadequacy in the supply of these resources. Thus it is clear from the previous discussion that blacks in general, and low-income black males in particular, encounter difficulty in establishing and maintaining certain attachments. While the larger society tends to view food and money as the primary resources that should be available to all persons, it is the severing of some of the other attachments or the failure to establish them that often proves most troublesome to segments of the ghetto population. This seems particularly true with reference to the attachments to a concept of self-identity, roles, persons, and a comprehensive system of meaning.

When one is unable to attach oneself to a meaning system such as a religion, there is a breakdown in social control. Human interaction which revolves around the severing of one or more of these basic attachments often leads to aggressive behavior. It has been demonstrated that most homicidal offenders, at least in the United States are neither neurotic nor psychotic but are persons whose life circumstances are such that adaptive patterns are acquired which tend to support the employment of violence when specific situations arise. It is generally agreed that both offenders and victims are persons not presently suffering from a psychiatric malfunction. The condition of the offender is logically better understood than that of the victim, but in a recent study of the psychiatric history of victims of homicide in the city of St. Louis, it was found that only 15% had previously been treated for a psychiatric diagnosis (Herjanic and Meyer, 1976). The authors were careful to point out that their search of psychiatric histories was confined to two public institutions, and errors occur to the extent that some of these persons might have been treated by psychiatrists in private practice. The most common diagnosis was alcoholism, which afflicted more than one-third of the total.

Ego needs and learned methods of resolving interpersonal conflict tend to trigger behavior which often leads to death. In order to better understand microspatial patterns of urban homicide, the geographer should be able to integrate the variables of culture, and the characteristics of persons and place in such a way as to lend more insight in understanding the role of the individual variables than now exists. Such an integration could lead to a process orientation that could assist in the development of tactics for intervention. Admittedly the difficult task associated with data collection, and the development of familiarity with the appropriate techniques works against extensive disciplinary involvement.

Previously brief descriptions of homicide patterns in segments of the Detroit and St. Louis black communities were analyzed employing a standard technique of ecological analysis. Because we are suggesting that both environmental data and personal data be integrated a data source focusing on the individual must be acquired. A standard information source which might be utilized in such a study is the death certificate. The death certificate, unlike health reports and police reports, provides information at the level of the individual. Place of residence, address of location of the homicidal event, place of employment, marital status, weapon used, and other important socio-demographic information can be secured from this data source. Thus, at least some information is available to support the kind of study suggested here. After the identification of the victim has been secured from death certificates, it is possible to secure information identifying the offender from police or court records. It is then possible to determine the spatial configuration which describes the location of the place of residence of the victim and the offender and the site of the homicidal event.

The relationship between the victim and offender will strongly influence the spatial configuration of homicide. Bullock, as well as Pokorny, have demonstated that victim and offender in Houston tended to reside very near one another. In the latter two studies, lovers' quarrels and family arguments contributed significantly to the homicide pattern and correspondingly to the spatial pattern of the event. With changes in the motive for homicide there should also occur changes in the spatial pattern of the event. The rise in the incidence of instrumental homicide should lead one to expect the emergence of a modified spatial pattern describing the place of residence of the victim and offender and the location of the event. Likewise, one would expect a proportionate decline in the number of events occurring in the house and an increase in the number of events occurring on the streets or in other public spaces.

SUPPORT FOR THE GEOGRAPHY OF VIOLENT BEHAVIOR

The geography of urban violence has only recently begun to receive attention from members of the discipline and the act of homicide has received even less attention than other dimensions of urban violence. This is regrettable when one considers the seriousness of the act and its growing importance as a major killer in large urban environments. Homicide has become increasingly associated with blacks, but there is no single source of information which documents the actual racial makeup of homicidal acts on a city-by-city basis.

The previous description of the changing pattern and magnitude of homicide stems from the desire to aid in curbing the frequency of the event. In order to do this, one must gain a better grasp of the homicide process, both in terms of its people and its place component. Homicides occur most often in the younger segment of the age structure, resulting in large losses in the average expected remaining years of life. With these losses, go losses in expected earnings, losses in sources of emotional support rendered to others, and the need to provide both economic and psychological support for those who are bereaved by the growing incidence. It is obvious that a real improvement in eliminating this behavior is not likely to occur until there exist major changes in the nature of society. Bouza, an inspector in the New York City Police Department, makes this point clear when he states

> "one of the learnings of a policeman involves the proposition that crime is a function of class. The lower economic classes resort to violence, escape into alcoholism and drug addiction and generally respond to the condition of their situation in a predictably violent mode." (Bouza, 1975)

Even given the existing societal structure, a geography of violent environments does not yet exist. It has been suggested here that there exists an opportunity for the discipline to aid in the contribution to an understanding of a phenomenon which is growing in importance as a killer of persons and a blighting element in the environment.

REFERENCES

Abrahamson, Mark. "The Social Dimensions of Urbanism." *Social Forces,* 1974, pp. 376-383.

Barnett, Arnold; Keitman, Richard J.; and Larson, Richard C. "On Urban Homicide: A Statistical Analysis." *Journal of Criminal Justice* 3 (1975):85-110.

Block, Richard. "Homicide in Chicago: A Nine-year Study (1965-1973)." *The Journal of Criminal Law and Criminology* 66 (1976):496-510.

Block, Richard, and Zimring, Franklin. "Homicide in Chicago, 1965-1970." *Journal of Research in Crime and Delinquency* 10 (1973):1-12.

Bouza, Anthony V. "Policing the Bronx." *Journal of Police Science and Administration,* 1975, pp. 55-58.

Bullock, Henry Allen. "Urban Homicide in Theory and Fact." *Journal of Criminal Law and Criminology* 4 (1955):565-575.

Cash, W.J. *The Mind of the South.* Garden City, New York: Doubleday, 1941.

Crain, Robert L., and Weisman, Carol S. *Discrimination, Personality, and Achievement.* New York: Seminar Press, 1972.

Curtis, Lynn A. *Violence, Race, and Culture.* Lexington, Massachusetts: Lexington Books, 1975.

Erlanger, Howard. "The Empirical Status of the Subculture of Violence Thesis." *Social Problems* 22 (1974):281-292.

———. "Is There a 'Subculture of Violence' in the South." *The Journal of Criminal Law & Criminology* 66 (1976):483-490.

Gorwitz, Kurt, and Dennis, Ruth. "On the Decrease in the Life Expectancy of Black Males in Michigan." *Public Health Reports* March-April, 1976, pp. 41-145.

Hackney, Sheldon. "Southern Violence." *The History of Violence in America.* Graham, Hugh D., and Gurr, Red R. (eds.). New York: Bantam Books, 1969.

Hansell, Norris. *The Person-in-Distress, on the Biosocial Dynamics of Adaptation.* New York: Human Services Press, 1976.

Harburg, Ernest and others. "Socio-ecological Stressor and Black-White Blood Pressure: Detroit." *Journal of Chronic Disease* 26 (1973):595-611.

Herjanic, Marijan, and Meyer, David A. "Psychiatric Illness in Homicide Victims." *American Journal of Psychiatry,* June 1976, pp. 691-693.

James, Sherman A., and Kleinbaum, David G. "Socio-ecological Stress and Hypertension Related Mortality Rates in North Carolina." *American Journal of Public Health* 66 (1976):354-358.

Klebba, Joan. "Homicide Trends in the United States, 1970-1974." *Public Health Reports.* May-June 1975, pp. 195-204.

Megargee, Edwin I. "Undercontrolled and Overcontrolled Personality Types in Extreme Antisocial Agression." *Psychological Monographs,* no. 611, 1966.

Metropolitan Life Statistical Bulletin. November 1974.

Moos, Rudolf H. *The Human Context, Environmental Determinants of Behavior.* New York: John Wiley & Sons, Inc., 1976.

Munford, Robert S., and others. "Homicide Trends in Atlanta." Unpublished Paper, Bureau of Epidemiology, Center for Disease Control. Atlanta, 1975.

Pokorny, Alex. "A Comparison of Homicide in Two Cities." *Journal of Criminal Law, Crimology and Political Science* 56 (1965):479-487.

Robins, Lee N. "Negro Homicide Victims—Who Will They Be?" *Transaction,* June 1968, pp. 15-19.

Rossi, Peter, and others. "The Seriousness of Crimes: Normative Structure and Individual Differences." *American Sociological Review,* April 1974, pp. 225-237.

Savitz, Leonard D. *"Black Crime" Comparative Studies of Blacks and Whites in the United States.* In Kent S. Miller, and M. Dreger. New York: Seminar Press, 1973, pp. 467-516.

Social Indicators, 1973.

Waldron, Ingrid, and Eyer, Joseph. "Socio-economic Causes for the Recent Rise in Death Rates for 15-24 Year Olds." *Social Science and Medicine,* 1975, pp. 383-398.

Wolfgang, Marvin F. *Patterns in Criminal Homicide.* Philadelphia: University of Pennsylvania, 1958.

Wolfgang, Marvin F., and Ferracuti, Franco. *The Subculture of Violence.* London: Tavistock Publications, 1967.

Current Research Themes in Geography and the Problem of Poverty

Ernest H. Wohlenberg
Indiana University

Much soul searching has occupied many academic geographers over the last decade or so. This is not unique to geographers, but characteristic of scholars in many if not most disciplines. Partly as a result of student demands for relevance, but probably to a greater extent as a result of increased awareness of problems outside of their ivory towers, geographers were increasingly concerned with the social relevance of their research and teaching. There was a sense of urgency. The commotion emanating from the central city ghettos, from the polluted environment, and from the rural poor had gotten so loud as to penetrate even the insulation surrounding the campus. Some within academia were no longer content to work with problems merely esthetically pleasing or intellectually stimulating, but devoid of real world applicability. Geographers were questioning their own motives and the ultimate significance of what they were doing. Indeed, David Smith viewed this change as a new revolution equal in its impact, its implications, and its effects to the quantitative revolution from which geography was beginning to recover (Smith, 1971). While there is reason to doubt that the revolution of social responsibility will have ramifications for the profession comparable to the magnitude of those from the quantitative revolution, it is real and its effects have taken many forms.

Symptomatic of this change was the creation in 1970 by the AAG Committee on Development and Planning of task forces "to explore how the Association might respond to society's needs by helping to mobilize its membership in significant, large-scale research efforts that are addressed to man's overriding problems" (Association of American Geographers, 1971). One of the seven problems identified as an expression of this concern was poverty in an affluent society.

Poverty has not fared well as a rallying point for research by geographers during the intervening years. Indeed the highest point of enthusiasm and interest in studying the problems of poverty in the United States by geographers may well have been reached at the AAG national meeting in Boston in 1971 when a special session on the subject was held with several hundred people in attendance. Since that time there has been declining interest as judged by published research, although interest has been sustained in such allied fields as crime and justice, minority problems, and access to health care services. Concomitant with declining concern for the less fortunate on the part of the general public has been a waning of interest in poverty by geographers. Perhaps responsible for this slackening of interest is the operation of an issue-attention cycle (Downs, 1972). Social concern is essentially faddish. We all want a new social issue to occupy us every year or so. Poverty has simply slipped from its brief status as a priority issue with the government, the media, and hence the public. Although the problem of poverty has ridden the issue-attention cycle to the pinnacle of concern only to disappear from the center of the stage, no one should make the mistake of presuming that poverty has disappeared from the country. In spite of the unprecedented prosperity of a majority of our citizens and the spending of many billions of dollars in various governmental programs to aid directly and indirectly those with low incomes, very sizeable numbers of poor people are still with us.

According to Census Bureau estimates, there were almost 26 million poor Americans in 1975 which was an increase of 2.5 million over 1974 (U.S. Bureau of the Census, 1976). More than one-third of the nation's statistical poverty was abolished during the 1960s, but the proportion of the population which is considered poor has remained nearly constant since 1970.

These figures are vigorously disputed by some who say they seriously underestimate the real dimensions of poverty in America. For example, the Campaign for Human Development, the antipoverty agency of the Catholic Church, estimated that there were 40,000,000 poor persons in the United States, circa 1974, or 65% higher than the government figures. They say, "The government's yardstick for measuring the number of poor Americans is radically unfair, given any of the variables listed—real cost-of-living, provision of adequate diet and hospitable housing, decent participation as a member of society" and "The relative capability of a poor person to participate in society, if the government's poverty standard is used, is far lower now than in 1959" (Procopio and Perella, 1976, p. 16). They suggest as an alternative criterion for determination of poverty status an income level of one-half the median. This emphasizes the relative nature of poverty in an affluent society rather than the subsistence definition adopted by government agencies.

An interesting aspect of the government poverty standards from the geographer's viewpoint is that they assume that the same level of income will provide the same level of subsistence everywhere in the country. There is no adjustment of the standard to allow for differences in the cost of living from place to place. Thus government poverty statistics for cities and regions may overstate the amount of poverty in a relative sense in low cost of living areas and understate the magnitude of the problem in high cost of living areas. Very fragmentary cost of living data for 39 metropolitan areas from the Bureau of Labor Statistics indicate that the 1970 cost of living for 4-person families with low incomes was 10% higher than the mean in San Francisco, the mainland city with the highest cost of living, and 11% lower than the mean in Austin, the city with the lowest living costs (U.S. Bureau of Labor Statistics, 1972). Considering the 1969 poverty threshold for an average 4-person family with nonfarm residence ($3,743) to be most appropriate for estimating the number of poor families in cities with cost of living near the mean for all urban areas, it would be reasonable to use a sliding poverty threshold. Then a value of $3,743 would have been appropriate for Philadelphia (a city with cost of living at the mean), $4,133 for San Francisco, $3,333 for Austin, and intermediate proportional values for other cities between these extremes.

Two meaningful methods of cartographic portrayal of poverty involve the concepts of absolute and relative poverty. The former concept is concerned with the actual number of poor families or persons residing in some geographical unit and can be represented on a map by proportional symbols. Relative poverty, on the other hand, pertains to the proportion of all families or persons residing in a geographic unit who are poor. A choroplethic technique would appear to be the most appropriate for map portrayal of relative poverty.

When county units are used for mapping poverty data on a national scale, the Four Corners area, southern Texas, the lower Mississippi Valley, eastern Kentucky, a continuous band across the south from Arkansas and Louisiana through the Carolinas, and scattered counties in the upper Midwest stand out as regions of high relative poverty (U.S. Bureau of the Census, 1974). Metropolitan counties are almost always freer of relative poverty than the surrounding hinterlands, the rare exceptions occurring where affluent major suburbs and the associated and declining central city are in different counties. Shifting to the states as units of analysis, the highest absolute poverty is found in our most populous states; California, Texas, New York, Pennsylvania, and Illinois (Fig. 1). With the exception of Texas, these states all rank thirtieth or below in relative poverty. One could summarize these data by saying that poverty defined in relative terms is largely a rural problem whereas poverty defined in absolute terms is mainly an urban one.

What geographic factors can help us understand something about the prevalence and distribution of poverty? Are there particular kinds of physical or natural environments associated with high relative poverty? Can we suggest geographic reasons why relative poverty

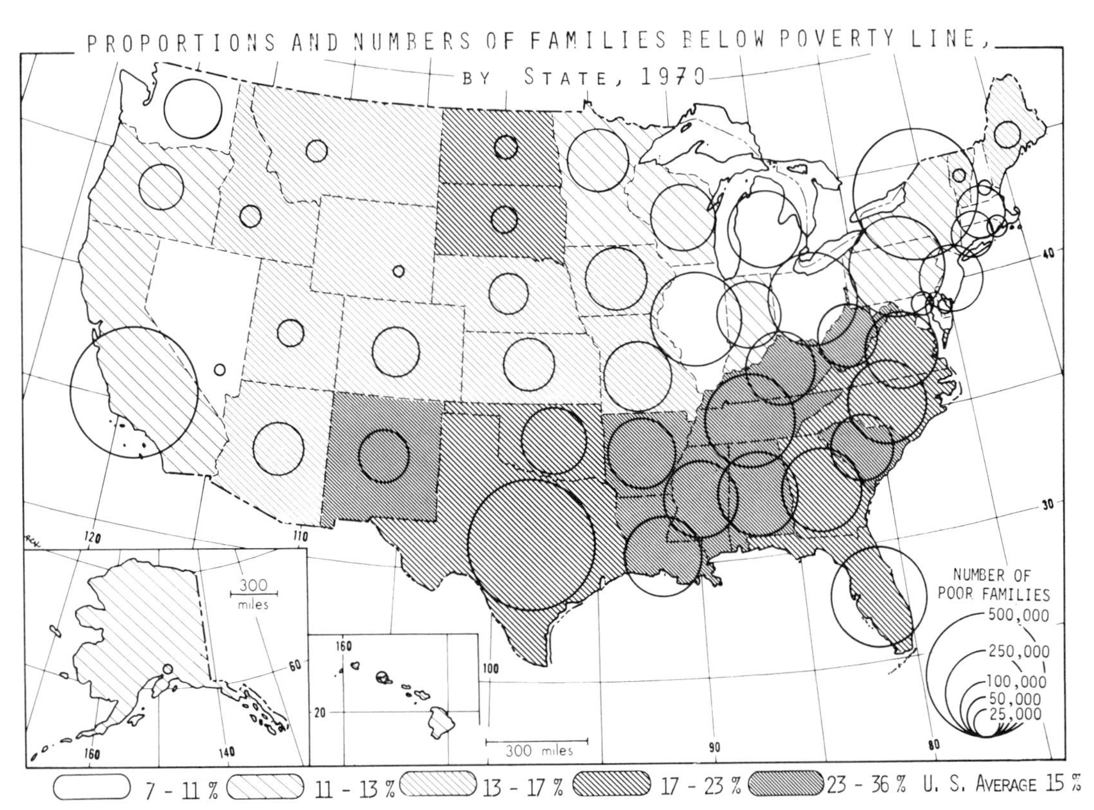

Figure 1.

is high here but low there? Clearly, in areas strongly dependent on agriculture or other forms of resource exploitation for their economic base there will be some connection between the quality and quantity of the environmental resource and the level of income of the population directly dependent on that resource. Agricultural population density may be a critical factor in determining the prevalence of poverty in a farming community. Back found some evidence that farm incomes in the United States were significantly higher in counties in which a relatively low density farm population was dependent on a relatively rich agricultural resource base than those in counties with opposite characteristics (Back, 1965). The rapid increases in labor productivity from mechanization in agriculture have left surplus labor in many farming areas. In spite of major adjustments in number and size of farms, there are still too many small farms in some parts of the country.

There also appears to be a connection between the diversity of the economic base of a community and its prosperity in the long term. Poverty in resource-dependent communities often results from lack of alternatives when faced with resource depletion, mechanization and/or automation of operations, or declining demand for the export staple (Fisher, 1966). Absentee ownership of valuable agricultural or mineral resources may also be a problem in some communities as is concentration of ownership. In these cases, economic benefits are exported from the local area or accrue to only a small minority of local residents.

Spatial factors may give some explanation why many resource-dependent communities have so few alternatives. Most such communities are relatively small and/or remote and economic activities if located there would be unable to take advantage of either economies of large-scale or agglomeration economies. Industries which gain significant economic advantage from close proximity to other related or linked industries, or which need elaborate or specialized business or transportation services, or which require large local markets or labor pools cannot operate profitably in small moderate sized communities remote from metropolitan centers.

The absence of alternative employment opportunities in smaller, more remote, one-industry communities tends to have an unfavorable impact on wages because of the lack of competition among employers for the available labor supply (Lineback, 1970). One firm controls the demand side of the local labor market in this situation. Incomes would be affected favorably in a smaller local community if it were within commuting distance of a larger city which can offer firms the benefits of agglomeration and large scale. Commuters living in smaller communities might participate in labor markets associated with the more prosperous larger cities if the distance of separation were not too great (Berry, 1968). Also, proximity to larger urban centers is expected to have a favorable impact on farm prices because delivery costs must be subtracted from city price to get actual realized price at the farm. This relationship works to the disadvantage of more remote farms (Nicholls, 1961; Ruttan, 1955; Lang, 1958).

More remote resource-dependent areas are less likely to benefit as much from out-migration as more accessible ones. For a number of reasons, a majority of migrants move relatively short distances. If perceived economic opportunities in nearby potential destinations are only marginally better, there is little incentive to move. When centers of opportunity are distant, the economic and social costs of migration are likely to be great. Out-migration from depressed areas may have positive effects on those areas by tending to bring population back into better balance with resource base, but it also has negative impacts. Migration is selective in nature, siphoning off the younger, more productive segments of the community leaving it less able to provide for those who remain (Parr, 1966).

Rural poverty may be transferred to an urban setting through the process of rural-urban migration (Kain and Persky, 1968). Davies, Elgie, Fowler, and other geographers have joined in the study of the problems encountered by poor rural migrants after they moved to the city and factors involved in the adjustment process which will ultimately control their success in escaping from poverty (Davies and Fowler, 1972; Elgie, 1970; Fowler, et al., 1973; Hyland, 1970). Studies have shown that many unskilled rural migrants are only marginally competitive in the urban job market. Other investigators have focused on deficiencies in the urban public transportation systems upon which the poor rural migrant is especially dependent for his journey to work and the trends toward suburbanization of workplace which have greatly increased the spatial separation of home and workplace for low income residents trapped in the central city (Davies, 1970, Davies and Huff, 1972; Davies and Albaum, 1972; Wheeler, 1968).

Roseman found that existing patterns of information flows lead to "channelized" migration from particular origins in the South to particular destinations in Midwestern cities (Roseman, 1971). These channelized migrant flows provide evidence that interpersonal communication between family members and friends plays a critical role in migrants' location decisions. They also provide evidence that a need exists for improvement in the wider public dissemination of information concerning opportunities in various potential destinations in rural areas of surplus population. Greater efficiency in migration flows might be achieved if information about cities outside in the information field of the potential migrant could be made available.

Much of the research on the geography of central cities where most of the urban poor live has been concerned with ghetto formation and evolution (Morrill, 1965; Morrill, 1972; Rose, 1970; Rose, 1971), and the fiscal disparities between central city and suburban areas which are manifested, for example, in unequal educational opportunities (Cox, 1973; Harvey, 1972). Concern for the metropolitan poor has spawned research on optimal location of day care centers (Brown, et al., 1972), spatial and social accessibility of quality health care services for the poor (DeVise, 1971; Earickson, 1971; Shannon and Dever, 1974), and availability and condition of low cost housing (Hartshorn, 1971; Mercer, 1972) which comes into being through a filtering down process starting when well-to-do residents desert older housing near the center for newer housing in the suburbs.

Incidence of relative poverty in a community is a function of accessibility and relative location at two scales: on a large national scale, and on a smaller, metropolitan hinterland scale. Economic history of the United States shows the extent to which the nation's economic development has been spatially concentrated in a relatively small core area or heartland. It is easy to draw contrasts between this core and the remainder of the country which may be called the periphery. In 1950, 65,000,000 people living on only 8% of the land earned 52% of the income, held 70% of its manufacturing jobs, and constituted 70% of those talented enough to be listed in *Who's Who* (Ullman, 1958). Not surprisingly, relative poverty is dependent on population potential, a measure of generalized accessibility to the nation's urbanized, industrialized heartland (Duncan et al., 1960). At the metropolitan hinterland scale, increasing distance from the outskirts of the city is associated with fairly steady increase in relative poverty, decreasing rates of participation in metropolitan labor markets, and decreasing efficiency of agriculture

(Berry, 1968; Berry, 1970). The degree of rise in incidence of poverty in intermetropolitan peripheries is related to the size and spacing of urban centers.

As a final consideration, one of the most relevant aspects of the problem of poverty is how to eradicate it. To that end, various programs of governmental assistance may be employed. First, we will examine programs that involve aid to regions or communities rather than individuals directly or programs which are place-oriented rather than people-oriented.

Appalachia was the poverty region upon which most attention was lavished during the 1960s. Among the geographers whose published research contributed to our edification regarding the problems of that region were Gauthier on the Appalachian development highway system (Gauthier, 1973); Blome on the dispersed Appalachian settlement patterns which contribute to the region's isolation and the outward orientation of the transportation system which severely curtails internal circulation (Blome, 1970), and the more general contributions of Estall (Estall, 1968), and Van Royen and Moryadas (Van Royen and Moryadas, 1966).

The issue of the proper approach to regional development is tied up with controversial questions. What is the optimal size for a city from the standpoint of economic efficiency or quality of life? Does an American citizen have the right to meaningful employment and a decent income regardless of where he or she chooses to live? Three approaches to the problem of depressed areas will be reviewed.

The first policy approach would promote rural industrialization in order that those who choose to live in rural areas remote from larger urban centers should not be denied this choice (Erickson, 1976; Lineback, 1972; Lonsdale, 1969). This approach places a priority on preserving the continuity of life in the distressed smaller community and discourages out-migration, especially of the young and productive from rural areas.

The second approach would encourage and augment the relocation of people from the declining rural areas to existing metropolitan centers with financial inducements and programs to ease the process of adjustment to life in the metropolis. The rationale for this approach is that the optimal size for cities is large and the best place to provide economic opportunities for the poor is in the metropolitan setting because of its special economic and social advantages.

A third alternative would create or encourage the development of a system of urban growth centers in lagging areas within commuting distance of as many as possible of the unemployed poor living outside metropolitan areas (Berry, 1973; Morrill, 1973; Ryan, 1970). Making urban work opportunities available to the rural poor without recourse to migration would be the goal of this approach. It would concentrate government investment and incentives in a carefully selected set of intermediate-sized cities with significant growth potential. Migration to such growth centers by those beyond commuting range would be assisted in lieu of migration to more distant metropolises. Intermediate-sized cities share many of the economic and cultural advantages of larger cities without the economic and cultural disadvantages of smaller places. It is not necessary to elaborate the details of the various approaches, but little has been done to evaluate and compare their merits in a systematic and comprehensive way. It is sad to say that no coherent and active regional development policy or program exists today in the United States.

Another class of government programs to fight poverty is the various income transfers to individuals directly, such as social security, veterans' benefits, public assistance, unemployment compensation, food stamps, and medicare and medicaid. Several of these programs provide fertile soil for geographic inquiry because of the role played by the states in determining eligibility standards and level of benefits and the consequent wide variation in the programs from state to state (Wohlenberg, 1976).

Much broader and more comprehensive income transfer programs to aid the poor such as guaranteed annual income or negative income taxes have been proposed and will probably be implemented in some form in the future (Kain and Schafer, 1971; Wilbanks and Huang, 1975). Geographers are well equipped to study the expected economic impact of different proposals and should play a role in the evaluation of plans to be adopted. As yet little has been done.

BIBLIOGRAPHY

Association of American Geographers. "AAG Task Forces Submit Proposals." *AAG Newsletter* 5 (August-September 1971):1.

Back, W.B. "Effects of the Natural Resource Base on Chronically Depressed Rural Areas." In *Workshop on Problems of Chronically Depressed Rural Areas.* Agricultural Policy Institute, North Carolina State University, 1965

Berry, Brian. "Commuting Patterns: Labor Market Participation and Regional Potential." *Growth and Change* 1 (1970):3-10.

———. *Spatial Organization and Levels of Welfare: Degree of Metropolitan Labor Market Participation as a Variable in Economic Development.* Paper prepared for the Economic Development Administration Conference, October 9-13, 1967.

———. *Growth Centers in the American Urban System.* Cambridge, Mass.: Ballinger, 1973.

Blome, Donald A. "A Spatial Model of the Urban Structure of Appalachia." *Proceedings of the Association of American Geographers* 2 (1970):12-16.

Brunn, Stanley D., and Wheeler, James O. "Spatial Dimensions of Poverty in the United States." *Geografiska Annaler.* Series B, 53 (1971):6-15.

Brown, Lawrence A. et al. *Day Care Centers in Columbus: A Locational Strategy.* Discussion Paper no. 26, Ohio State University, Department of Geography, 1972.

Colenutt, Bob. "Poverty and Inequality in American Cities." *Antipode* 2 (December 1970):55-60.

Cox, Kevin R. *Conflict, Power, and Politics in the City: A Geographic View,* New York: McGraw-Hill, 1973.

Davies, Shane. *The Reverse Commuter Transit Problem in Indianapolis.* Unpublished Ph.D. dissertation, Indiana University, 1970.

Davies, Shane, and Melvin Albaum. "Mobility Problems of the Poor in Indianapolis." In Richard Peet (ed.) *Geographical Perspectives on American Poverty.* Worcester, Mass.: *Antipode,* 1972.

Davies, Shane, and Fowler, Gary L. "The Disadvantaged Urban Migrant in Indianapolis." *Economic Geography* 48 (1972):153-167.

Davies, Shane, and Huff, David L. "Impact of Ghettoization on Black Employment." *Economic Geography* 48 (1972):421-427.

DeVise, Pierre. "Cook County Hospital: Bulwark of Chicago's Apartheid Health System and Prototype of the Nation's Public Hospitals." In *Antipode,* November 1971, pp. 9-30.

Downs, Anthony. "Up and Down With Ecology—the 'Issue Attention Cycle.' " *The Public Interest* 28 (September 1972):38-50.

Duncan, Otis; Scott, W.R.; Lieberson, S.; and others. *Metropolis and Region.* Baltimore: John Hopkins Press, 1960.

Earickson, Robert. "Poverty and Race: The Bane of Access to Essential Public Services." *Antipode* 3 (November 1971):1-8.

Elgie, Robert. "Rural Immigration, Urban Ghettoization and Their Consequences." *Antipode* 2 (December 1970):35-54.

Erickson, Rodney. "The Filtering Down Process: Industrial Location in a Nonmetropolitan Area." *The Professional Geographer* 28 (1976):254-260.

Estall, R.C. "Appalachian State: West Virginia as a Case Study in the Appalachian Regional Development Problem." *Geography* 53 (1968):1-24.

Fisher, Joseph L. "Poverty and Resource Utilization." In L. Fishman (ed.) *Poverty Amid Affluence.* New Haven: Yale University Press, 1966, pp. 150-162.

Fowler, Gary et al. "The Residential Location of Disadvantaged Urban Migrants: White Migrants to Indianapolis." In Melvin Albaum (ed.) *Geography and Contemporary Issues.* New York: Wiley, 1973.

Gauthier, Howard L. "The Appalachian Development Highway System: Development for Whom? *Economic Geography* 49 (1973):103-108.

Hartshorn, Truman A. "Inner City Residential Structure and Decline." *Annals of the Association of American Geographers* 61 (1971):72-96.

Harvey, David. *Society, the City and the Space Economy of Urbanism.* Commission on College Geography Resource Paper no. 18. Washington, D.C.: Association of American Geographers, 1972.

Hyland, Gerard A. "Social Interaction and Urban Opportunity: The Appalachian In-Migrant in the Cincinati Central City." *Antipode* 2 (December 1970):68-83.

Jonish, James E., and Kau, James B. "Locational and Racial Variations in Poverty Incidence." *Growth and Change* 7 (1976):24-27.

Kain, John F., and Persky, J. "The North's Stake in Southern Rural Poverty." In President's National Advisory Commission on Rural Poverty. *Rural Poverty in the United States.* Washington, D.C.: Government Printing Office, 1968, pp. 288-308.

Kain, John F., and Schafer, Robert. *Regional Impacts of the Family Assistance Plan: Some Revised Estimates.* Cambridge, Mass.: Harvard University Program on Regional and Urban Economics Discussion Paper no. 69, 1971.

Kiang, Ying-cheng. "Recent Changes in the Distribution of Urban Poverty in Chicago." *The Professional Geographer* 28 (1976):57-60.

Lanegran, David A., and Snowfield, John G. "Families with Low Incomes in Rural America." In D.A. Lanegran and R. Palm *An Invitation to Geography.* New York: McGraw-Hill, 1973, pp. 140-151.

Lineback, Neal G. *The Developing Spatial Patterns of Low Wage Manufacturing: A Case Study in East Tennessee.* Unpublished Ph.D. dissertation, University of Tennessee, 1970.

———. "Low-wage Industrialization and Town Size in Rural Appalachia." *Southeastern Geographer* 12 (1972):1-13.

Lonsdale, Richard. "Barriers to Rural Industrialization in the South." *Proceedings* Association of American Geography 1 (1969):84-88.

Mercer, John. "Housing Quality and the Ghetto." In Harold Rose (ed.) *Geography of the Ghetto.* Dekalb: Northern Illinois University Press, 1972, pp. 144-167.

Morrill, Richard L. "The Negro Ghetto: Problems and Alternatives." *Geographical Review* 55 (1965):339-361.

———. "Geographical Aspects of Poverty in the United States." *Proceedings* Association of American Geographers 1 (1969):117-121.

———. "A Geographic Perspective of the Black Ghetto." Harold M. Rose (ed.) *Geography of the Ghetto.* Dekalb, Ill.: Northern Illinois University Press, 1972.

———. "On the Size and Spacing of Growth Centers." *Growth and Change* 4 (1973):21-24.

Morrill, Richard L., and Wohlenberg, Ernest H. *The Geography of Poverty in the United States.* New York: McGraw-Hill, 1971.

Nicholls, W.H. "Industrialization, Factor Markets, and Agricultural Development." *Journal of Political Economy* 69 (1961):319-340.

Parr, John B. "Outmigration and the Depressed Area Problem." *Land Economics* 42 (1966):149-159.

Peet, Richard. "Poor Hungry America." *The Professional Geographer* 23 (1971):99-104.

———. "Inequality and Poverty: A Marxist-Geographic Theory." *Annuals Association of American Geographers* 65 (1975):564-571.

Procopio, Mariellen, and Perella, Frederick J., Jr. *Poverty Profile USA.* New York: Paulist Press, 1976.

Reckord, Gordon E. "The Geography of Poverty in the United States." In Saul Cohen et al. *Problems and Trends in American Geography.* New York: Basic Books, 1967, pp. 92-112.

Rose, Harold M. "Development of an Urban Subsystem: The Case of the Negro Ghetto." *Annals of the Association of American Geographers* 60 (1970):1-17.

———. *The Black Ghetto: A Spatial Perspective.* New York: McGraw-Hill, 1971.

Roseman, C.C. "Channelization of Migration Flows from the Rural South to the Industrial Midwest." *Proceedings* Association of American Geographers 3 (1971):140-146.

Ruttan, Vernon W. "The Impact of Urban Industrial Development on Agriculture in the Tennessee Valley and the Southeast." *Journal of Farm Economics* 37 (1955):38-56.

Ryan, Bruce. "The Criteria for Selecting Growth Centers in Appalachia." *Proceedings* Association of American Geographers 2 (1970):118-123.

Shannon, Gary W., and Dever, Alan G.E. *Health Care Delivery Spatial Perspectives.* New York: McGraw-Hill, 1974.

Smith, David M. "Radical Geography—the Next Revolution?" *Area* 3 (1971):153-157.

Tang, Anthony M. *Economic Development in the Southern Piedmont.* Chapel Hill, N.C.: University of North Carolina Press, 1958.

Thompson, Gary L. "The Spatial Convergence of Environmental and Demographic Variables in Poverty Landscapes." *Southeastern Geographer* 12 (1972):14-22.

Thompson, John H. et al. "Toward a Geography of Economic Health: The Case of New York State." *Annals* Association of American Geographers 52 (1962):1-20.

Ullman, Edward L. "Regional Development and the Geography of Concentration." *Papers* Regional Science Association 4 (1958):179-216.

U.S. Bureau of the Census. Map of "Families Below the Low Income Level in 1969." GE-50, no. 58, 1974.

———. "Money Income and Poverty Status of Families and Persons in the United States: 1975 and 1974 Revisions." (Advanced Report) *Current Population Reports.* Series P-60, no. 103, September 1976.

U.S. Bureau of Labor Statistics. *Three Budgets for an Urban Family of Four Persons,* 1969-1970. Supplement of Bulletin 1570-5. Washington: Government Printing Office, 1972.

Van Royen, W., and Moryadas, S. "The Economic Geographic Basis of Appalachia's Problems." *Tidschrift Voor Economische en Sociale Geografie* 57 (1966):185-193.

Wheeler, James O. "Work-Trip Length and the Ghetto." *Land Economics* 44 (1968):107-111.

Wilbanks, Thomas J., and Huang, Hsung. "The Regional Impact of the Proposed Family Assistance Plan." *Proceedings* Association of American Geographers 7 (1975):283-288.

A Geography of Education in the United States, Some Preliminary Considerations

Alan L. Backler
Indiana University

Recently several geographers have called upon their colleagues to give more attention to the geography of education, by which they mean the examination of educational phenomena from the spatial point of view.[1] These arguments are based upon the involvement of more geographers in education than in any other single professional pursuit, the systematic examination of so many other social and cultural phenomena by geographers, and the growing research of several other social science disciplines on educational problems. More significantly, they contend that the application of geographic theories, concepts, and research strategies to such topics as the availability of educational opportunities, the spread of educational innovations, and the creation of efficient school attendance districts would provide insights into these topics not possible using the perspectives of other disciplines.

A review of the geographic literature in English since 1960 indicates that many geographers have in fact applied their perspective and skills to the consideration of educational phenomena. Several major geographic themes can be differentiated within the educationally-oriented geographic literature.[2] Certain aspects of one of these themes, spatial arrangements, will be considered in this paper.

In a recent study dealing with spatial dimensions of education in England and Wales, Coates and Rawstron concluded that regional variations existed both in the provision of educational facilities and in educational attainment.[3] Intuitively, one might decide that a similar conclusion is appropriate for the United States. However, the paucity of research on the topic by geographers makes it difficult to draw an empirically based conclusion. It is therefore the primary purpose of this paper to begin to explore such a geography of education for the United States by bringing together existing geographic literature and other relevant spatial data which deal with educational attainment.

EDUCATIONAL ATTAINMENT

Numerous indicators, especially those measuring aptitude and achievement might be useful in examining educational attainment. Unfortunately, data on variables such as these are not sufficiently available through time and over space. For example, the author originally considered using Scholastic Aptitude Test (S.A.T.) scores as an indication of educational attainment. Investigation indicated, however, that results by state have only been available since 1972.[4] Furthermore, these data are not always comparable since the proportion of students taking the tests varies considerably. In this regard, a study done by Humm for midwestern states in 1975 indicated that the higher the proportion of high school graduates taking the S.A.T. in a state, the lower the average S.A.T. scores.[5] Proportions of high school graduates taking the S.A.T. in midwestern states varied from less than 3% to 45%. Similar variations exist nationally.

In order to use an indicator for which comparable spatial oriented data are available over time, and which reflects the educational experience of the United States population and its major social, economic, and racial subgroups, the official Census Bureau definition of educational attainment is measured by the U.S. Census as median school years completed by persons 25 years old and over, on the assumption that most people of 25 have completed their formal education. The data have been collected in this manner since 1940, and therefore generally reflect the educational experiences of Americans born during the first half of the twentieth century.[6]

The U.S. Census definition of educational attainment provides information on the number of school years completed by the population, which may not be synonymous with the quality of the education received. In fact, while the United States is one of the world's best-educated nations as measured by median school years completed, a recent study published by the University of Texas indicated that about one-fifth of the adult population has difficulty coping with everyday chores such as shopping, obtaining a driver's license, or reading an insurance policy.[7] Nevertheless, the U.S. Census definition of educational attainment does provide data which can be used to begin an exploration of the geography of education in the United States.

In the last thirty-five years, there has been a steady rise in the number of years people spend in school.[8] Most people, male and female, black and white, are staying in school longer than their parents. Changes over time in the educational levels of individuals 25 to 34 years old, shown in Table 1, illustrate this trend. For example, the proportion of white males of this age group who have four years of high school or more has risen from 36.1% to 82.3% between 1940 and 1970. In 1940 more women than men 25 to 34 years old had completed at least four years of high school. By 1974, the situation had changed so that about the same percentage of males and females of this age group had at least a high school diploma.

A sizable difference in educational level still exists between blacks and whites, however. For example, 82% of white males and 67% of black males 25 to 34 years old in 1974 had completed four years of high school or more, a difference of 15 percentage points. Yet the differential between the races has decreased substantially. In 1940, 27 percentage points separated black and white males. At present, the annual increase in the proportion of those 25 to 34 years old who have at least a high school diploma is two times higher for black males than for white males.

REGIONAL DIFFERENCE IN EDUCATIONAL ATTAINMENT

Some interesting regional variations exist with respect to educational attainment. The 1970 data (see Map 1) indicate that educational levels are generally lowest in the South and highest in the West with very little variation elsewhere. Kariel's plotting of the 1960 Census data, using slightly different class intervals, reveals the same pattern.[9] Rhode Island is unusual for a

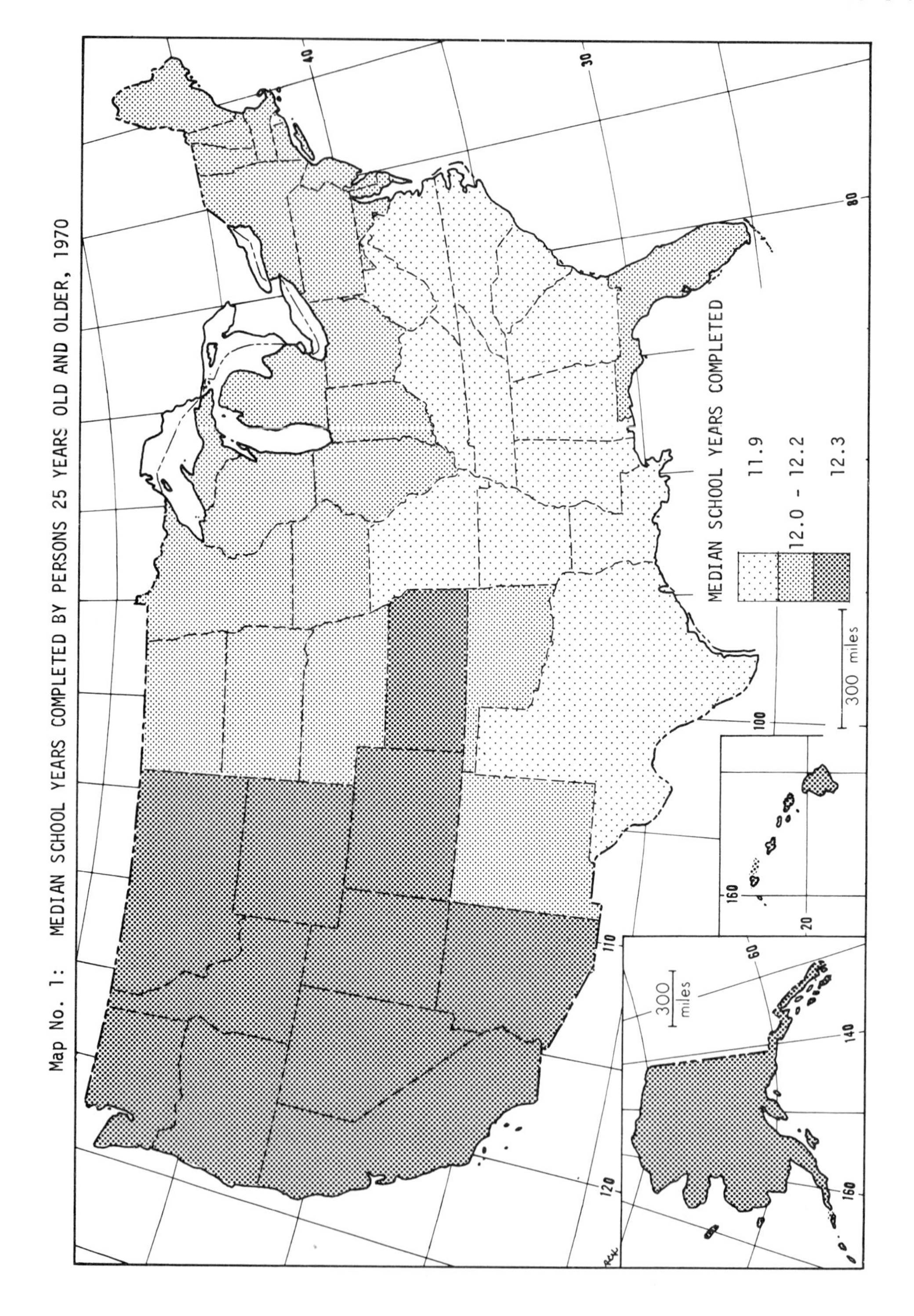

Map No. 1: MEDIAN SCHOOL YEARS COMPLETED BY PERSONS 25 YEARS OLD AND OLDER, 1970

Table 1
Percent of Persons 25 to 34 Years Old Who Have Completed Four Years of High School or More, by Race and Sex: 1940-1974

Year	White		Black		Black-White Differential	
	Males	Females	Males	Female	Males	Females
1974	82.3	81.0	67.0	63.9	15.3	17.1
1973	80.2	79.7	62.3	60.5	17.9	19.2
1972	79.7	78.3	59.1	61.6	20.6	16.7
1971	78.4	76.5	52.6	58.8	23.1	17.7
1970	77.0	75.3	49.4	57.0	27.6	18.3
1969	75.2	74.7	53.9	52.8	21.3	21.9
1968	73.4	73.6	52.0	50.0	21.4	23.6
1967	72.9	72.3	49.9	54.5	23.0	17.8
1966	72.5	71.6	44.3	46.4	28.6	25.2
1965	71.0	70.5	45.2	45.8	25.8	24.7
1960	59.3	62.8	30.1	35.8	29.2	27.0
1950	51.5	55.4	18.4	22.2	33.1	32.8
1940	36.1	40.9	8.9	12.3	27.2	28.6

Source: U.S. Bureau of the Census, *Current Population Reports,* Series P-20, No. 274, "Educational Attainment in the U.S.: March 1973 and 1974." U.S. Government Printing Office, Washington, D.C., 1974.

northern state in that its educational level in both 1960 and 1970 is more than 0.5 years below the national median. Goldstein and Mayer feel that this (referring specifically to the 1960 score) "may be a function of the older age structure of the population (in Rhode Island) as well as a high proportion of foreign born."[10] A similar argument may explain North Dakota's relatively low educational levels in 1950 and 1960.

An examination of the data contained in Table 2 reveals that the regional variations in educational levels have remained generally unchanged, at least since 1940. Educational attainment always has been lowest in the South and highest in the West. Elsewhere, little variation has existed.

On the basis of the evidence presented in this section, it is possible to conclude that regional variations do in fact exist with respect to educational attainment, as the term is defined here. As a logical next step in exploring the geography of education in the United States, we should begin to examine the factors areally associated with educational attainment. In the next section of this paper, therefore, three factors (race, income, and migration) are considered as they relate spatially to educational attainment.

EDUCATIONAL ATTAINMENT AND RACE

Kariel hypothesized that an inverse relationship would exist between the percentage of blacks in a state and median school years completed in that state.[11] He based his hypothesis on the argument that job and educational discrimination, especially in the South, result in blacks dropping out of school. Such a relationship did in fact exist for both 1960 and 1970.

Table 2
Median School Years Completed by Persons 25 Years Old and Over,
by Region and State: 1940, 1950, 1960, 1970

	1940	1950	1960	1970
UNITED STATES	8.4	9.3	10.6	12.1
N EAST	8.8	10.7	11.2	12.1
ME	8.9	10.2	11.0	12.1
NH	8.7	9.8	10.9	12.2
VT	8.8	10.0	10.9	12.2
MA	9.0	10.9	11.6	12.2
RI	8.8	9.3	10.0	11.5
CT	8.5	9.8	11.0	12.2
MID ATLANTIC	8.4	9.3	10.5	12.1
NY	8.4	9.6	10.7	12.1
NJ	8.4	9.3	10.6	12.1
PA	8.2	9.0	10.2	12.0
E N CENTRAL	8.5	9.6	10.7	12.1
OH	8.6	9.9	10.9	12.1
IN	8.5	9.6	10.8	12.1
IL	8.5	9.3	10.5	12.1
MI	8.6	9.9	10.8	12.1
WI	8.3	8.9	10.4	12.1
W N CENTRAL	8.5	9.0	10.7	12.1
MN	8.5	9.0	10.8	12.2
IA	8.7	9.8	11.3	12.2
MO	8.3	8.8	9.6	11.8
ND	8.3	8.7	9.3	12.0
SD	8.5	8.9	10.4	12.1
NE	8.8	10.1	11.6	12.2
KS	8.7	10.2	11.7	12.3
S ATLANTIC	7.8	8.6	9.8	11.4
DE	8.5	9.8	11.1	12.1
MD	8.0	8.9	10.4	12.1
DC	10.3	12.0	11.7	12.2
VA	7.7	8.5	9.4	11.7
WV	7.8	8.5	8.8	10.6
NC	7.4	7.9	8.9	10.6
SC	6.7	7.6	8.7	10.5
GA	7.1	7.8	9.0	10.8
FL	8.3	9.6	10.9	12.1
E S CENTRAL	7.5	8.3	8.8	10.5
KY	7.7	8.4	8.7	9.9
TN	7.7	8.4	8.8	10.6
AL	7.1	7.9	9.1	10.8
MS	7.1	8.1	8.9	10.7
W S CENTRAL	8.1	8.8	9.9	11.3
AR	7.3	8.3	8.8	10.5
LA	6.6	7.6	8.8	10.8
OK	8.4	9.1	10.4	12.1
TX	8.5	9.3	10.8	11.6
MOUNTAIN	8.9	10.7	12.0	12.4
MT	8.7	10.2	11.6	12.3
ID	8.9	11.0	11.8	12.3
WY	9.2	11.1	12.1	12.4
CO	8.9	10.9	12.1	12.4
NM	7.9	9.3	11.1	12.2
AZ	8.6	10.0	11.3	12.3
UT	10.2	12.0	12.2	12.5
NV	9.6	11.5	12.1	12.4
PACIFIC	9.7	11.5	12.0	12.4
WA	9.1	11.2	12.1	12.4
OR	9.1	10.9	11.8	12.3
CA	9.9	11.6	12.1	12.4
AK	N/A	11.3	12.1	12.4
HI	N/A	8.7	11.3	12.3

Sources: U.S. Bureau of the Census, *U.S. Census of Population: 1940.* Vol. 11, *Characteristics of the Population,* Pt. 1, U.S. Summary and Alabama-District of Columbia, Table 31. U.S.G.P.O., Washington D.C., 1943.

U.S. Bureau of the Census, *U.S. Census of Population:* 1950. Vol. 11, *Characteristics of the Population,* Part 1, United States Summary. Table 67. U.S.G.P.O., Washington D.C., 1953.

U.S. Department of Health, Education, and Welfare, *Digest of Education Statistics: 1966.* Table 12. U.S.G.P.O., Washington, D.C., 1966.

U.S. Department of Health, Education, and Welfare, *Digest of Education Statistics: 1975.* Table 12. U.S. Government Printing Office, Washington, D.C., 1976.

Table 3
Median School Years Completed by Persons 25 Years Old and Over, by Age and Race: March 1975

Age	Total	Black	White
25-29	12.8	12.5	12.8
30-34	12.7	12.4	12.7
35-44	12.5	11.9	12.5
45-54	12.3	10.4	12.4
55-64	12.1	8.5	12.1
65-74	9.8	7.0	10.3
75+	8.6	5.3	8.7

Source: U.S. Bureau of the Census, *Current Population Reports,* Series P-20, No. 295, "Educational Attainment in the United States: March 1975," U.S. Government Printing Office, Washington, D.C., 1976.

As was mentioned in the previous section, educational attainment is related to both race and age. A 1975 Census Bureau study indicates further that when these variables are considered together, very little difference in educational attainment exists between the races, among young adults. Dramatic differences do exist, however, for older age groups (see Table 3). This evidence suggests that the educational and economic climate is improving for blacks. It also suggests that a modified version of Kariel's hypothesis[12] should be tested, specifically, that an inverse relationship exists between the percentage of older Blacks in a state and median school years completed. Using blacks 65 years old and over as an approximation of "older blacks," an inverse relationship much stronger than Kariel's was established for 1970.

There is evidence to indicate that the presence of blacks contributes to low-median educational attainment levels more in some southern states than in others. Anderson in a study done during the 1950s concluded that levels of attainment and differences in attainment between the races varied significantly from state to state in the South.[13] He argued that these differences reflected a combination of the stage of general educational progress and the extent to which the two races are members of distinct subcultures. In 1970, in this regard, it is interesting to observe that in Alabama, Georgia, Louisiana, Mississippi, South Carolina, and Virginia at least 4 years of schooling separate black and white males. In contrast, less than 2 years separate these groups in Kentucky and West Virginia, where both have low-attainment levels (see Table 4). The patterns are somewhat less pronounced for females.

Before concluding this section one further observation must be made. Map 2 shows the distribution of white male educational attainment. It is interesting to note that when educational attainment is viewed in this way, regional differences are similar to those produced using state totals (see Map 1). This indicates that white southerners are educationally deprived as compared to whites in other regions.

The evidence presented in this segment of the paper indicates that the distribution of blacks is a major variable associated with regional variations in educational attainment. However, the

Table 4
Median School Years Completed by Persons 25 Years Old and Over,
by Sex, Race, and Selected States: 1970

	Men			Women		
Region and State	White	Black	Difference	White	Black	Difference
United States	12.1	9.4	2.7	12.1	10.0	2.1
South Atlantic						
Delaware	12.3	9.6	2.7	12.2	10.2	2.0
Maryland	12.3	9.5	2.8	12.2	10.3	1.9
D.C.	15.1	11.1	4.0	12.8	11.5	1.3
Virginia	12.1	8.0	4.1	12.1	9.0	3.1
W. Virginia	10.3	8.9	1.4	10.8	10.0	0.8
N. Carolina	10.8	7.9	2.9	11.2	9.0	2.2
S. Carolina	11.4	7.1	4.3	11.4	8.1	3.3
Georgia	11.6	7.3	4.3	11.5	8.3	3.2
Florida	12.2	8.3	3.9	12.2	9.2	3.0
E. South Central						
Kentucky	9.6	8.9	0.7	10.3	9.7	0.6
Tennessee	10.9	8.4	2.5	11.2	9.0	2.2
Alabama	11.6	7.4	4.2	11.6	8.5	3.1
Mississippi	12.1	6.5	5.6	12.2	8.1	4.0
W. South Central						
Arkansas	10.8	7.1	3.7	11.3	8.3	3.0
Louisiana	12.0	7.4	4.6	11.9	8.3	3.6
Oklahoma	12.1	10.0	2.1	12.1	10.4	1.7
Texas	12.0	9.3	2.7	11.9	9.0	2.9

Source: U.S. Department of Health, Education, and Welfare, *Digest of Education Statistics: 1975.* Table 12. U.S. Government Printing Office, Washington, D.C., 1976.

improved inverse relationship produced by testing the spatial association between the arrangement of older blacks and educational attainment, the differential impact of blacks on median attainment levels among southern states, as well as the spatial pattern of white male educational attainment suggest further that race operates together with other variables areally correlated with regional variations in educational attainment. One of these variables, income level, suggested by what has already been said, will be discussed in the next segment of the paper.

EDUCATIONAL ATTAINMENT AND INCOME LEVELS

Kariel hypothesized that a direct relationship exists between median family income and median educational attainment. He used the 1960 census data for states to test his hypothesis.

Map No. 2: MEDIAN SCHOOL YEARS COMPLETED BY WHITE MALES 25 YEARS OLD AND OLDER, 1970

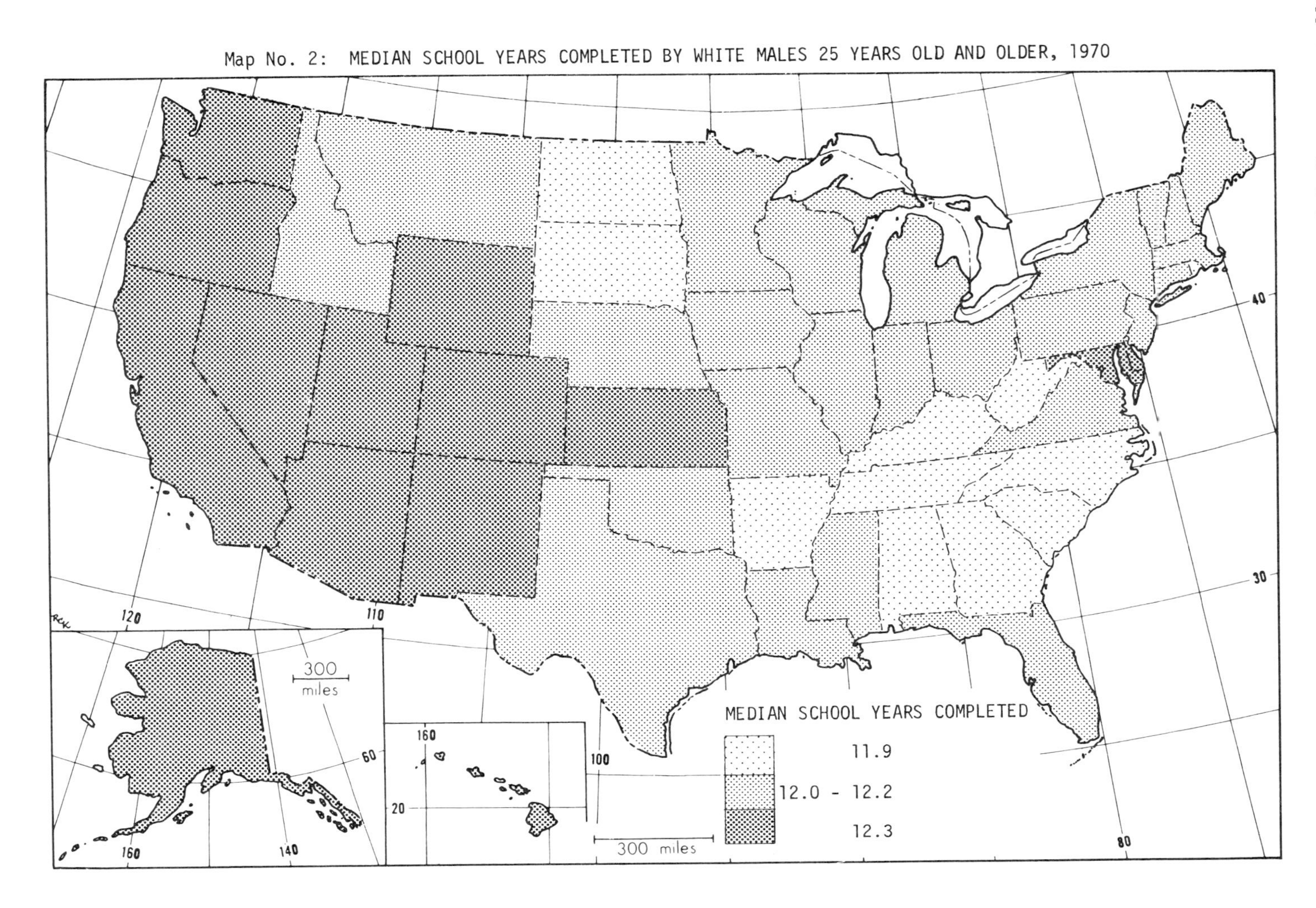

Table 5
Relationship Between Adult Illiteracy and per Capita National Income, Around 1950, in 41 Selected Countries

Adult Illiteracy Around 1950	Per Capita National Income in 1950: Less than $300 (U.S.)	$300 (U.S.) or More
High (> 50%)	Bolivia, Brazil, Dominican Rep., Egypt, El Salvador, Guatemala, Haiti, Honduras, India, Malaya, Nicaragua, Turkey	
Medium (20-49%)	Ceylon, Chile, Costa Rica, Ecuador, Greece, Panama, Paraguay, Philippines, Portugal, Thailand, Yugoslavia	Puerto Rico, Venezuela
Low (< 20%)	Japan	Argentina, Australia, Belgium, Canada, Denmark, Finland, France, Netherlands, New Zealand, Norway, Spain, Sweden, Switzerland, United Kingdom, United States

Source: UNESCO, *World Illiteracy at Mid-Century: A Statistical Study,* Monograph on Fundamental Education, No. 11, Paris, 1957.

His finding (r = 0.73) is consistent with one of the observations made in a UNESCO study of world literacy (literary, like educational attainment, may be viewed as a measure of educational level).[14] The study indicated that a direct relationship existed between spatial variations in literacy rate and per capita national income. This relationship is shown in Table 5. The UNESCO study further suggested that educational attainment is probably related to the distribution of income within a country even more closely than to the level of per capita income. The study argued that where income is highly concentrated in a small proportion of the population, a situation which could not be detected using per capita income, education tends to be the privilege of the few, and a large part of the people will remain illiterate. It was probably this argument which led Kariel to use median family income rather than per capita income figures in his hypothesis. Interestingly, the coefficients of correlation produced using median family income (1960, 0.73, and 1970, 0.64) are only slightly higher than those produced using per capita income—(1960, 0.68, and 1970, 0.57).

A clue to the difficulty involved in interpreting Kariel's findings is found in the UNESCO study mentioned previously. The study concludes that educational levels and income levels influence each other. On the one hand a "higher level of literacy in the population tends to accelerate the economic development of a country; and conversely, a higher rate of illiteracy tends to slow down the increase of prosperity."[15] On the other hand, as the productivity of a country increases, demand for workers with basic education and a variety of technical skills also increases. More people will therefore attend school to attain these skills. Kariel established the existence of a strong positive relationship between family income and educational attainment in the United States. However, his study does not tell us whether the spatial distribution of attainment is influenced by the distribution of income levels, whether the spatial distribution of income levels is influenced by the distribution of educational attainment, or both.

A potentially fruitful way of approaching the relationship between income and educational attainment, in which the direction of the relationship is clearer, involves looking at income as an indicator of the family environment of the person being educated. The general attainment literature indicates that father's education and occupation influence children's educational attainment directly. For example, a U.S. Census Bureau study using data collected in 1962 concluded that "among men 20 to 64 years old, those whose fathers lacked an eighth-grade education, those whose fathers had farm occupations, were less likely than those whose fathers were in nonfarm occupations to complete eight or more years of schooling, and those from families where the father had a white-collar job were more likely than others to have completed some college. Where a father was better educated than average, the fact that he was a white-collar worker also made a difference in the chances of his son's educational level exceeding his own."[16] If median family income is used as an indicator of education and occupation, and therefore of family environment, then it too should be directly related to educational attainment. However, the relationship cannot be tested by using income level data and educational data from the same year in a manner similar to Kariel. Since median educational attainment figures are available only for persons 25 years old and older, using income and attainment data from the same year does not reflect the family environment existing while those persons were attending school. In a preliminary effort to test the home environment-educational attainment relationship in a spatial sense, *1960* median family income data were correlated with *1970* educational attainment figures. A coefficient of correlation of 0.76 was found. Admittedly, using 1960 data as an indicator of the home environment of *all* persons 25 and over in 1970 is somewhat gross. Nevertheless, in using a procedure such as this, in contrast to Kariel's, one is at least confident in the direction of the relationship.

Kariel also tested an hypothesis which considered the influence of income levels on educational attainment at a different scale.[17] At this scale he was interested in the extent to which a state devoted its income to education. He proposed that a positive relationship existed between median school years completed in an area and expenditures on education. Relating 1960 data on median school years completed to per capita revenue of state and local governments for the same year (Kariel's indicator of educational expenditures) produced a coefficient of correlation of 0.75.

It can be argued that Kariel's indicator of educational expenditure is inappropriate for two reasons: (1) there is no reason to assume that all states appropriate the same proportion of revenues for educational purposes; and (2) if, as seems reasonable, the hypothesis is based on the assumption that state educational expenditures are being used as a measure of commitment to education, and that commitment has an impact on attainment, then commitment and attainment should not be examined for the same year.

With these points in mind, educational expenditures per capita for 1960 were correlated with educational attainment for 1970, in an effort to approximate the state levels of commit-

Table 6
Relationship Between Adult Illiteracy and Government Expenditure on Education as Percentage of National Income, Around 1950, in 39 Selected Countries

	Government Expenditure on Education as Percentage of National Income	
Adult Illiteracy Around 1950	**Less than 2%**	**2% or More**
High (> 50%)	Bolivia, Brazil, Dominican Rep., El Salvador, Guatemala, Haiti, Honduras, India, Malaya, Turkey	Egypt
Medium (20-49%)	Ecuador, Paraguay, Portugal, Thailand, Venezuela	British Guiana, Ceylon, Chile, Cyprus, Panama, Philippines, Puerto Rico, Yugoslavia
Low (< 20%)	Argentina, France, Spain	Australia, Belgium, Canada, Denmark, Finland, Japan, Netherlands, New Zealand, Norway, Sweden, United Kingdom, United States

Source: UNESCO, *World Illiteracy at Mid-Century: A Statistical Study,* Monograph on Fundamental Education, No. 11, Paris, 1957.

ment to education during which the attainment being measured occurred. This yielded a coefficient of correlation of 0.63.

It can be argued that per capita expenditures on education may simply reflect the economic situation in a state and not its commitment to education. In an effort to measure degree of commitment, expenditures considered as a percentage of per capita income might be more appropriate. Although the UNESCO study on world literacy concluded that the percentage of a nation's income devoted to educational expenditures and literacy rates were directly related,[18] (see Table 6) no relationship between educational expenditures as a percentage of per capita income and educational attainment could be established for the United States. One should not use this finding to conclude that state commitment to education is not related to educational attainment. It can be reasonably argued that the problem is with the indicator of commitment being used and not with the underlying relationship. That is to say, there are at least two factors which can influence levels of educational expenditures as a percent of per capita income aside from variations in commitment. A state's level of income can be such that it may spend only a small proportion of its income on education and yet be spending more than a state allocating a similar or greater percentage of per capita income to education (see Table 7). The size of a school population of a state may influence levels of expenditures more than does commitment. Given these possibilities, expenditures per student might provide a more useful measure of commitment than do expenditures as a percentage of per capita income (correlating 1960 expend-

Table 7

Direct Expenditures of State and Local Governments for Education, by per Capita Amount and Percent of per Capita Income: 1972-73

State	Amount per Capita	% of per Capita Income
UNITED STATES	$331.53	7.3
Alabama	240.49	6.9
Alaska	867.29	16.7
Arizona	385.98	8.9
Arkansas	209.89	6.3
California	353.58	7.0
Colorado	382.86	8.3
Connecticut	340.60	6.3
Delaware	504.11	9.7
District of Columbia	338.14	5.7
Florida	248.81	5.5
Georgia	265.88	6.7
Hawaii	418.07	8.2
Idaho	274.65	7.3
Illinois	357.37	7.0
Indiana	331.98	7.6
Iowa	345.81	8.0
Kansas	314.18	6.9
Kentucky	261.37	7.2
Louisiana	273.16	7.6
Maine	276.46	7.5
Maryland	374.66	7.5
Massachusetts	342.24	7.1
Michigan	392.07	7.9
Minnesota	424.09	9.8
Mississippi	241.86	7.6
Missouri	283.65	6.6
Montana	348.05	8.6
Nebraska	300.58	6.8
Nevada	375.55	7.3
New Hampshire	272.60	6.5
New Jersey	331.02	6.2
New Mexico	367.59	10.5
New York	421.69	8.0
North Carolina	273.18	7.1
North Dakota	332.96	8.3
Ohio	294.25	6.4
Oklahoma	268.36	7.0
Oregon	358.23	8.3
Pennsylvania	332.74	7.3
Rhode Island	288.02	6.4
South Carolina	273.18	7.8
South Dakota	376.72	9.9
Tennessee	239.63	6.5
Texas	286.99	7.0
Utah	393.08	10.5
Vermont	346.01	8.9
Virginia	297.91	6.8
Washington	391.23	8.6
West Virginia	260.99	7.2
Wisconsin	373.29	8.7
Wyoming	500.72	11.7

Source: U.S. Department of Health, Education, and Welfare, *Digest of Educational Statistics: 1975.* U.S.G.P.O., Washington, D.C., 1976.

itures per student with 1970 median school years completed, by states, results in a coefficient of correlation of 0.70).

In concluding this segment of the paper, the evidence suggests that to understand regional variation in educational attainment, income variables on at least two scales must be considered. It appears that family environment, measured here by median family income, is directly related to educational attainment, a relationship which expresses itself both spatially and aspatially. Furthermore, educational attainment appears to be associated areally with state commitment to education, measured in terms of educational expenditures, despite the difficulty involved in establishing the appropriate operationalization of expenditures and therefore of commitment. The evidence suggests finally that the income variables at both scales may be related causally to attainment. At this point, one can conclude only that research at both scales should be pursued by geographers interested in the geography of education.

EDUCATION AND MIGRATION

Educational attainment is a characteristic of people living in a particular area, regardless of where they were raised. The areal associations examined to this point have assumed the existence of a stationary population living where it was raised and educated. It is conceivable, however, that regional variations in educational attainment are influenced by factors associated with the selective movement of the population. That is to say, for example, that educational levels in some areas may be inflated by in-migration of well-educated people, while levels in other areas are deflated by the out-migration of well-educated persons. There is empirical evidence to indicate that a relationship between migration and attainment does exist. For example, a 1975 Census Bureau study of mobility concluded that educational attainment influences the likelihood of migration. Specifically "college graduates are more likely to move between counties or states than are high school graduates who, in turn, migrate more often than persons with only a grade school education."[19] Also, a 1968 Census study revealed that the level of schooling completed by persons 25 to 44 years old, who moved to another region of the country, was higher than that of the general population of that age group in that region (see Table 8).[20]

For selective migration to have had an influence on 1970 spatial variations in educational attainment, the South would have had to experience substantial net out-migration. Until the mid-1950s, the South did experience a pattern of net out-migration. Since that time, however, a change has occurred. In the late 1950s, this change was brought about as more whites moved into the area than moved out. The pattern continues today as the result of return migration, decreased out-migration, and increased in-migration of persons not born in the South. In the period 1965-1975, the South and the West have experienced net in-migration and the Northeast and North Central Regions have witnessed net out-migration.

This segment of the paper may be concluded by pointing out that while migration and educational attainment are clearly related, the nature and extent of the areal association between them is unclear. Thus, when migration experiences of states for the period 1950-1970, as measured by net migration, are correlated with educational attainment for 1970, a coefficient of correlation of only 0.36 is produced. Despite this finding, pursuit of this areal association, perhaps in the form of a historical study, is intuitively appealing.

Table 8
Level of School Completed by All Residents and
In-migrants 25 to 44 Years Old, for Regions: March 1968

(Percent)

Level of School and Migration Status	Total	Northeast	North Central	South	West
ALL RESIDENTS					
4 Years of High School or More	66.8	69.4	69.1	58.6	73.4
1 Year of College or More	25.4	25.1	23.8	22.5	33.3
IN-MIGRANTS FROM OTHER REGIONS					
4 Years of High School or More	76.3	85.4	77.0	73.1	75.2
1 Year of College or More	42.3	52.9	41.0	40.7	40.1

Source: U.S. Census Bureau, *Current Population Reports,* Series P-20, No. 182, "Educational Attainment: March, 1968," U.S. Government Printing Office, Washington, D.C., 1969.

SUMMARY AND CONCLUSIONS

The purpose of this paper is to explore the geography of education in the United States, in a preliminary way, with special reference to educational attainment. Using median years of school completed by persons 25 years old and over as an indicator, regional variations in educational attainment were found to exist, at least since 1940. Attainment has been lowest in the South and highest in the West. While the Census Bureau definition may be considered less than a perfect indication of educational attainment, other measures also show the South to be a "disadvantaged region." This is true, for example, of the latest spatial data available on literacy in the United States (see Table 9). Similarly, a study dealing with the "functional competency" of adult Americans concluded that the South had the highest proportion of functional incompetents.[21] Finally, data showing the percent of draftees who failed to meet the mental requirements for induction into the Armed Forces in 1972, at a time when the lottery system and minimal deferments were in effect, indicated that failure rates were highest for southern states (see Table 10).[22]

In an effort to begin to examine factors areally associated with educational attainment, race, income levels, and migration were considered. The list of factors is by no means exhaustive, but is, however, suggestive of the types of variables which are related to the spatial arrangement of educational attainment. In future studies, the relative influence of the variables already discussed and the additional influence of other variables on the distribution of attainment levels should be considered.

Educational attainment was found to be inversely related to the distribution of blacks. Additional evidence was presented which indicated that race operates together with other variables. As to what these variables are, the evidence suggested that family environment and

Table 9
Illiteracy of the Population by Region and State: 1960 (percent)

Region/State	Percent
UNITED STATES	2.4
N EAST	1.8
Maine	1.3
New Hampshire	1.4
Vermont	1.1
Massachusetts	2.2
Rhode Island	2.4
Connecticut	2.2
MID ATLANTIC	2.4
New York	2.9
New Jersey	2.2
Pennsylvania	2.0
E N CENTRAL	1.5
Ohio	1.5
Indiana	1.2
Illinois	1.8
Michigan	1.6
Wisconsin	1.2
W N CENTRAL	1.1
Minnesota	1.0
Iowa	0.7
Missouri	1.7
N Dakota	1.4
S Dakota	0.9
Nebraska	0.9
Kansas	0.9
S ATLANTIC	3.2
Delaware	1.9
Maryland	1.9
D.C.	1.9
Virginia	3.4
W Virginia	2.7
N Carolina	4.0
S Carolina	5.5
Georgia	4.5
Florida	2.6
E S CENTRAL	4.0
Kentucky	3.3
Tennessee	3.5
Alabama	4.6
Mississippi	4.9
W S CENTRAL	4.0
Arkansas	3.6
Louisiana	6.3
Oklahoma	1.9
Texas	4.1
MOUNTAIN	1.7
Montana	1.0
Idaho	0.8
Wyoming	0.9
Colorado	1.3
N Mexico	4.0
Arizona	3.8
Utah	0.9
Nevada	1.1
PACIFIC	2.3
Washington	0.9
Oregon	0.8
California	1.8
Alaska	3.0
Hawaii	5.0

Source: U.S. Census Bureau, *Current Population Reports,* Series P-23, No. 8, "Estimates of Illiteracy, by States: 1960." U.S. Government Printing Office, Washington, D.C., 1963.

Table 10

Percent of Draftees Who Failed to Meet the Mental Requirements for Induction into the Armed Services, by Region and State: 1972

Region / State	Percent
UNITED STATES	6.7
N EAST	2.8
Maine	2.9
New Hampshire	2.0
Vermont	2.1
Massachusetts	2.4
Rhode Island	5.1
Connecticut	2.1
MID ATLANTIC	4.7
New York	5.3
New Jersey	5.8
Pennsylvania	3.7
E N CENTRAL	4.0
Ohio	4.2
Indiana	4.6
Illinois	4.8
Michigan	3.2
Wisconsin	3.2
W N CENTRAL	1.8
Minnesota	1.1
Iowa	2.1
Missouri	2.2
N Dakota	2.1
S Dakota	1.6
Nebraska	1.8
Kansas	1.7
S ATLANTIC	12.6
Delaware	6.5
Maryland	4.8
D.C.	12.5
Virginia	11.5
W Virginia	9.4
N Carolina	15.2
S Carolina	25.5
Georgia	19.8
Florida	8.0
E S CENTRAL	17.1
Kentucky	10.0
Tennessee	13.5
Alabama	18.3
Mississippi	26.6
W S CENTRAL	10.7
Arkansas	10.8
Louisiana	17.2
Oklahoma	6.3
Texas	8.3
MOUNTAIN	3.5
Montana	1.2
Idaho	2.3
Wyoming	1.8
Colorado	3.0
N Mexico	8.6
Arizona	4.4
Utah	2.9
Nevada	4.1
PACIFIC	4.5
Washington	1.8
Oregon	2.0
California	4.4
Alaska	3.8
Hawaii	10.6

Source: U.S. Department of Health, Education, and Welfare, *Digest of Education Statistics: 1975.* Table 15. U.S. Government Printing Office, Washington, D.C., 1976.

state commitment to education both of which vary spatially, are associated with regional variations in attainment. Their relative influence on the areal arrangement of attainment and the extent to which they are related to race have not yet been established.

Finally, recognizing that educational attainment figures for a state do not only reflect the activities of people reared there, an effort was made to briefly explore the role of migration on educational attainment. It was concluded that while a relationship between educational attainment and migration certainly existed, the spatial manifestation of the relationship has yet to be clarified.

NOTES

1. See for example: W.T.S. Gould, "Geography and Educational Opportunity in Tropical Africa," *Tijdschrift voor Economische, en Sociale Geografie* 62 (1971), 2, pp. 82-89; Gerald H. Hones, and Raymond Ryba, "Why Not a Geography of Education?" *Journal of Geography* 71 (1972), 3, pp. 135-139; R.H. Ryba, "The Geography of Education and Educational Planning," summary of paper presented at the International Geographical Congress (22d, Quebec), Symposium CA7, "Geography in Education" (1971).
2. These include spatial arrangements, functional regions, spatial interaction, location analysis, and diffusion. For examples see: Alice C. Andrews, "Some Demographic and Geographic Aspects of Community Colleges," *Journal of Geography* 73 (1974), 2, pp. 10-16; Stanley Brunn, and Wayne Hoffman, "The Geography of Federal Grants-in-Aid to States," *Economic Geography* 45 (1969), 3, pp. 226-238.; Donald Maxfield, "Social Desegregation and Overcrowding: A Mathematically Based Solution," *Bulletin of the Georgia Academy of Science* 28 (1970); Maurice H. Yeates, "Hinterland Delimitation: A Distance Minimizing Approach," *Professional Geographer* 15 (1963), 6, pp. 7-10; Herbert Kariel, "Student Enrollment and Spatial Interaction," *The Annals of Regional Science* 2 (1968), 2, pp. 114-127; Harold McConnell, "Spatial Variability of College Enrollment As a Function of Migration Potential," *Professional Geographer* 27 (1965), 6, pp. 29-37; Paul Scipione, A Computer Solution for Determining Study Migration," *Professional Geographer* 25 (1973), 3, pp. 249-254; Fred L. Hall, *Location Criteria for High Schools: Student Transportation and Racial Integration* (University of Chicago, Department of Geography, Research Paper no. 150, 1973); John Florin, "The Diffusion of the Decision to Integrate: Southern School Desegregation, 1954-1964," *Southeastern Geographer* 11 (1971), 2, pp. 139-144; Judith W. Meyer, *Diffusion of an American Montessori Education* (University of Chicago, Department of Geography, Research Paper no. 160, 1975).
3. B.E. Coates, and E.M. Rawston, *Regional Variations in Britain* (London: B.T. Batsford, Ltd., 1971).
4. Personal correspondence with L.J. Abernathy, Assistant Regional Director, College Entrance Examination Board, Evanston, Ill.
5. William L. Humm, "An Investigation of the Decline in College Entrance Examination Board Scores," Research and Statistics Section, Illionois Office of Education, Springfield (1975), 6.
6. U.S. Bureau of the Census, *Statistical Abstract of the United States: 1972,* Washington, D.C. (1972), 101.
7. Norvell Northcutt, *Adult Functional Competency: A Summary* (University of Texas, Austin, Division of Extention, 1975).
8. This discussion is based on: U.S. Bureau of the Census, *Current Population Reports,* Series P-20, no. 274, "Educational Attainment in the U.S.: March, 1973 and 1974." (U.S. Government Printing Office, Washington, D.C., 1974).
9. Herbert Kariel, and Patricia Kariel, *Exploration in Social Geography* (Menlo Park: Addison-Wesley 1972), p. 157.
10. Sidney Goldstein, and Kurt Mayor, *The People of Rhode Island: 1960,* State Planning Section Publication no. 8 (Providence: Planning Division, Rhode Island Development Council, 1963).
11. Kariel, *op. cit.,*p. 158.
12. Kariel, *op. cit.,* p. 158.
13. Arnold C. Anderson, "Patterns and Variability in the Distribution and Diffusion of Schooling," *Education and Economic Development,* C.A. Anderson and M.J. Bowman, eds. (Chicago: Aldine Publishing Company, 1965), p. 332.
14. UNESCO, *World Illiteracy at Mid-Century: A Statistical Study,* Monographs on Fundamental Education no. 11 (Paris, 1957), p. 172.
15. *Ibid.,* p. 176.
16. U.S. Bureau of the Census, *Current Population Reports,* Series P-20, no. 132, "Educational Change in a Generation, March 1962." (U.S. Government Printing Office, Washington, D.C., 1964), p. 2.

17. Kariel, *op. cit.,* p. 156.
18. UNESCO, *op. cit.,* pp. 175-176.
19. U.S. Bureau of the Census, *Current Population Reports,* Series P-20, no. 285, "Mobility of the Population of the U.S.: March, 1970 to March, 1975" (U.S. Government Printing Office, Washington, D.C., 1975), p. 4.
20. U.S. Bureau of the Census, *Current Population Reports,* Series P-20, no. 182, "Educational Attainment: March, 1968" (U.S. Government Printing Office, Washington, D.C., 1969).
21. Northcutt, *op. cit.*
22. Selective Service System, *Draft: Past, Present, and Future,* Draft Information Series (U.S. Government Printing Office, ashington, D.C., 1972).

Recent Themes in Ethnic Geography

Don C. Bennett

Ethnic geography has not yet formally come of age as a subfield of the discipline. However, a rather substantial literature has been produced which has an important ethnicity component and there are many geographers whose works are solely or primarily concerned with ethnicity in some way. But there has been little discussion which attempts to develop ethnic studies as a type within the discipline. The literature reviews of ethnic geography focus on limited themes or the cultural biases of work done on particular ethnic groups. None, to my knowledge, has attempted a broad view across groups, countries, and time.

The purpose of this study is to identify some major themes that have occurred in the ethnic geographic literature from 1963 to 1975 as a step towards sorting out the topics and relationships of greatest interest to geographers. The 1963 date was selected as a convenient division between the relatively few and scattered prior studies and the much greater emphasis given ethnicity by geographers subsequently. It is patent that the enlarged emphasis in the United States resulted from the combined civil rights and poverty issues which the urban riots in the black ghettoes made so visible. It was a time for change, for renewed recognition of ethnic diversity and persistence, and the association of ethnicity and inequity in the United States. Many geographers responded with an increased attention to the circumstances of disadvantaged ethnic minority groups.

Somewhat fortuitously, the 1963 date can also be used as a rough division between the pre- and post-methodological and philosophical revolution which geography has experienced. Ethnic geographic studies after that date are increasingly set in a more nomothetic framework.

Our classification of ethnic geographic works differentiates them according to their objectives. Eight types of objectives have been determined and will be discussed later. These types were arrived at by the following procedure. First, the relevant literature was defined as that which had an important racial and/or ethnic component and was either published in geographic journals or was written by geographers in English between 1963 and 1975. Next, a sample of about one hundred works was examined for their stated or implied objectives. Finally, these objectives were subjectively grouped according to general themes which are frequently discussed in the literature. As the work progressed, new themes were added and some former ones were collapsed or changed. Our categorization of ethnic geographic studies is a progress report and further work will likely suggest additional themes and/or changes. This method of classification will be recognized as a "special purpose" type, using a grouping-from-below procedure (Harvey, 1969 p. 337). Its advantage is that it does not force individual studies into predetermined and perhaps inappropriate slots. Its disadvantage is that there is no "inherent logic" to the set of categories which results.

This review of the ethnic-geographic literature is extremely selective and will not mention many worthwhile studies. A relatively few works have been chosen to illustrate the richness and variety of the literature.

A REVIEW OF REVIEWS

Before initiating our own review of the recent ethnogeographic literature, it is useful to see what other geographers have said about the kind of work their colleagues were doing or not doing in this area. The 1960s had produced a sufficient literature so that by the early 1970s the first reviews and exhortations began to appear. Taken as a whole, the eight reviews identified show a very strong orientation towards an interest in racial groups, especially blacks. Birdsall (1971), Donaldson (1971), Leach (1973), Lewis (1974), and Ritter (1971) essentially confine their evaluations to the geographic work dealing with blacks. Carlson (1972) briefly reviews the literature on American Indians, while only Thompson and Agocs (1973) argue for a more general approach which would embrace all ethnic groups.

Two studies, those of Ritter and Thompson-Agocs, were oriented towards a teaching framework. Ritter suggested an outline for a course on urban black Americans, thus indirectly recognizing the main group of interest of existing work and implicitly recommending particular topics. Thompson and Agocs evaluated a number of problems the teacher should be aware of in teaching ethnicity and commented on some teaching approaches which have proved successful. Further, they made a plea for greater attention to studies at the microscale (individuals, families, neighborhoods), and more exchange among the social science disciplines that study ethnicity.

Birdsall, Lewis, and Carlson have sorted the ethnogeographic literature in terms of perceived themes. Lewis and Carlson, writing about the work on United States blacks and Native Americans respectively, arrived independently at themes which were very similar. Carlson's five themes (regional orientation; urban setting; rural setting; migrations; and general) are the same as Lewis' except that Lewis, writing later, has an additional "historical" theme. The formulation of these themes focuses attention to the significance of scale to geographers, a view supported by Thompson and Agocs above.

Another, rather polemic perspective is taken by both Donaldson and Leach. These authors take their colleagues to task for the biased way they perceive geographers who have dealt with the issues of race, racial groups, and racism. Donaldson's review, the earlier of the two, purports ". . . to show that geographers in their writings have contributed to the maintenance of white supremacy over Afro-Americans." Leach is basically concerned with the inequities that exist between blacks and whites and has evaluated the ethnogeographic literature in terms of its approaches to this issue.

With this brief review of reviews we turn to the eight themes which have been identified. There is no logic to the order of discussion.

ETHNIC GROUP DISTRIBUTIONS, CHARACTERISTICS, AND CHANGES

How is an ethnic group distributed in this country, region, or city? How is some characteristic (e.g., age, income, education) of this group distributed among them? What is the nature of the changes which have occurred in either of the above during a certain period of time? These three are the most popular types of questions geographers have formulated in their ethnogeographic work. They are perhaps the most obvious questions for geographers to ask and, in part, the emphasis given to them reflects the rather recent geographic interest in ethnicity. To be sure, most researchers go beyond these simple questions to probe, with varying levels of sophistication, the causes and/or consequences of these distributional patterns.

NATIONAL AND REGIONAL SCALE STUDIES

Blacks have been the major group of interest for American geographers at every scale of inquiry. The significant relocation of blacks nationally since World War I has unquestionably influenced the focus on them. Their long identification as a southern subgroup which rather rapidly changed to that of a national subgroup, concurrent with their changeover from rural to metropolitan communities, has generated considerable research interest. The greater part of this interest has been concerned with clarifying various aspects of socioeconomic-related conditions. For example: Fisher (1973) has looked at black farm ownership in the South; Henderson and Hart (1971) examined the distribution of black colleges; Katzman (1969) analyzed the relationship between changing national distribution and changing relative incomes; Lowry (1971) focused on Mississippi, and assessed some demographic consequences of the very divergent distributional patterns of the blacks and whites there; Smith and Raitz (1974) investigated the black hamlets associated with Kentucky's Bluegrass (former slave plantations) region; Wheeler and Brunn (1969, 1968) studied atypical northern black rural farm and nonfarm settlements; and Meyer (1972) factor-analyzed 1960 SMSAs in an effort to identify the main dimensions of black variation among the metropolitan areas and to regionalize them. Lewis (1969) regionalized the black population into domain, spheres, periphery, and exclaves by employing the concepts of numbers, density, and black/white ratios.

Studies of ethnic distributions of United States groups other than blacks have varied widely but are few considering the range of possibilities. American Indians have received some attention, as for example Neils' (1971) dissertation, which provides perhaps the best historical explanation for the distribution of that group in the mid-sixties and Meinig's (1971) analysis of the Southwest in which the residual Indian pattern is understood more fully as one component in a three-way competition. Neils describes the distributional effects of the vacillating policies of the U.S. Government in dealing with Indian sometimes as an individual, and sometimes as a member of a group. Additionally, the government assumed the conflicting roles of acting to satisfy white demands for Indian resources while at the same time protecting Indians from frequently ruthless and unequal competition with whites. The spatial patterns and socioeconomic conditions of the Indians are clarified against this historical background.

Another American ethnic group which has been examined on the national or regional scale is the numerically and regionally important Mexican-American population. Nostrand (1970) has regionalized this group on the basis of minimum numbers and the group/nongroup ratio, providing a county-based nationwide cartographic distribution. Meinig (1971) also devotes considerable attention to Mexican-Americans in his study of the Southwest in which he embraces the entire Texas-to-California area, the predominant location of the group.

It is still not possible to identify either a significant focus of attention or a cluster of scholars who have concentrated on any other American ethnic groups. Indeed, for many groups which have received rather wide attention in sociology and psychology, such as the Jews, Puerto Ricans, Italian-Americans, Polish-Americans, Japanese-Americans, and Chinese-Americans, only a very few studies, mostly of a preliminary type, have been done. A significant part of these are in unpublished masters' thesis or doctoral dissertations and are thus not readily available. Fellows (1972) provides a general distributive picture of the nonwhite groups mentioned as well as the Mexicans and Puerto Ricans.

NON-UNITED STATES AREAS

In general, the geographers of each country have looked at the ethnicity conditions within their own nation, clearly reflecting their greater awareness of those minority groups and the accessibility of data. Exceptions of this would primarily be a few studies by Europeans and Americans abroad such as: Doeppers' (1972, 1974) research on indigenous language and/or religious minorities in the Philippines; Fair and Shaffer's (1964) and Sabbagh's (1968) work on the Republic of South Africa; Lowenthal's (1972) study of West Indian societies; and Neville's (1965) and McTaggart's (1968) work on Malaysia and Singapore.

The settling of many thousands of New Commonwealth immigrants in the United Kingdom since World War II has generated a moderate geographic literature. Among the best of that work at the national/regional scale is Peach's (1968) study of West Indians. In this, he explained the West Indian distribution as a replacement population which had gone to those regions having heavy labor demands at the time of their immigration but which had failed to attract a sufficient native white population.

INTRAURBAN SCALE STUDIES

Geographers have been far more interested in examining ethnic patterns within urban environments than at either national or global scales. The urban riots associated with the black ghettos in the 1960s undoubtedly played the major role in promoting the urban focus as well as kindling the interest in ethnicity in general and blacks in particular. A substantial literature exists which focuses on the nature, causes, and consequences of urban ethnic locations and distributions. Because the greater part of this literature has to do with the black population, we turn to them first.

Harvey (1972) calls into question the entire perspective by which ghetto origins and maintenance are examined in the West. He views the work done by scholars in the West as fundamentally reflecting the biases of the middle class within a capitalist framework.

Most students of the causes of black ghettoisation assume the continuance of the existing socioeconomic framework in their analyses. A variety of causal factors have been suggested and a smaller number tested. Within this notable literature the work of Harold Rose is preeminent. However, many geographers have contributed to unraveling the subtleties and complexities of the phenomenon. For example, Groves and Muller (1975) have looked at the border cities of

Washington and Baltimore in the 1880s because these cities expressed both northern and southern patterns of black residence and had already experienced a substantial immigration of southern blacks by that time. Thus, analyzing them would help, they thought, to identify the earliest historical patterns which may have been models for later developments elsewhere. In other work which employed a substitute historical method, both Morrill (1965) and Rose (1972) used simulation models to examine some factors they believed were relevant to the growth of black ghettos such as (1) the rate of black population growth in the city, (2) the degree of white neighborhood resistance to black entry, (3) the role of hilly and attractive terrain, (4) the distance to the ghetto center, and (5) the level and extent of economic differentiation within the black population. Extensive discussions of the causes of black ghettos are found in Rose (1969) and Davis and Donaldson (1975:127-45).

The externally imposed restrictive residential pattern that defines ghettos results in a number of disadvantages to all concerned, but especially to those so restricted. The migration of blacks to the central cities of the larger SMSAs has been an important factor in the creation of heavily populated ghetto clusters. These massive numbers of people combined with the inevitable variations in human abilities and motivations and the length of time lived in a given city have produced a type of community-within-a-community. The spatial variations within these subcommunities have been the subject of some attention. Rose (1971) discussed several of the major types: housing, income, jobs, and retail activities. In an earlier work (Rose, 1964) he examined in detail the variations among Miami's ghetto clusters and the changes they had recently experienced. Deskins' (1971) study of Atlanta provides another example of the spatial differentiation of several socioeconomic variables within the black population clusters of a large southern city. The age structures within the black ghetto clusters of Cleveland were analyzed by Sanders and Adams (1971) in order to evaluate the specific age patterns in the zones of ghetto expansion. They also looked at age structure changes through time.

The spatially separated nature of an ethnic group in cities, with blacks experiencing the most extreme circumstances for the longest periods of time in communities of all sizes and regions, has some rather direct consequences on travel-distance interaction, and other behavioral forms with the larger society. Some examples are: Bennett's (1973) examination of consequences for friendship formation for the school-age population; Wheeler's (1968) study of the effects of work trips in Pittsburgh; and his later (Wheeler, 1971) investigation of all trips within the ghetto and between the ghetto and other areas of the city. Further, focusing on the consequences of black occupancy of a formerly white area, Lewis (1965) analyzed the resulting changes in the voting patterns within the area.

The geographic literature on groups other than blacks in urban settings is still very meager. Ward's studies of immigrants in the pre-World War I era are the most prominent, and among these the 1968 article and the 1971 volume focused most directly on the question of the causes of immigrant ghettos. In the earlier study, Ward examined the immigrants' role in competing with commerce for central locations and the relationship between varying types of commercial activities and certain ethnic immigrant locations. Later (1971:105-43) in the larger study, he discussed living conditions in terms of the housing, density, and health characteristics of immigrant groups in the central areas of cities and how central business expansion affected immigrant ethnic concentration patterns. In another study, Jakle and Wheeler (1969) placed the changing degrees of clustering of the Dutch in Kalamazoo over a one hundred year period within the acculturation theory format in one of the few examples of ethnic research structured to relate to a well-defined body of theory. A final example of the ethnic location/distribution theme is Ojo's (1968) study in which he examines the reasons for the high degree of clustering of the Hausa within the Yoruba towns in Nigeria.

MIGRATION

Because migration is one of the major processes which creates ethnic minorities, it is not surprising that there is an easily identifiable migration theme within ethnic geographic studies. It is moreover, an obviously spatial process and a theme that is strongly represented within geography in general. In spite of this, there is a notable dearth of studies at the global scale. The fascinating and wide-ranging intercontinental movements of the Chinese, Indians, Jews, Lebanese, Italians, and Greeks, to mention a few, have received very slight attention by geographers (at least by those who publish in English). An example at this scale is work by Kosinski (1969), in which he discussed the ethnic international relocations and resulting changes in group distribution in the East European shatter belt region that occurred during and after World War II. At the national scale, Ward (1971), Davies and Donaldson (1975), and Neils (1971) looked at interregional and rural-to-urban migration experiences of European immigrants, southern blacks, and Native Americans respectively. Each of these three gives extensive coverage of its topic. Davies and Donaldson devote half of their book to examining black interregional and interstate migration patterns from the colonial period to the present. Neils' major thrust is to assess the reservation-to-city relocation of Native Americans, particularly the role played by the U.S. Government in this activity.

At the intraurban scale are studies such as those of Roseman and Knight (1975) or Davies and Fowler (1972). Roseman and Knight's purpose was to test the validity of the adjustment process hypothesis and the mover-stayer hypothesis as they related to blacks who had recently moved into a city. This is another good example of ethnic research done within a clearly defined theory-developing context, although the particular theoretical structures are not directly related to ethnicity. Davies and and Fowler examined the source areas and migration streams of disadvantaged Southern whites and blacks who had moved to Indianapolis and related their selection of different intracity destinations to employment opportunities.

INTERGROUP INTERACTION

Another theme within ethnic geography focuses on the spatial patterns associated with the interactions among two or more ethnic groups. In pursuing their goals to maintain their identity, to assimilate, or to gain some greater reward in the socioeconomic sphere, minority group members inevitably interact with members of other groups, especially with members of the dominant group, and these interactions have a variety of spatially patterned contents and consequences. Studies of this sort may be conveniently divided into those at an intraurban scale and those at a larger scale. We look first at some examples at the intraurban scale.

School integration has been among the more popular topics. For example, Christian, et al. (1971) briefly examined the concepts of 'action space,' 'territoriality,' and 'neighboring' as they pertain to potential manipulations and assessments of school assignment strategies. More concretely, both Florin (1971) and Lowry (1973) analyzed school integration practices: Florin for the 1954-1964 period over a 10-state southern region; and, Lowry for a longer period, 1940-1970, but for just one state, Mississippi.

When a minority group's distribution at local scale is evaluated relative to that of the dominant and/or total population distribution, then the concepts of integration or segregation are appropriate. Integration and segregation are concepts that necessarily imply an interaction relationship. Therefore much of the literature that examines residential integration or segregation has some relevance to the intergroup interaction theme. Illustrative of this literature are the aforementioned ghetto growth simulation studies by Morrill (1965) and Rose (1970) both of which included interaction concepts or processes such as Morrills' use of the distance decay model when he reduced white neighborhood resistance to black entry as the distance to the ghetto center lengthens, or Rose's producer component which related the rate of white neighborhood exit and white avoidance of mixed neighborhoods, to the black-white ratio continuum. At a slightly larger scale, Dudas (1971) analyzed the impact of federally funded programs on housing integration in Dade County (Miami) and compared the differing effects of relocation on native whites, blacks, and Cubans. At a smaller scale, Bennett (1973) analyzed the relationship between integration in the neighborhood and school contexts with friendship formation and friendship activities. And finally, Salisbury (1971) saw a rather marked similarity between black ghetttos and insurgent states in their responses to hostile environments.

Studies of other minority groups as well as studies at larger scales also display the intergroup interaction theme. For example, Bounds (1971) examined the consequences of specific white-Indian interactions on subsequent Indian relocation and changes in the Indian economy. And Meinig's (1971) study of the Southwest carefully related the changing spatial relationships of the Anglos, Mexicans, and Indians in part to their dynamic interactions over a period of more than 200 years.

Turning to other areas, Lowenthals' (1972) book on West Indian societies has as a major theme the intricate and varying roles that color and ethnicity have for social interaction. Still another example is Sabbagh's (1968) study of the impact of the Apartheid policy on the interaction of the several ethnic groups in a number of contexts in South Africa.

INTERETHNIC DIFFERENTIALS

An ethnic minority group always differs in some ways from other ethnic groups and from the total population. These differences may be so many, so great, and so reinforcing that they persist over generations. Such intergroup differentials may reflect either the superior or inferior circumstance of the minority relative to the others. Minorities having a relatively inferior standing are frequently characterized by one or more of the following: visible distinctions from the dominants, recent other-language immigrants, a recent peasant or low status urban background, recent immigrants from a mass-exodus experience, and participants or descendents from a simple-technology tribal culture.

Whatever the circumstances that initiate intergroup differentials, the persistance of conditions that are perceived to be the foundation for the continuing dominant-subordinate relationship may well become a focus of resentment and hostility among the minority members. These also become the differentials that are most frequently selected for research analysis. The following examples illustrate geographers' concerns with them.

Studies of interethnic differentials in the United States primarily examine intraurban conditions. For example, Rose (1971) compared blacks with whites in terms of occupational structure, income, education, population growth, and other characteristics over a period of several decades. Zonn (1974) wished to determine if blacks and whites had similar patterns of association between occupational structure and places of residence. Wheeler (1968) examined the association between occupations and commuting distances in Pittsburgh for blacks and whites. Brunn and Hoffman (1970) and Salter and Mings (1972) focused on black-white voting preference differentials. Brunn and Hoffman standardized for income, education, and housing value and then compared the two racial groups in terms of their vote on an open housing issue. Salter and Mings, looking forward, wished to project the likely voting pattern changes that will result if Cubans replace Anglos in sections of Miami. Looking backward, Ward (1969) compared several immigrant ethnic groups in New York City in 1890 in terms of their population densities and mortality rates in order to develop a model of the internal spatial structure of immigrant residential areas. Looking inward, Bennett (1974) examined black and white differences in attitudes towards territoriality, residential space-sharing, residential proximity, and ethnic residential clustering. An example beyond the United States is Neville's (1966) study of Singapore in which he compared the Chinese, Indians, and Malays in terms of age-dependency welfare, educational costs and equality, and unemployment rates and the difficulties of changing these.

Illustrative of the work done at the regional and/or national scale are Katzman's (1969) study in which he analyzed trends in the average incomes of several ethnic groups within an eight-region division of the United States for the 1880-1960 period; and Lowry's (1970) article which compared living standards (education, income, occupation) of blacks and whites in Mississippi and related the differentials to the levels of urbanization, economic activities, and the black-white ratios in different areas of the state.

ACCULTURATION

Acculturation is only a moderately popular topic among geographers. This is surprising in view of the strong interest within geography in diffusion and migration processes. Within the literature which has a well-defined acculturation component there is, moreover, a notable dearth of both studies of American blacks and studies at the intraurban scale. The above characteristics combine to give acculturative studies an exceptional orientation within ethnic geography.

Acculturation, or the adoption of certain cultural attributes be they mentifacts, sociofacts, or artifacts, from another group occurs to some degree whenever two culturally different groups are in communication. Because communication is facilitated by proximity and because there are many advantages for minority group members if they adopt some cultural traits of the dominant group the process of minority acculturation proceeds in varying ways at varying rates in different places. We turn now to some of the kinds of topics geographers have considered within this fundamental interethnic process.

One of the more philosophical and thorough studies is that by Bjorklund (1964) in which she examined the broad relationships between a group's ideology and their acculturation. Her analysis of the components of ideology and their linked effects were stated in detail for the general case as well as being applied specifically to a Dutch Reformed group in Michigan. Both

Johnston (1971), and Jakle and Wheeler (1969) present especially good examples of research which explicitly recognizes and contributes to acculturation theory. The earlier work by Jakle and Wheeler examined the relationship of acculturation to a group's distributional pattern. In this case, they followed a Dutch group in Kalamazoo through several historical periods, identifying major factors which affected rates and levels of acculturation and the association of those with the group's tendencies to concentrate or disperse. Johnston has written an excellent summary of intraurban ethnic spatial models, including the models of acculturation.

Illustrative of work that examines acculturation changes for a variety of groups and at differing scales are: Jayawardena (1968) who investigated several factors associated with changes in a number of overseas Indian communities, especially those related to family structure, caste, and religious practices; Fellows (1972) who was particularly concerned with acculturation in the religious sphere of six major ethnic groups in the United States; and Nostrand (1970) who focused on Mexican-American acculturation in the Southwest and pondered the causes for what he considered a slow rate of change. As a final example, McIntire (1971) studied how Hopi housetypes and settlement patterns changed in response to contact with other, especially Anglo, cultural groups.

LAND USE PATTERNS

The profound concern which geographers have for land use variations is incorporated into their research on ethnicity. Prior to 1963, much of the work on this theme examined rural contrasts at a regional scale, such as the land surveys associated with the French settlements of Quebec, Louisiana and the Upper Mississippi-Ohio River areas, or Spanish patterns of the Southwest. Post-1963 ethnic-related research, in contrast, has a strong tendency to focus on intraurban phenomena and it is to examples of that that we now turn.

For the most part ethnic groups in urban settings experience, and expect to experience, a slow relocation with time. The aging and deterioration of neighborhoods combined with the improving economic circumstances of most groups with time are major factors in the ethnic succession process. Because succession is an expected pattern and housing production is usually unrelated to specific ethnic groups, it follows that visible land use characteristics which are associated with specific ethnic groups are quite muted. Ethnic distinctiveness appears to be much more related to demographic, cultural, and socioeconomic variations among the populations than visible differences in land use. Still, ethnic groups inevitably have some impacts on site use, style, and pattern.

Several studies have focused on ethnic differences in commercial areas. Pred's (1963) early work in Chicago compared a black lower income neighborhood commercial street with a lower income Polish street and a middle income WASP commercial street in an effort to identify the distinctive "black" cultural attributes. Later Harries (1971) examined Mexican-American, black, and Anglo commercial districts in Los Angeles for interethnic land use differences. Rose (1970) charted the changes that occur in commercial activities when a neighborhood changed from white to black occupancy. Still at the intraurban scale, Fellows (1972) discussed some of the visible landscape features associated with six ethnic groups, especially the Japanese.

In a rural setting, a recent type of study is that of Raitz and Mather (1971) in which they trace the distribution of tobacco and tobacco barns in southern and western Wisconsin specifically to the settlements of Norwegians.

ETHNIC NEIGHBORHOODS

The reality of ethnic clustering in urban areas, whether by voluntary or involuntary processes, inevitably creates a kind of homemade environment to which the ethnic group further responds. Thus, cycles of creation-response, propelled by the vehicle of communication produce ethnic neighborhoods: areas where at least one minority group lives with a proximity which definitely facilitates group interaction and cohesion. Such neighborhoods may persist or alternatively may be subjected to pressures which can result in their dissolution such as the invasion of another ethnic group, the expansion of commerce, or the deterioration and removal (renewal) of neighborhood sections. While they exist, such neighborhoods can be recognized as multifeatured cultural regions. They give rise to a number of interesting questions in which geographers have been interested.

Doeppers' (1967) study, for example, of the German-Polish Globeville neighborhood in Denver sought answers to several questions: (1) What is the spatial expression of this ethnic neighborhood? (2) How do spatial patterns reinforce the identity of the ethnic groups? (3) How have subneighborhood social groupings responded (adjusted) to the Mexican-American invasion? (4) What impact has the construction of a major elevated interstate highway, creating a land use barrier, had on the functioning of the neighborhood?

An exceptionally provocative and extensive study of an ethnic neighborhood has recently been done by Ley (1974). In this book he has attempted to identify the salient forces of the "totality of reality" of a black inner city neighborhood in Philadelphia. In this effort, he has abstracted the behavioral and cognitive processes which operate in this environment of stress and uncertainty. Moreover, his evaluation of this neighborhood did not support two widely accepted views of inner city black populations: (1) the mass media popular view, which promotes the image of black homogeneity combined with a hostility to all whites, and (2) the social science view whose image is of a black population culturally separate and distinct from that of whites.

PARTICULAR PROBLEMS

One category of themes in ethnic geographic studies can be conveniently described as representing a variety of particular problems or circumstances associated with an ethnic group. Under this heading are assigned such topics as lynchings, riots, housing standards, job specializations, and antisocial behaviors. While it would be possible to assign each of these studies to one or more of the other themes, it seemed more desirable to recognize their special-interest nature. We turn now to some examples of the kind of studies identified in this manner.

The ghetto riots of the late 1960s not only generated a national commission to enquire into their nature and causes but likewise set in motion individual efforts with the same objectives. The studies of Adams (1972) and Rose (1971) are especially notable. Rose devoted a chapter to what he terms, "collective violence" and posed the question of whether these actions should be considered race riots or rebellions. Adams examined several hypotheses about ghetto rioting at both interurban and intraurban scales and couched his work neatly within a theory-building framework. Salter and Mings (1969) paid particular attention to the role of police-community relations in their study of the Miami riot of 1968.

Others have focused on socioeconomic conditions. Davies and Fowler (1971), for example, examined the circumstances surrounding black females in Indianapolis as they related to employment opportunites. At a wider scale, Meyer (1973) investigated the variations in the housing quality of blacks throughout the intermetropolitan system. Rose (1971) has considered a number of items relevant to blacks such as the economic development of the ghetto, the provision of health service and public/personal safety services.

As a last example of this theme there is the study of ethnic separatism by Burghardt (1971) in which he analyzed the role of the French-speaking population in Canada, compared their situation to that of the Hungarians in the Austro-Hungarian empire, and finally considered the possible consequences of a separate Quebec.

SUMMARY AND CONCLUSIONS

In brief summary three observations can be made about the recent geographic literature which has an ethnic component. First, while the very strong focus on blacks is understandable, still the virtual absence of study of several ethnic groups gives the literature a decided group bias. The negative consequences of this bias for an understanding of both the omitted groups and blacks is self-evident. Second, there is a notable dearth of studies at the global scale and these too could provide a comparative perspective that would most likely be very beneficial to theory testing and/or development. Finally, too few of the recent research works have been set in a theory-testing mold. Too many have attempted to answer interesting questions but said little about how those questions link up with other more general questions.

BIBLIOGRAPHY

Articles, Monographs, Books

Adams, John S. "The Geography of Riots and Civil Disorders in the 1960s." *Economic Geography* 48 (1972):24-42.

Afolabi, Ojo F.J. "Hausa Quarters of Yoruba Towns, With Special Reference to Ile-Ife." *Journal of Tropical Geography* 27 (1968):40-49.

Allen, James P. "Migration Fields of French Canadian Immigrants to Southern Maine." *Geographical Review* 62 (1972):366-383.

Ballas, Donald. "Geography and the American Indian." *Journal of Geography,* 1960, pp. 156-168.

Bennett, Don C. "Segregation and Racial Interaction." *Annals, Association of American Geographers* 63 (1973):48-57.

———. "Southeast Asian Indigenous Minorities." *Journal of Geography* 69 (1970):428-433.

———. "Interracial Ratios and Proximity in Dormitories: Attitudes of University Students." *Environment and Behavior* 6 (1974):212-232.

Birdsall, Stephen. "An Introduction to Research on Black America: Prospects and Preview." *Southeastern Geographer* 11 (1971):85-89.

Bjorklund, E. "Ideology and Culture Exemplified in Southwestern Michigan." *Annals, Association of American Geographers* 54 (1964):227-241.

Blaut, J.M. "The Ghetto as an Internal Neo-Colony." *Antipode* 6 (1974):37-41.

Boal, F.W. "Territoriality on the Shankhill-Falls Divide, Belfast." *Irish Geography* 6 (1969):30-50.

Bounds, John H. "The Alabama-Coushatta Indians of Texas." *Journal of Geography* 70 (1971):175-182.

Brunn, Stanley D., and Hoffman, W.L. "The Spatial Response of Negroes and Whites Toward Open Housing: The Flint Referendum." *Annals, Association of American Geographers* 60 (1970):18-36.

Bunge, W. *Fitzgerald: Geography of a Revolution.* Cambridge, Mass.: Schenkman Publ. Co., 1971.

Burghardt, A.F. "Quebec Separatism and the Future of Canada." Chapter in L. Gentilicore,

Geographical Approaches to Canadian Problems. Scarborough, Ont.: Prentice-Hall, 1971, pp. 229-235.
Burrill, Robert M. "The Establishment of Ranching in the Osage Indian Reservation." *Geographical Review* 62 (1972):524-543.
Carey, G.W.; Macomber, L.; and Greenberg, M. "Educational and Demographic Factors in the Urban Geography of Washington." *Geographical Review* 58 (1968):515-537.
Carlson, A.W. "A Bibliography of the Geographical Literation on the American Indian, 1920-1971." *Professional Geographer* 24 (1972):258-263.
Chan, Kok E. "The Distribution of the Portugese-Eurasian Population of Malacca: A Study of Spatial Continuity and Change." *Geographica* 6 (1970):56-64.
Chang, Sen-dou. "The Distribution and Occupations of Overseas Chinese." *Geographical Review* 58 (1968):89-107.
Christian, Chas.; Jakle, J.A.; and Roseman, C.C. "The Prejudicial Use of Space: School Assignment Strategies in the United States." *Journal of Geography* 70 (1971):105-109.
Clarke, Colen G. "Residential Segregation and Intermarriage in San Fernando, Trinidad." *Geographical Review* 61 (1971):198-218.
Clark, W.A. "Patterns of Black Interaction Mobility and Restricted Relocation Opportunities." In H. Rose (ed.) *Geography of the Ghetto.* DeKalb, Ill.: Northern Illinois Univ. Press. *Perspectives in Geography* 2 (1972):111-127.
Dagodag, Tim. "Non-White Residence and the Location of Freeways: An Evaluation of Locational Policies and Procedures." *Oregon Geographer* 6 (1972):25-32.
Dalton, M., and Seaman, J.M. "The Distribution of New Commonwealth Immigrants in the London Borough of Ealing, 1961-1966." *Transactions, Institute of British Geographers* 58 (1973):21-39.
Davies, S., and Fowler, G. "The Disadvantaged Black Female Household Head: Migrants to Indianapolis." *Southeastern Geographer* 11 (1971):113-120.
Davies, C.S., and Fowler, G.L. "The Disadvantaged Urban Migrant in Indianapolis." *Economic Geography* 48 (1972):153-167.
Davies, C. Shane, and Huff, D. "Impact of Ghettoisation on Black Employment." *Economic Geography* 48 (1972):421-427.
Davis, George A., and Donaldson, O.F. *Blacks in the United States: A Geographic Perspective.* Boston: Houghton Mifflin Co., 1975.
Deskins, D. "Race as an Element in the Intracity Regionalization of Atlanta's Population." *Southeastern Geographer* 11 (1971):90-100.
Deskins, Donald R., Jr. "Race, Residence, and Workplace in Detroit, 1880-1965." *Economic Geography* 48 (1972):79-94.
Doeppers, Daniel. "The Globeville Neighborhood in Denver." *Geographical Review* 57 (1967):506-522.
———. "The Development of Philippine Cities Before 1900." *Journal of Asian Studies* 31 (1972):769-792.
———. "Ethnic Urbanism and Philippine Cities." *Annals, Association of American Geographers* 64 (1974):549-559.
Donaldson, Fred. "The Geography of Black America: Three Approaches." *Journal of Geography* 71 (1972):414-420.
Dudas, J., and Longbrake, D. "Problems and Future Directions of Residential Integration: The Local Application of Federally Funded Programs in Dade County, Florida." *Southeastern Geography* 11 (1971):151-158.
Eidt, R.C. "Japanese Agricultural Colonization: A New Attempt at Land Opening in Argentina." *Economic Geography* 44 (1968):1-20.
Ernst, Robt. T., and Hugg (eds.). *Black America: Geographic Perspective.* New York: Anchor-Doubleday, 1976.
Fair, T.J.D., and Shaffer, N.M. "Population Patterns and Policies in South Africa, 1951-1960." *Economic Geographer* 40 (1964):261-274.
Fellows, Donald K. *A Mosaic of America's Ethnic Minorities.* New York: John Wiley & Sons, Inc., 1972.

Fisher, James S. "Negro Farm Ownership in the South." *Annals, Association of American Geographers* 63 (1973):478-489.

Forin, John W. "The Diffusion of the Decision to Integrate: Southern School Desegregation, 1954-1964." *Southeastern Geographer* 11 (1971):139-144.

Gale, Donald. "The Impact of Canadian Italians on Retail Functions and Facades in Vancouver, 1921-1961." In J. Minghi (ed.) *Peoples of the Living Land: Geography of Cultural Diversity in British Columbia.* Vancouver: University of British Columbia, Geographical Series no. 15 (1972), pp. 107-124.

Groves, Paul A., and Muller, E.K. "The Evolution of Black Residential Areas in Late Nineteenth-Century Cities." *Journal of Historical Geography* 1 (1975):169-191.

Hansell, C.R., and Clark, W.A.V. "The Expansion of the Negro Ghetto in Milwaukee: A Description and Simulation Model." *Tijdschrift voor Economische en Sociale Geographie* 61 (1970):267-277.

Harries, Keith D. "Ethnic Variations in Los Angeles Business Patterns." *Annals, Association of American Geographers* 61 (1971):736-743.

Harvey, David. *Explanation in Geography.* New York: St. Martin's Press, 1969.

———. "Revolutionary Theory in Geography and the Problem of Ghetto Formation." In H. Rose (ed.) Geography of the Ghetto. DeKalb, Ill.: Northern Illinois University Press, *Perspectives in Geography* 2 (1972):1-25.

Harvey, Melton E. "Social Change and Ethnic Relocation in Developing Africa: The Sierra Leone Example." *Geografiska Annaler, Series B, Human Geography* 53 (1971):94-106.

Heller, Chas. F., and Redenti, A.L. "Residential Location and White Attitude Toward Mixed-Race Neighborhoods in Kalamazoo, Mich." *Journal of Geography* 72 (1973):15-25.

Henderson, Janet, and Hart, J.F. "The Development and Spatial Patterns of Black Colleges." *Southeastern Geographer* 11 (1971):133-138.

Hill, A.G. "Segregation in Kuwait." Institute of British Geographers, Special Publication no. 5 (1972), pp. 123-142.

Jakle, John, and Wheeler, J.O. "The Changing Residential Structure of the Dutch Poulation in Kalamazoo, Michigan." *Annals, Association of American Geographers* 59 (1969):441-460.

Jayawardena, Chandra. "Migration and Social Change: A Survey of Indian Communities Overseas." *Geographical Review* 58 (1968):426-449.

Johnston, R.J. *Urban Residential Patterns.* London: G Bell & Sons, Ltd., 1971.

Jones, H.R. "The Pakistani Community in Dundee." *Scottish Geographical Magazine* 88 (1972):75-85.

Jones, P.N. "Some Aspects of the Changing Distribution Of Coloured Immigrants in Birmingham." *Transactions, Institute of British Geographers* 50 (1970):199-219.

Jordan, Terry. *German Seed in Texas Soil: Immigrant Farmers in Nineteenth Century Texas.* Austin: University of Texas Press, 1966.

Kantrowitz, Nathan. *Negro and Puerto Rican Population of New York City in the Twentieth Century.* New York: American Geographical Society, Studies in Urban Geography, no. 1, 1969.

Katzman, Martin T. "Ethnic Geography and Regional Economies, 1880-1960." *Economic Geography* 45 (1969):45-52.

Kaups, M., and Mather, C. "Eben: Thirty Years Later in a Finnish Community in the Upper Peninsula of Michigan." *Economic Geography* 44 (1968):57-70.

Kosinski, L.A. "Changes in the Ethnic Structure in East-Central Europe, 1930-1960." *Geographical Review* 59 (1969):388-402.

Landing, James E. "Geographic Models of Old Order Amish Settlements." *Professional Geographer* 21 (1969):238-243.

Leach, Bridget. "The Social Geographer and Black People: Can Geography Contribute to Race Relations?" *Race* 15 (1973):230-241.

Lee, Trevor, R. "Immigrants in London: Trends in Distribution and Concentration, 1961-71." *New Community* 2 (1973):145-158.

———. "The Role of the Ethnic Community as a Reception Area for Italian Immigrants in Melbourne, Australia." *Internal Migration* 8 (1970):50-64.

Lemon, James T. "The Agricultural Practices of National Groups in Eighteenth-Century Southeastern Pennsylvania." *Geographical Review* 56 (1966):467-496.

Lewis, G.M. "The Distribution of the Negro in the Conterminous United States." *Geography* 54 (1969):410-418.

Lewis, Lawrence T. "The Geography of Black America: The Growth of a Sub-Discipline." *Journal of Geography* 73 (1974):38-43.

Lewis, Peirce F. "Impact of Negro Migration on the Electoral Geography of Flint, Michigan, 1932-1962: A Cartographic Analysis." *Annals, Association of American Geographers* 55 (1965):1-25.

Ley, David. *The Black Inner City as Frontier Outpost.* Association of American Geographers, Monograph Series, no. 7, Washington, D.C.

Lowry, Mark II. "Population and Race in Mississippi." *Annals, Association of American Geographers* 61 (1971):576-588.

———. "Race and Socioeconomic Well-Being: A Geographical Analysis of the Mississippi Case." *Geographical Review* 60 (1970):511-528.

———. "Racial Segregation: A Geographical Adaptation and Analysis." *Journal of Geography* 71 (1972):28-40.

Lung, Julie. "The Termination of the Klamath Indian Reservation." *Oregon Geographer* 5 (1970):25-31.

McArthur, N., and Garland, M.E. "The Spread and Migration of French Canadians." *Tijdschrift voor Economische en Sociale Geographie* 52 (1961):141-147.

McIntire, Elliot G. "Changing Patterns of Hopi Indian Settlement." *Annals, Association of American Geographers* 61 (1971):510-521.

McPhail, I.R. "The Vote for Mayor of Los Angeles in 1969." *Annals, Association of American Geographers* 61 (1971):744-758.

McTaggart, W.D. "The Distribution of Ethnic Groups in Malaya, 1947-1959." *Journal of Tropical Geography* 26 (1968):69-81.

Meinig, Donald W. *Southwest: Three Peoples in Geographical Change, 1600-1970,* New York: Oxford University Press, 1971.

Mercer, John. "Housing Quality and the Ghetto." H. Rose (ed.) *Geography of the Ghetto.* DeKalb, Ill.: Nothern Illinois University Press, *Perspectives in Geography* 2 (1972):143-167.

Meyer, David R. "Classification of U.S. Metropolitan Areas by Characteristics of Their Non-White Populations." In B.J.L. Berry (ed.) *City Classification Handbook: Methods and Applications.* New York: Wiley-Interscience, 1972, pp. 61-93.

———. "Implications of Some Recommended Alternative Urban Strategies for Black Residential Choice." In H. Rose (ed.) *Geography of the Ghetto.* DeKalb, Ill.: Northern Illinois University Press, *Perspectives in Geography* 2 (1972):129-142.

———. "Blacks in Slum Housing: A Distorted Theme." *Journal of Black Studies* 4 (1973):139-152.

Morrill, Richard L. "A Geographic Perspective of the Black Ghetto." In H. Rose (ed.) *Geography of the Ghetto.* DeKalb, Ill.: Northern Illinois University Press, *Perspectives in Geography* 2 (1972):27-58.

———. "The Negro Ghetto: Problems and Alternatives." *Geographical Review* 55 (1965):339-361.

———. "The Persistence of the Black Ghetto as Spatial Separation." *Southeastern Geographer* 11 (1971):149-156.

Morrill, R., and Donaldson, O.F. "Geographical Perspectives in the History of Black America." *Economic Geography* 48 (1972):1-23.

Neville, R.J.W. "The Areal Distribution of Population in Singapore." *Journal of Tropical Geography* 20 (1965):16-25.

———. "Singapore: Ethnic Diversity and Its Implications." *Annals, Association of American Geographers* 56 (1966):236-253.

Nostrand, Richard L. "The Hispanic-American Borderland: Delimitation of an American Culture Region." *Annals, Association of American Geographers* 60 (1970):638-661.

———. "Mexican Americans Circa 1850." *Annals, Association of American Geographers* 65 (1975):378-390.

Peach, G.C.K. "Factors Affecting the Distribution of West Indians in Great Britain." *Transactions, Institute of British Geographers* 38 (1966):151-163.

———. *Urban Social Segregation.* New York: Longman, 1975.

Peach, Ceri. *West Indian Migration to Britain: A Social Geography.* Oxford University Press for the Institute of Race Relations, 1968.

Pred, Allan. "Business Thoroughfares as an Expression of Urban Negro Culture." *Economic Geography* 39 (1963):217-233.
Raby, Stewart. "Indian Land Surrenders in Southern Saskatchewan." *Canadian Geographer* 17 (1973):36-52.
Raitz, Karl, and Mather, C. "Norwegians and Tobacco in Western Wisconsin." *Annals, Association of American Geographers* 61 (1971):684-696.
Ray, Arthur J., Jr., "Indian Adaptations to the Forest-Grassland Boundary of Manitoba and Saskatchewan." *Candian Geographer* 16 (1972):103-118.
Ray, Michael. "Cultural Differences in Consumer Travel Behavior in Eastern Canada." *Canadian Geographer* 11 (1967):143-156.
Ritter, Frederic A. "Toward a Geography of the Negro in the City." *Journal of Geography* 70 (1971):150-156.
Rose, Harold, M. "Metropolitan Miami's Changing Negro Population, 1950-1960." *Economic Geography* 40 (1964):221-238.
———. "The All-Negro Town: Its Evolution and Function." *Geographical Review* 55 (1965):362-381.
———. *Social Processes in the City: Race and Urban Residential Choice.* Washington, D.C.: Commission on College Geography, Resource Paper no. 6, Association of American Geographers, 1969.
———. "The Development of an Urban Subsystem: The Case of the Negro Ghetto." *Annals, Association of American Geographers* 60 (1970):1-17.
———. "The Structure of Retail Trade in a Racially Changing Trade Area." *Geographical Analysis* 2 (1970):135-148.
———. *The Black Ghetto: A Spatial Behavioral Perspective.* New York: McGraw-Hill, 1971.
———. "The Spatial Development of Black Residential Subsystems." *Economic Geography* 48 (1972):43-65.
Roseman, C.; Christian, C.M.; and Bullamore, H.W. "Factorial Ecologies of Urban Black Communities." Rose, H. (ed.) *Geography of the Ghetto.* DeKalb, Ill.: Northern Illinois University Press, *Perspectives in Geography* 2 (1972):240-255.
Roseman, C., and Knight P.L. III. "Residential Environment and Migration Behavior of Urban Blacks." *Professional Geographer* 27 (1975):160-165.
Sabbagh, M. Ernest. "Some Geographical Characteristics of a Plural Society: Apartheid in South Africa." *Geographical Review* 58 (1968):1-28.
Salisbury, Howard G. "The State Within a State: Some Comparisons Between the Urban Ghetto and the Insurgent State." *Professional Geographer* 23 (1971):105-112.
Salter, Paul S., and Mings, R. "The Projected Impact of Cuban Settlement on Voting Patterns in Metropolitan Miami, Florida." *Professional Geographer* 24 (1972):123-131.
Sanders, Ralph, and Adams, J.S. "Age Structure in Expanding Ghetto Space: Cleveland, Ohio, 1940-1965." *Southeastern Geographer* 11 (1971):121-132.
Sandhu, Kernial S. *Indians in Malaya: Immigration and Settlement.* Cambridge: Cambridge University Press, 1969.
Smith, Peter C., and Raitz, K.B. "Negro Hamlets and Agricultural Estates in Kentucky's Inner Bluegrass." *Geographical Review* 64 (1974):217-234.
Stimson, R.J. "Patterns of European Immigrant Settlement in Melbourne, 1947-1961." *Tijdschrift voor Economische en Sociale Geographie* 61 (1970):114-126.
Tata, R., et al. "Defensible Space in a Housing Project: A Case Study from a South Florida Ghetto." *Professional Geographer* 27 (1975):297-303.
Thompson, Bryan, and Agocs, C. "Ethnic Studies: Teaching and Research Needs." *Journal of Geography* 72 (1973):13-23.
Uhlig, H. "Hill Tribes and Rice Farmers in the Himalayas and South-East Asia." *Transactions, Institute of British Geographers* 47 (1969):1-24.
Wagner, Philip L. "The Perisistence of Native Settlement in Coastal British Columbia." J. Mighi (ed.) *Peoples of the Living Land: Geography of Cultural Diversity in British Columbia.* Vancouver: University of British Columbia, Geographical Series no. 15 (1972), pp. 13-27.
Ward, David. *Cities and Immigrants: A Geography of Change in Nineteenth Century America.* New York: Oxford University Press, 1971.

———. "The Emergence of Central Immigrant Ghettoes in American Cities: 1840-1920." *Annals, Association of American Geographers* 58 (1968):343-359.

———. "The Internal Spatial Structure of Immigrant Residential Districts in the Late Nineteenth Century." *Geographical Analysis* 1 (1969):337-353.

———. "Some Locational Attributes of the Ethnic Division of Labor in Mid-Nineteenth Century American Cities." *Proceedings of the Conference in National Archives and Historical Geography*. Washington, D.C.: Howard University Press, 1975, pp. 258-270.

Western, John. "Social Groups and Activity Patterns in Houma, Louisiana." *Geographical Review* 63 (1973):301-321.

Wheeler, J.O. "The Spatial Interaction of Blacks in Metropolitan Areas." *Southeastern Geographer* 11 (1971):101-112.

———. "Work-Trip Length and the Ghetto." *Land Economics* 44 (1968):107-112.

Wheeler, J.O., and Brunn, S.D. "Negro Migration into Rural Southwestern Michigan." *Geographical Review* 58 (1968):214-230.

———. "An Agricultural Ghetto: Negroes in Cass County, Michigan, 1845-1968." *Geographical Review* 59 (1969):317-329.

Winsberg, Morton D. "Jewish Agricultural Colonization in Argentina." *Geographical Review* 54 (1964):487-501.

Zonn, Lev E. "Residential Distribution and Occupational Stratification of Blacks and Whites: The Milwaukee Example." Paper presented at the 1974 annual meeting of the West Lakes Division of the Association of American Geographers.

Ph.D. Dissertations

Brown, Wm. H., Jr., "Class Aspects of Residential Development and Choice in the Oakland Black Community." Unpublished Ph.D. dissertation (Geography), University of California, Berkeley, 1970.

Carlson, Alvar W. "The Rio Ariba" A Geographic Appraisal of the Spanish-American Homeland (Upper Rio Grande Valley, New Mexico)." Unpublished Ph.D. dissertation (Geography), University of Minnesota, 1971.

Chammon, Eliezis. "Migration and Adjustment: The Case of Sephardic Jews in Los Angeles." Unpublished Ph.D. dissertation (Geography), UCLA, 1976.

Cho, Tai Jun. "Commercial Structures in Three Ethnically Different Areas: Compton, East Los Angeles, and Riverside." Unpublished Ph.D. dissertation (Geography), UCLA, 1975.

Craig, W.W. "Weekend and Vacation Recreational Behavior of a Negro Community in Louisiana: A Spatial Study." Unpublished Ph.D. dissertation (Geography), University of Michigan, 1968.

Cybriwsky, R.A. "Social Relations and the Spatial Order in the Urban Environment: A Study of Life in a Neighborhood in Central Philadelphia." Unpublished Ph.D. dissertation (Geography), The Pennsylvania State University, 1972.

Dagodag, Wm. T. "Public Policy and the Housing Patterns of Urban Mexican-Americans in Selected Cities of the Central Valley." Unpublished Ph.D. dissertation (Geography), University of Oregon, 1972.

Darden, Jos. T. "The Spatial Dynamics of Residential Segregation of Afro-Americans in Pittsburgh." Unpublished Ph.D. dissertation (Geography), University of Pittsburgh, 1972.

Davies, C.S. "The Reverse Commuter Transit Problem in Indianapolis." Unpublished Ph.D. dissertation (Geography), Indiana University, 1970.

Deskins, D.R. "Residential Mobility of Negro Occupational Groups in Detroit, 1837-1965." Unpublished Ph.D. dissertation (Geography), University of Michigan, 1971.

Doherty, Jos. Michael. "Immigrants in London: A Study of the Relationship Between Spatial Structure and Social Structure." Unpublished Ph.D. dissertation (Geography), 1973, University of London, United Kingdom.

Ernst, Robt. T. "Factors of Isolation and Interaction in an All Black City: Kinlock, Missouri." Unpublished Ph.D. dissertation (Geography), University of Florida, 1973.

Hall, Frederick T. "High Schools, Integration and Student Transportation: A Case Study of Location Criteria for the Public Sector." Ph.D. dissertation, University of Chicago, 1972.

Harries, Keith D. "An Analysis of Inter-Ethnic Variations in Commercial Land Use in Los Angeles." Unpublished Ph.D. dissertation (Geography), 1969.

Hopple, Lee. "Spatial Development and Internal Spatial Organization of the Southwestern Pennsylvania Plain Dutch Community." Unpublished Ph.D. dissertation (Geography), Pennsylvania State University, 1971.

Jordan, Terry. "A Geographical Appraisal of the Influence of German Settlement on Nineteenth-Century Texas Agriculture." Unpublished Ph.D. dissertation (Geography), University of Wisconsin, 1965.

Landing, James E. "Organization of an Old Order Amish-Beachy Amish Settlement: Nappanee, Indiana." Unpublished Ph.D. dissertation (Geography), Pennsylvania State University, 1967.

Lee, Trevor Ross. "Concentration and Dispersal: A Study of West Indian Residential Patterns in London, 1961-1971." Unpublished Ph.D. dissertation (Geography), University of London, United Kingdom, 1973.

Ley, David F. "The Black Inner City as Frontier Outpost: Images and Behavior of a Philadelphia Neighborhood." Unpublished Ph.D. dissertation (Geography), Pennsylvania State University, 1972.

Lowry, Mark III. "Geographical Characteristics of a Bi-Racial Society: The Mississippi Case." Unpublished Ph.D. dissertation (Geography), Syracuse University, 1973.

McHenry, Stewart. "The Syrians of Upstate New York." Unpublished Ph.D. dissertation (Geography), Syracuse University, 1973.

McIntire, Elliot G. "The Impact of Cultural Change on the Land Use Patterns of the Hopi Indians." Unpublished Ph.D. dissertation (Geography), University of Oregon, 1968.

Mackum, Stanley. "The Changing Pattern of Polish Settlements in the Greater Detroit Areas: Geographic Study of the Assimilation of an Ethnic Group." Unpublished Ph.D. dissertation (Geography), University of Michigan, 1964.

McLennan, Marshall. "The Ilocano Occupanee of Northern Nueva Ecija, Philippines." Unpublished Ph.D. dissertation (Geography), University of California, Berkeley, 1973.

Meyer, David R. *Spatial Variation of Black Urban Households.* Chicago, Ill.: University of Chicago, Department of Geography, Research Paper no. 129, 1970.

Meyers, D.K. "The Changing Negro Residential Patterns of Lansing,Michigan, 1850-1969." Unpublished Ph.D. dissertation (Geography), Michigan State University, 1970.

Neils, Elaine M. "Reservation to City: Indian Migration and Federal Relocation." Unpublished Ph.D. dissertation (Geography), University of Chicago, 1971.

Nostrand, Richard L. "The Hispanic-American Borderland: A Regional, Historical Geography." Unpublished Ph.D. disseration (Geography), UCLA, 1968.

Peach, G.C.K. "Socio-Geographic Aspects of West Indian Migration to Great Britain." Unpublished Ph.D. dissertation (Geography), Merton College, Oxford University, United Kingdom, 1965.

Rechlin, Alice T. "The Utilization of Space in the Nappanee, Indiana Old Order Amish Community: Minority Group Study." Unpublished Ph.D. dissertation (Geography), University of Michigan, 1970.

Ropka, Gerald W. "The Cultural Impact of the Spanish Speaking People in the City of Chicago." Unpublished Ph.D. dissertation (Geography), Michigan State University, 1973.

Smith, Peter C. "Negro Hamlets and Gentlemen Farms: A Dichotomous Rural Settlement Pattern in Kentucky Bluegrass Region." Unpublished Ph.D. dissertation (Geography), University of Kentucky, Lexington, 1972.

Stafford, John W. "Cross Cultural Change: A Geographical Analysis." Unpublished Ph.D. dissertation (Geography), Michigan State University, 1971.

Sutton, I. "Land Tenure and Changing Occupance on Indian Reservations in Southern California." Unpublished Ph.D. dissertation (Geography), University of California, 1965.

Varady, D.P. "The Houshold Migration Decision in Racially Changing Neighborhoods." Unpublished Ph.D. dissertation (Geography), University of Pennsylvania, 1971.

Vicero, Ralph. "A Study of the Changing Pattern of French-Canadian Migration to New England 1840-191, and Its Regional Significance." Unpublished Ph.D. dissertation (Geography), University of Wisconsin, Madison, 1968.

Villenueve, Paul Y. "The Spatial Adjustment of Ethnic Minorities in the Urban Environment." Unpublished Ph.D. dissertation (Geography), University of Washington, 1971.

Physical Geography and Environmental Conservation

Introduction

The recent emergence of a broadly based environmentalist movement in the United States has awakened the public and the scholarly community to the necessity for a scientifically sound awareness of environmental problems as a prerequisite for the formulation of effective policies in this area. Case studies of several important types of environmental disruption are provided by papers in this section.

A persistent theme in the paper by Thomas Barton on environmental problems in Indiana is the existence of serious conflicts between the welfare of the public, the interests of private sectors, and the performance of legislative and regulatory agencies at the local, state, and national levels. The federal and state policies toward land use, water resources, air pollution, energy, and toxic substances in Indiana are often ineffectual and contradictory. For example, certain agencies of the federal government actively promote the development of single-family residences on the fringes of metropolitan areas while other federal offices issue reports citing the undesirability of urban sprawl. Barton documents many other examples of the unfulfilled role of government in protecting public interests against environmental disruption.

The participation of scientists in the analysis of environmental issues and the broad dissemination of their findings are indispensable for the formulation and implementation of effective public policies which contribute to societal welfare. This is the basic premise in the paper of Peter Kakela. He illustrates this theme by presenting an analysis of the problems of recycling bottles and throwaway beverage containers. These case studies focus on two of the most serious examples of industrial and commercial practices which have a negative environmental effect but which hitherto have escaped close scrutiny in geographic literature. Kakela makes a careful examination of the technological and economic aspects of these activities and discusses reasons for their persistence even though they do not conform to the interests of society as a whole.

The question of the complex and often delicate balance between man and his physical environment is treated in the paper by John Hidore. This study examines the problems of drought and starvation in the Sahel region of Africa, which has been one of the worst problem areas in the world in recent years. The ecological disaster in the Sahel serves as an example of the environmental difficulties induced by the excessive expansion of population and marginal agricultural land in an area subject to serious physical constraints on development. This dilemma has been intensified by the inflexibility and unresponsiveness of social, political, and cultural values and institutions. Hidore notes that the Sahel experience provides support for the notion of regional limits to growth.

Much of the environmental research carried on by geographers involves the application of basic scientific findings to societal problems. An example of basic research in biogeography can be found in the paper by Jerry Davis. This study is a review and analysis of the massive outpouring of literature in recent years devoted to the precise definition and measurement of two of the most important elements of the fundamental equation developed by Richard Gates for the energy budget of a leaf. The first part of this paper deals with alternative forms of measuring the connective heat transfer coefficient and the second section is concerned with the response of the plant stomata to environmental variables. This is an innovative and promising area of research with important geographic applications.

The effective cartographic representation of quantitative data is a pervasive problem in geographic research and teaching. In his paper on variable-sized point symbols, Robert Kingsbury discusses the pitfalls of employing continuously graduated circles to depict quantitative information and suggests diverse approaches to enhance user perception.

Environmental Conservation Problems of Indiana

Thomas Frank Barton
Indiana University

In Indiana, conservation problems are found in all three spheres of the environment—the lithosphere, the hydrosphere, and the atmosphere. Problems of inefficient and detrimental use of land, water, and air abound and are increasing in number and severity. Many of these have existed for decades and some, if not most, are more serious today than when they first appeared, such as floods and pollution.

Unfortunately, Indiana has relatively few public leaders and elected officials either in the state or on the national level who have concerned themselves with these ecological problems. What U.S. Senator from Indiana has received national recognition for promoting environmental conservation? About a century ago, President Benjamin Harrison received national recognition as a conservationist, but at that time many of our present-day problems either did not exist or were not so serious.

Apparently the social, political, and economic environment of Indiana produces a greater number of opponents to environmental conservation problem solving than proponents. It seems that most of these opponents with wealth and influence favor unbridled exploitation of resources or only token regulations. These people, whether they outnumber conservationists or not, have successfully blocked, sidetracked and in other ways delayed or made ineffectual programs intended to promote and implement wise and efficient use of land, water, and air. In the struggle for better and more efficient use of resources by the general public, a little progress has been made—some battles won—but are not the conservationists losing the war? The conservationists face resistance on at least three fronts. One is that of developers whose apparent attitude is "full steam ahead" with exploitative development and the environment "be damned." The second front is that of the conservation extremists lacking background in the sciences who take undefensible, illogical positions apparently because they are against the establishment. These people unintentionally do the environmental conservation cause more harm than good. And on the third front are the problem solvers and planners faced with problems for which only new and tentative solutions exist, and these potential solutions may take a decade or two to try out in pilot projects. Certainly environmental conservationists need to be well educated in physical, biological, and social sciences and be stout-hearted persons.

Although one wishes it were not so, one must admit that without federal aid and influence, the environment of Indiana would be in far worse shape than it is now, and environmental conservation problems far more serious and grim. But it is also true that some of the problems we now face exist because some federal agencies have been working at cross purposes and creating problems or making them harder to solve.

However, there is always hope. During the 1960s and 1970s the youth of Indiana and the news media contributed toward making the public aware of environmental problems of Indiana. Becoming aware of the fact that problems do exist and recognizing that something should be done is perhaps the first step leading to better management. Moreover, pilot projects are now in operation both in Indiana and in other states. Hopefully, these projects may contribute toward providing alternative solutions. Furthermore, some proven constructive environmental controls are being enforced, if at times only half-heartedly.

LAND

Since Indiana became a state in 1816, the overwhelming goal for land use has been to bring the highest percentage possible of the area under cultivation. Reclamation stressed the removal of the forests, plowing of the grasslands, and draining of the wetlands. Before a hundred years of reclamation had passed, some people became aware that all of the land in Indiana not used for cities and mining was not suitable for twentieth-century crop production. This was true especially in the southern one-third of the state. The Clark County State Forest was established as early as 1904, five years after the largest lumber cut in the history of Indiana.[1] After 1,037 million board feet were cut in 1899, lumber production rapidly declined.[2] By 1954, lumber production had declined to 159 million board feet which accounted for only 2.1% of the nation's hardwood production, and Indiana had to buy 85% of its lumber beyond its boundaries.[3] But perhaps more significantly, Indiana with only somewhat more than 23 million acres had a 5 million acre land-use problem. These 5 million acres consisted of approximately 4 million acres of commercial forest land and 1 million acres in cultivation more suited to grass and hay production or reforestation.

During the depression of the 1930s, ten of Indiana's state forests were established between 1930-1937 when it became obvious that millions of acres were not suitable for cultivation. The Hoosier National Forest was established by Indiana's General Assembly in 1935.[4]

Unfortunately, Indiana's State Forests contain only about 138,563 acres and the Hoosier National Forest owned only 177,289 acres in 1975 so that the combined State and Federal managed forestland comes to approximately 316,000 acres while 5 million acres should be reforested and properly managed or placed in grass and hay production.

During the 1930s there was a serious discussion in the Ohio River basin as to whether the hilly land should be reforested or used as pasture and hayland, and the production of sheep or cattle. Many recommended sheep because their production was more suited to the small farms typical of the region. Because the southern parts of Ohio, Indiana, and Illinois as well as parts of Kentucky and the Ozarks of Missouri have rough land with somewhat similar climatic and relief conditions, in the 1930s the Dixon Springs Pasture and Erosion Control Project was established in Southern Illinois. Grasses from all parts of the world were collected, brought to the project and grown under various conditions and in various combinations to determine how to produce the greatest yields of feed per acre for sheep. Pilot sheep herds were established but dogs destroyed the potential sheep industry. And the proper management of the commercial forest lands owned by the farmers and others, the state forests leave much to be desired. The problem of efficient wise use of over 5 million acres remains one of the largest land-use problems in the state. Ohio and Illinois located on the eastern and western sides of Indiana apparently have superior reforestation programs. For example, the Shawnee National Forest in Southern Illinois is being developed more rapidly than the Hoosier National Forest.

In the 1930s another serious land-use problem was recognized. Land depletion by soil erosion on farms was attacked with vigor and substantial gains were made in bringing the problem under control.

However, before the soil erosion problem was completely solved another type of land depletion accelerated. Urban sprawl is doing more damage than soil erosion. This new threat to cultivated land takes land permanently out of production by using it for other purposes such as highways, airfields, factories and businesses, public buildings, and residences. Land taken out of cultivation for residential uses has not only accelerated because of population increase but also in part due to a higher standard of living which results in increased lot sizes. Today many rural nonfarm dwellers, who in 1970 outnumbered the combined city and farm dwellers in 32 of Indiana's counties, are building houses on parcels of land 3 or 5 or 10 acres in size, or even larger. Unfortunately, in the center one-half or more of the state these acreages are taking the most productive Indiana cropland (land with number 1 and 2 capabilities) out of cultivation. There is only an average of 4.6 acres per person in the state.

Russell W. Peterson, chairman of the President's Council on Environmental Quality, reports that in the United States between 1960 and 1970, rural land was converted to urban use at the rate of 2,000 acres per day.[5] Consequently, in that ten-year period, 7.3+ million acres of rural land were changed to urban use.[6]

In 62 of Indiana's counties the rural nonfarm dwellers outnumber the people living in the largest city and their numbers are accelerating as the definition of a farm changed in 1975, and as the federal government subsidizes their growth. Two of the agencies promoting rural nonfarm development are the Farmers Home Administration and the Federal Commission of Rural Water. One of the most influential federal laws encouraging the growth of rural nonfarm dwellers and the copious detrimental use of farmer-cultivated land for urban uses is the Rural Development Act of 1972, Public Law 92-419. Later, in 1974, the Housing and Community Development Act consolidated a number of other federal programs so communities can request federal money for a wide range of activities. Guaranteed federal loans at relatively low-interest rates and federal matching funds are available for airports, industrial parks and industries, businesses, libraries, hospitals, parks, and other community activities, especially public services of water, fire control, water, and waste disposal, planning grants, retirement homes, recreational complexes, housing demolition and housing rehabilitation and other community services and activities. In 1970, 70% of the federal Housing and Community Development funds went to settlements with fewer than 5,000 residents. One-third of the 1974 fund or 33% went to settlements with populations of 1,000 inhabitants or less.[7]

This type of federal subsidizing has had its influence. The national growth of nonmetropolitan counties (counties without at least one city of at least 50,000 population) was more than 4% during the years of 1970-1975 whereas metropolitan counties grew by less than 3%.

It is estimated that by the year 2000, some 19.7 million acres in the United States will be gobbled up by urban sprawl. Two million of these acres will be taken for right-of-ways for power lines.

At the same time that the federal government was financially promoting and subsidizing urban land-use sprawl outside the city limits in the countryside and/or in towns up to 10,000 population or communities not within cities up to 50,000 population,[8] other federal money was appropriated to support the Council on Environmental Quality, the Office of Policy Development and Research of the Department of Housing and Urban Development, and the Office of Planning and Management of the Environmental Protective Agency so these agencies might hire the Real Estate Research Corporation to prepare a detailed cost analysis of the costs of sprawl. In 1974, a volume entitled *The Cost of the Sprawl* was published by the Superintendent of Documents, U.S. Government Printing Office. According to this study, the traditional pattern of scattering single-family residences and subdivisions around a city's fringe is the most expensive method of community development.

Although the state is aware that it has a land-use problem, as indicated by a land-use conference held in Indianapolis in September, 1973, the leadership representing farmers, elected officials, conservationists, urbanologists and university professors did not agree on what action should be taken to study or solve the land-use problem. Lt. Governor Robert Orr feels the state should not get involved in land-use control. Moreover, he has been critical of a state land-use planning study made prior to the start of the present administration.[9] In contrast, Professor Charles Bonser, Dean of Indiana University's School of Public and Environmental Affairs suggested:

> . . . the formation of state coordinating agencies to work with regional and planning boards, the creation of state review boards which could inspect and pass judgement on large development proposals and state or regional land-use commissions with the power to designate areas for conservation, agriculture, and limited development.[10]

With the present stalemate, Indiana's land-use problems are becoming more serious. In contrast, other states

> have passed innovative legislation giving the state government a direct role in approving important changes in land use. The Congress has been considering legislation to encourage these state efforts and is at the same time trying to delineate the appropriate role if any of the national government in land-use decisions. . . .[11]

Some of the states which have adopted land-use legislation are California, Maine, Vermont, Delaware, Florida, Oregon, and Hawaii.

WATER

Although the amount of water is limited and the annual rainfall is relatively steady in Indiana, its state population grows and the per capita use of water soars. With the removal of forests and grasslands during the nineteenth century and the drainage of wetlands during the past 100 years, the water table has dropped, springs have dried up, and streams and rivers flow more erratically than in pioneer days. Streams, rivers, and lakes have become more and more polluted during the twentieth century as more and a greater variety of nonorganic pollutants and toxic substances such as polychlorinated biphenyls were added to the growing amount of human wastes of small and large settlements. Most of these villages and cities only give primary treatment (the lowest stage) to their sewage and only a few settlements apply secondary treatment.

As stream and river flow became more erratic and polluted, the habitat for both water and land wildlife decreased and some recreational facilities declined and other recreational potentialities were destroyed.

Some water problems in Indiana today are:

1. reducing the number, size, and damage of floods
2. lowering the pollution of streams, rivers, and lakes
3. building reservoirs to provide:
   a. domestic and industrial water
   b. a habitat for wildlife and/or
   c. recreational facilities
4. designating primary and secondary uses of water in our rivers and lakes for
   a. domestic and industrial use
   b. cooling of electrical generating machinery
   c. navigation
   d. recreation
   e. restoration of wildlife
   f. sewers of storm water and partially treated sewage
5. coordinating present and future water transportation with other major forms of transportation such as railways, highways, airlines, and major line and pipe systems
6. Integrating settlements and countryside drainage systems

Many state, interstate and federal agencies are now engaged in helping solve these problems. Some of these are: (1) the Great Lakes Basin Commission; (2) Wabash Valley Interstate Commission; (3) Ohio River Basin Commission; (4) Indiana's Soil Conservation Service; (5) Army Corps of Engineers; (6) Water Resources Division of the Indiana Department of Natural Resources; (7) The Indiana Stream Pollution Control Board and others too numerous to list here. There is no dearth of governmental agencies involved in conservation problem solving but an almost complete lack of coordinating the activities of the various agencies, who do not agree on solutions to the existing problems. One agency advocates construction of small dams and reservoirs on minor tributaries upstream from the major stream while another agency recommends much larger dams and reservoirs on the larger rivers for flood control and other water uses. At the same time, some conservationists are strongly opposed to building dams and reservoirs of any size for flood control and advocate settlement control by planning and zoning of land use on the flood plains as the primary solution for reducing the number, size, and damage of floods. At least one agency has been teaching that reforestation is a major way of reducing flooding while university texts on the subject stress that reforestation is of only secondary importance. Unfortunately, the flood control chapters in some university texts on the subject make no effort to distinguish between primary and secondary solutions to flood control. With different governmental agencies advocating different solutions to the same problems and failing to coordinate their activities, it is a little wonder that the public is confused, resentful and rebellious against bureaucracies and the water problem has not been solved.

With some justification, the public has started to rebel against a proliferation of federal and state agencies and this soaring number of bureaucrats. It is reported that at least 25 authorities in a dozen different national departments and agencies are conducting research on water pollution.[12]

AIR

In part because Indiana is a state with few large cities, air pollution is not as serious as in many states. This means that Hoosiers have a golden opportunity to keep the problem from becoming serious if they act quickly and firmly and cooperate with other states.

Of the 151 cities with populations of over 2,500, only nine had a population of over 50,000 in 1970 and only six of these had over 100,000. Most of the air pollution is associated with these nine cities and the pollution blown into the state from metropolitan areas adjacent to its boundaries. These metropolitan areas are Chicago, Detroit, Toledo, Cincinnati, and Louisville. The Louisville *Courier Journal* reports on stagnant air and its pollution, thereby alerting people to serious conditions not only in Jefferson County in which Louisville is located but also to those in counties north of the Ohio River in Indiana such as Clark, Floyd and Harrison.[13] During the alerts persons with respiratory conditions are advised to limit their activities and stay indoors as much as possible and take precautions as advised by family physicians.

Indiana's six largest cities have air pollution control programs but the programs are too new to be evaluated at this time. Indiana is in the Great Lakes Northeast atmospheric area as delineated by the Secretary of Health, Education and Welfare and subject to its regulations.

There are pockets of air pollution associated with the production of minerals and their manufactured products and with the generation of electricity. In March, 1976, Public Service of Indiana announced that air-quality monitoring systems worth $600,000 have been installed in at least six plants. These systems will provide readings on sulfur dioxide concentrations. Public Service of Indiana, the state's largest utility with a half-million customers in sixty-nine counties, hopes the systems eventually will save consumers millions of dollars. The utility company is confident that all of their coal-fired stations are already meeting state air quality standards. But under current Environmental Protection Agency standards, the burden of proof is on the electric utilities. If electric generating companies do not comply with EPA regulations, they must install scrubber systems on their large coal-fired stations. Unfortunately, Indiana coal has relatively large amounts of sulfur in it.

Although air pollution problems are not as serious or of as long duration as land and water problems, various polluting establishments and state agencies should cooperate now with each other and agencies in adjacent states to solve air pollution problems before they become more serious.

ENERGY

Although Americans were forewarned of the coming energy crisis by the Paley Report about a quarter of a century ago, the influence of this report was lost during the Korean crisis. The public was not prepared for the energy crisis which came in the 1970s and the energy problem has not yet been faced realistically. According to unchallenged studies, ". . . Americans consume twice as much energy as they need to maintain their high standard of living and waste more than two-thirds of what the rest of the world consumes."[14] The Federal Energy Administration hired a Washington-based group called Worldwatch Institute to make an in-depth study of the energy problem. This group recommends that it is desirable to search for innovative and conventional new sources of energy, but that the "meeting of new energy needs through conservation would be cheaper, safer, more reliable, less polluting and would create more jobs than obtaining energy from any other sources."[15] Yet it is difficult if not impossible in Indiana and other states to enforce the 55 mile an hour maximum speed limit. This same Worldwatch Institute maintains that Americans "could lead lives as rich, healthy, and fulfilling with as much comfort and with more employment merely by using less than half the energy now used." In Switzerland, Sweden, and Germany, where the per capita gross national product is about the same as in the United States, the per capita use of energy is 40% less. In contrast with the United States, these countries have launched stringent energy conservation measures. Are Hoosiers and Americans in general facing up to the reality of the energy problem?

TOXIC SUBSTANCES

Another problem of the 1970s which hit the general public in Indiana and the nation with the suddenness of a clap of thunder was the pollution of land, water, and air with polychlorinated biphenyls (PCBs) and other toxic substances. Some people who have respected chemistry for decades as almost a god are now depicting industrial chemists and the companies for which they work as the devil and his cohorts. Unfortunately, at least one industry notified state regulatory agencies but not the local sewage management plant about PCBs. The local sewage plant gave the sludge to gardeners and farmers. Suddenly, these people learned that their gardens are polluted, the grass in their pastures is polluted, the milk is polluted and the fish in creeks and rivers are so polluted that warnings are posted not to eat the fish caught in these waters.

To use the Bloomington-Bedford area as an example, the people are becoming disturbed and seeking help. It is an understatement to say that help from state agencies is minimal, illusive, and reluctantly given. Since no one now wants sludge from the Bloomington sewage treatment plant to enrich soil, the sludge pile gets larger and larger by the week.

Indiana is now becoming aware of problems which had to be faced by states east of the Appalachian Mountains at an earlier date. In this region, some industries must treat their own sewage and it is not put into city sewage systems. Some questions raised are:

1. Should not industries or service establishments be required by law to notify city officials if they are placing toxic substances in a city sewer system, or otherwise be subject to fines and damages?
2. Should not industries that manufacture products of which less than 10% are consumed in the city be required to provide their own sewage treatment especially if their factory wastes are more difficult to treat than human wastes?
3. Should or do cities have the legal right to refuse sewage from any industry or service establishment when tests indicate that the wastes are placing a hardship on a city sewage treatment system?
4. Is it fair, just, legal or wise for city administrations to levy an indirect tax (an increased sewage rate) on its citizens so as to pay for the treatment of industrial and/or service waste when the industry or service establishment is not located in the city where it would be paying city taxes?

There is nothing so constant as change and nothing so necessary. One hundred years ago in Indiana it seemed wise to run storm water into the same sewers carrying human wastes. The commonly-held philosophy was that dilution was a partial solution to pollution. Now that primary, secondary, and tertiary sewage treatment plants are necessary, it is no longer efficient to have dual purpose sewers carrying both storm water and human wastes. Have we not now reached a new era when many chemical industrial wastes should not be permitted in city sewers and the producers of these wastes should be required to provide their own waste treatment or disposal systems?

NOTES

1. Thomas Frank Barton, "Indiana State Forests and Reclamation," *Proceedings of the Indiana Academy of Social Science* 6 (1961):56-61.
2. ———, "A Four Million Acre Problem: Commercial Forest Land in Indiana," *Proceedings of the Indiana Academy of Social Science* 5 (1960):8.
3. *Ibid.*
4. Thomas Frank Barton, "The Hoosier National Forest," *Proceedings of the Indiana Academy of Social Sciences* 3 (1958):82-93.
5. *Ibid.*
6. Jeff Stansbury, and Edward Flattau, "How the Traditional Economics Undervalues Natural Resources," *The Courier Journal and Times* (July 27, 1975), p. D-9.
7. Leslie Ellis, "Hoosier Communities Face Competition for U.S. Housing Development Funds," *The Courier Journal* (February 12, 1975), p. A-1.
8. Rural Development Act of 1972, *Congressional Record* (July 27, 1972).
9. Ric Manning, "A Land-Use Planning Meeting Concern over Unchecked Growth," *Bloomington-Bedford Sunday Herald Times* (September 9, 1975), p. 9.
10. *Ibid.*
11. Robert G. Healy, "Controlling the Use of Land," *Resources* (Resources for the Future, Inc.), Number 50 (October 1975), p. 1.
12. Finlay Lewis, "Cost and Size of Social Programs Challenged," *Daily Herald-Telephone* (January 21, 1976), p. 10.
13. "Stagnant Pollution Alert Persists; Weekend Relief Possible," *The Courier Journal* (June 10, 1976), p. A-1.
14. "Pollution Meters Installed," *The Courier Journal* (March 3, 1976), p. C-2.
15. Editorial, "Conservation Best Energy Source," *Bloomington and Bedford Sunday Herald Times* (February 29, 1976), p. 14.

Recycling Resistance
The Case of Cans

Peter Kakela
Sangamon State University

INTRODUCTION

The public is demanding that scientists account for the human consequences of what they are doing. Even H. Guyford Stever, director of the National Science Foundation, agrees.[1] To be accountable, scientists must deliberate about the human consequences of their work and must communicate perceived consequences to public decision makers and the general public. The process of making decisions in public arenas is based on different assumptions than the process of making scientific decisions. Effective communication in the public decision-making arena demands an understanding of these different assumptions. Thus, for scientists to accept their responsibility to the public requires new learning and development of new skills. We must ask our colleagues if we might not already have enough basic knowledge to solve many of the problems of today involving energy, food, population, pollution, poverty, etc. Is it not true that we lack more the will and the way than the basic knowledge? If that be the case, should not the application of existing knowledge to attempt solutions to societal problems be as important and as legitimate as digging up new knowledge?

Personally, I have drifted from pure physical geography to environmental issues and especially energy conservation as a result of that conviction. During the last several years, I have devoted much effort to communicating scientific and technical information about energy to state legislators and other policy decision makers. In this paper, I would like to relate the content basis of two case studies that I have been involved with in an effort to demonstrate a model of mutual environmental and economic benefit.

In 1976, environmentalists are backed into a dusty corner. The public today views environmentalists as obstructionists to economic development. With a tight job market, tight money, and restricted economic growth, there just does not seem to be the money or the will to fight for the environmental improvements called for by this same public during the environmental enlightenment period of the early 1970s.

In 1972, Dennis Meadows et al., told us in the arousing book *Limits to Growth* that environmental destruction would bring a catastrophic decline in population if mankind did not consciously and quickly change its ways and halt economic growth.[2] Ironically, reduced economic growth has done more to endanger the environment than enhance it. Job considerations and costs are taking priority over environmental protection and social benefit. We have presidential vetoes of strip mine reclamation bills and postponement of the enforcement of air quality standards for new cars. The push for an Alaskan pipeline was successful and offshore oil exploration was expanded in the midst of an "energy crisis." Increased sulfur oxide emissions are deemed necessary and quicker approval of nuclear power plants is urged. These issues involve some of the major fights environmentalists waged in the early 1970s. Hard times, in the economic sense, quickly eradicated environmental defense. On the day-to-day issues, there is great slippage in meeting the added costs of pollution abatement devices. Thus, the "obstructive" environmentalist is pushed aside and relegated to his or her dusty corner.

On the grander issue, when the environmentalist talks of "limits," as in *Limits to Growth* or limits to resources, the economist sees "scarcity." Economic theory is based on scarcity (and insatiable wants). In fact, economics as we know it must have scarcity to function. Interpreting limits as scarcity, economists are in their glory and reach for: (1) increased supply, (2) new technologies, or (3) substitution of materials, as solutions. Talk of limits can even pose a challenge to some and stimulate efforts to prove once more that we can overcome the physical environment. Action here takes such forms as mining leaner ores, striping more fragile lands, using more chemicals in farming, and pushing for the "ultimate" energy supply, nuclear power. For the most part, these economic adjustments cause increased environmental impacts.

This paper outlines an alternative production strategy in which economics and ecology work together. With this strategy:

1. It *is* possible to reduce energy consumption;
2. It *is* possible to reduce virgin material consumption;
3. It *is* possible to reduce air and water pollution;
4. It *is* possible to reduce solid waste production;

All this *is* possible while still maintaining the same quantity and quality of goods and services delivered. Further, jobs can be created and consumer costs reduced by more environmentally sound production systems. These alternatives do not come from an added on environmental cleanup device, which is an added expense, nor from drastically cutting production. They come from a more fundamental change in our production practices. I will develop two examples of this alternative production approach: (1) replacing throwaway beverage containers with refillable bottles, and (2) the recycling of material resources.

CASE STUDY OF BOTTLES AND CANS

The first case study is, in itself, a very small issue. Literally, it is garbage and specifically it is throwaway beer and soft drink bottles and cans. I do not apologize for the scale of this case study because there is much to learn about environmental problems and societal decisions from beer cans and pop bottles. In fact, a whole model of environmental and economic interaction and of public opinion and pressure politics can be generated from this seemingly small-scale example. It is a case study that is manageable in scope and fruitful in lessons.

Refillable beverage bottles represent reuse of resources. On the average, some fifteen reuses occur but one container may be reused more than 40 times. Recycling is a form of remaking products from once used products and, as the term implies, throwaways are thrown away after one use. They are the "one-way," or the "no deposit, no return" containers that were introduced some 25 years ago and now dominate to present-day beverage market. Throwaways can be recycled, but not reused.

To make the 15 cans or throwaway bottles requires much more energy, raw materials, and complex equipment. Few employees are required to run this highly automated system.

To be more specific, throwaway bottles use 3.11 times as much energy on the average to produce and distribute than refillable bottles do. Cans use 2.70 times as much energy as refillables.[3] By converting from present market practices to a complete refillable system, the beverage industry as a whole could reduce its energy consumption by 40% to 55% without reducing the amount of beer or soft drink delivered to consumers.[4] Assuming just ten trips for the refillable bottle but delivering the same quantity of beverage, the refillable system when compared to a wide variety of throwaway containers provides a 34% to 87% reduction in water borne wastes and 30% to 71% reduction in air effluents.[5] Beverage containers make up 20% to 32% of total roadside litter by item count. Over 90% of these littered beverage containers are throwaway containers.[6] Beverage containers now constitute 7% of our total municipal solid waste or more than 8.2 million tons per year.[7] Because of the trend towards throwaways, beverage containers are now the fastest growing segment of solid waste, growing at 8% per year, whereas overall solid waste is growing at only 3% to 3.5%.[8]

Throwaway containers are hard on the environment. They are also hard on the pocketbook. In a recent cost survey conducted by some of my students, it was found that the difference in price between refillable bottles and cans for one soft drink in multiple pack, individual serving size containers was 90%. That is, an ounce of soda in cans was 90% more expensive than an ounce of the same soda in refillable bottles. Additionally, soda in a 32 ounce throwaway bottle was 38% more expensive than the same soda in a 32 ounce refillable bottle. Actually, even if one were to buy the 32 ounce refillable bottle requiring a deposit and throw the refillable away, it still would be cheaper than buying the throwaway bottle.

Throwaway beverage containers adversely affect employment, too. The highly automated can and one-way bottle production requires much less labor than refillables. Refillables require more handling, sorting, and washing, but less resource consumption.

A number of studies have substantiated this. Charles Gudger and Jack Bailes of Oregon State University found that one year after an Oregon law requiring a deposit on all beverage containers had gone into effect, 95% or more of all beverage containers purchased were refillable. This shift caused a net increase of 315 jobs in Oregon's beverage industry. Employment projections of what is expected to occur have been made for other states. A 1972 study for Illinois was conducted by economist Hugh Folk of the University of Illinois, who estimated a net increase of 1,494 jobs in Illinois in the beverage industry with a return to the refillable bottle.[9] More recent studies have found that a return to refillable bottles would create 390 additional jobs in Minnesota, at least 1,403 additional jobs in Michigan[10] and a net increase of some 4,000 jobs in the beverage of industry of New York.[11] The U.S. Department of Commerce has studied the impacts of national mandatory deposit regulations. Even though other uncertainties prohibited the support of nationwide legislation, they calculated a net increase of from 13,000 jobs to 33,000 jobs in the beverage industry for the nation.[12]

These figures are for net increases. There would be some job relocation, but this might be eased by phasing in changes over time and allowing for normal attribution rates in factories to assimilate job decreases. The jobs that will be lost are in the can manufacturing and bottle manufacturing plants. The main increases in jobs come in the retail stores, in transportation, and in the bottling plants where washing and sorting are increased.

THE RECYCLING EXAMPLE

In the more general case of materials recycling, similar trends exist that provide evidence of energy and environmental savings. In the example of steel, 75% to 80% less energy is required to make a given quantity of finished steel from iron and steel scrap than from virgin iron ore.[13] Producing a given quantity of steel with 80% less energy is an important contribution to energy conservation. The same holds true for aluminum made from recycled scrap where there is approximately 96% energy conservation.[14] Copper made from scrap rather than virgin ores saves 94% of the energy.[15] Paper is another example. A given quantity of finished paper produced from waste paper instead of virgin pulp requires 70% less energy.

In addition to energy savings, there also are a number of environmental impacts which are reduced. According to the U.S. EPA *Report to Congress on Resource Recovery* (1973), the production of a given quantity of low-grade paper from waste paper as opposed to virgin pulp results in the following:

25% reduction in water pollution (suspended solids),
39% reduction in process solid wastes generated,
44% reduction in water pollution (BOD),
61% reduction in process water used,
70% reduction in energy consumed,
73% reduction in air pollution emissions,
100% reduction in virgin materials used, and
129% reduction in post-consumer solid waste generated.[16]

Recycling affords the production of a given quantity of product while reducing energy consumption as well as environmental impacts.

I have written at greater length elsewhere about the environmental savings from recycling.[17] An important point to be made here is that recycling represents a poor second best when reuse is a possibility. Consider the aluminum can and the 96% energy savings when aluminum is recycled (i.e., remelted and recast). Because of the high initial energy requirements of making aluminum, and the low rates of actual retrieval of aluminum beverage cans (about 17%), the aluminum can system requires some 3 1/2 times as much energy to deliver a given quantity of beverage as does the refillable glass bottle making just ten trips.[18] If aluminum can retrieval rates were to increase, then it would become more competitive in energy consumption with the ten-trip bottle. Approximately 95% reclamation of aluminum cans would be required before energy costs of recycling this container would be cut to match the refillable bottle. (See Figure 1.)

RESISTANCE TO REUSE AND RECYCLING

Why then do we not recycle materials at a larger rate than our present 25% overall? Scrap is available and furnace capacity could handle more scrap. In fact, we are exporting some of our clean scrap metals and importing dirtier, more energy intensive virgin ores.

Why do we not use products and packages that are easily reused? We used to. It saves energy and resources. It reduces pollution, solid waste, and litter. It saves consumers money and creates jobs.

To understand the more complicated economics that have led to these practices, one must look harder at the structure of the production systems. What are the industrial structures of the beverage industry and the materials production processes that would make it profitable to promote waste of resources and excessive pollution?

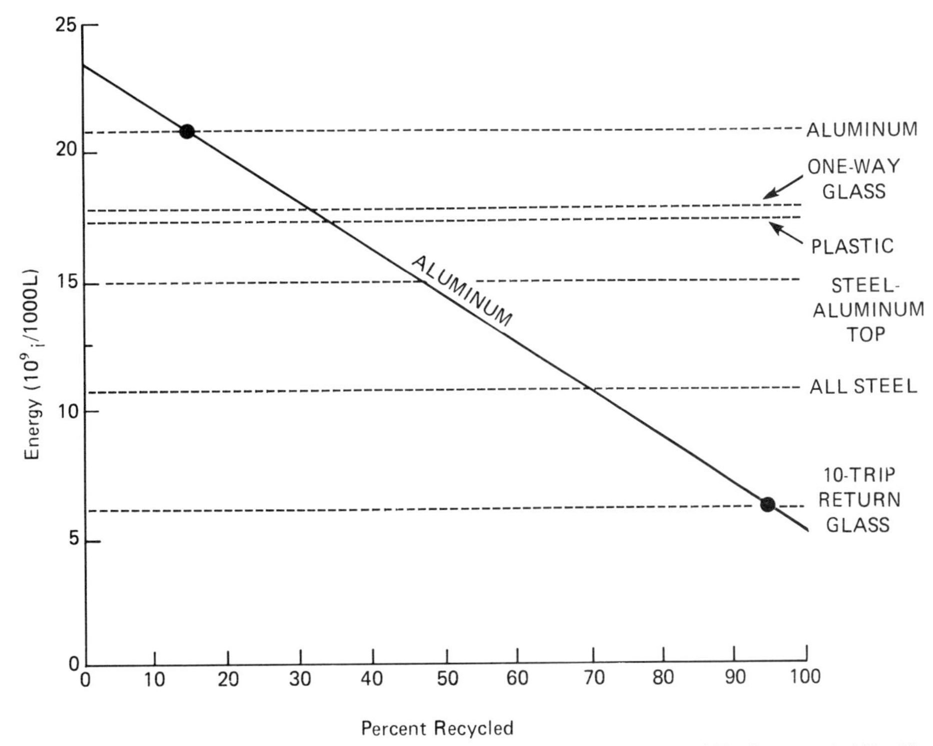

Source: U.S. EPA, *Resources and Environmental Profile Analysis of Nine Beverage Container Alternatives* (Washington, D.C., U.S. Government Printing Office, EPA/530/sw91c), 1974, p. 41.

Figure 1. Energy Costs and Recycling for Six Beverage Containers.

For the purpose of examining the beverage industry, I want to tell the tale of "The Beer that Jack Drank."

After prohibition, in the early 1930s Jack could have obtained his beer from any one of some 735 United States breweries. The beer that Jack drank was mostly from local, family operations. Today there are approximately 65 breweries left, but most likely the beer that Jack drinks comes from one of the five largest United States breweries: Anheuser-busch, Schlitz, Pabst, Coors, or Millers. These five control about 70% of the U.S. beer market.[19]

The metal can industry which supplies the throwaway containers that held the beer that Jack drinks is also highly centralized and exhibits monopolistic tendencies. Two companies, American Can and Continental Can control about 80% to 90% of the can market.[20]

The aluminum industry that supplies the materials to American Can and Continental Can, (who supply the containers, which contains the beer that Jack drinks), is made up primarily of three companies; Alcoa, Kaiser, and Reynolds. These three have another near monopoly in the beverage container system—80% to 90% of the U.S. aluminum market.[21]

The steel industry which supplies the other metal that American Can and Continental Can use to make the bi-metallic cans that hold the beer that Jack drinks is dominated by four manufacturers who control 50% to 60% of the total U.S. steel market. These companies are U.S. Steel, Bethlehem, Armco, and Republic Steel.[22]

If monopolies do exist, then prices can be fixed. Even short of functional monopolies, large corporations can influence policies. For example, regulated freight rates on the United States common carriers run some 2 1/2 times more to move scrap iron and steel than it costs to ship virgin iron ore. In defending these discriminatory rates recently, one found the regulator (the Interstate Commerce Commission), the regulatee (the railroads), and the customers (the major U.S. steel companies) all cooperating in lobby efforts to oppose federal attempts to equalize freight rates. They argued that "discrimination is not that rampant" while the recycling people complained that over the past seven years, freight rates for scrap increased 60% faster than those for virgin ores.[23]

Another example of political pressure from big corporations is in the realm of tax policies. Major virgin ore mining companies have received and retained corporate tax deductions, the most notable example is the depletion allowance, which encourages the exploration of virgin natural resources and provides an economic disincentive to recycle scrap. With beverage containers, one could argue that throwaways receive a tax subsidy too. Throwaways have 21 times greater chance of being littered than returnables, and throwaways add greatly to the volume of household garbage.[24] The cost of litter pick-up along state highways and usually the cost of municipal refuse collection and disposal are paid for with tax dollars. Illinois alone spends 2.6 million dollars annually on litter pick-up along state highways.[25] The nation's cities spend 6 billion dollars annually for solid waste treatment.[26] A portion of these costs is directly attributable to throwaway containers. These are hidden costs of throwaway containers.

There is another characteristic of corporate structure which causes further resistance to recycling. This is vertical integration. Major steel firms own and operate steel-making facilities, but they also own and operate virgin iron ore mines (in this country and abroad). They own and operate pelletizing plants, coal mines, transport ships, and even construction and erection companies. They do not own the many and widespread iron and steel scrap dealers.

In the beverage industry there is vertical integration too. The Coca-Cola Company, which makes the syrups for beverages has long operated its franchise system of local bottlers. The local Coca-Cola Bottling Company buys the syrups, mixes them to strict specifications with carbonated water, bottles the product, and markets it locally. More recently, with the introduction of the throwaway cans, vertical integration has been extended. For example, the Coca-Cola Company owns and operates a can manufacturing plant in Chicago to supply canned beverages to that market. The Coca-Cola Company not only owns and runs the can making plant, but they lease the canning equipment from Continental Can Company. The beer industry has similarly expanded into the throwaway container business. Budweiser makes many of its own cans and as one market analysis puts it, since 1969, "Schlitz began getting into the can making business in a big way."[27]

In the beverage example, there are two important results of extending vertical integration into the throwaway container business. One, the beverage companies created another profit center, the throwaway container. Therefore, they became not just beverage salesmen but container salesmen as well. Second, the throwaway container has become a link in the beverage production delivery system. With five links, instead of four, there is added total cost to the consumer and added total profits to industry. This is a waste link, however. It is unnecessary to the enjoyment of the product. As Barry Commoner stated, "lets take a long hard look at the (throwaway) container. Does it have any intrinsic value? Not as far as I can tell. I don't want to consume a container, what I want are the contents."[28]

CONCLUSIONS

Throwing away cans and bottles, and avoiding recycling are counter to ecology. Both vastly increase environmental impact for a given rate of consumer goods produced.[29]

These examples result in more costly products and, with throwaway containers, fewer jobs. This is countereconomic from the consumer's standpoint. Further, these throwaway and one-use trends go against public opinion.[30] When one reads economic theory, on the other hand, we find that restricted competition, inefficient technology, excess resource use, and higher prices are associated with industries that are monopoly controlled and vertically integrated.[31] Regardless, the results are greater environmental impact at the expense of consumers.

So it turns out that our "small" example involving beer cans has led us to issues involving energy waste, the underpinings of our throwaway society, promotion of a "no-deposit, no return" attitude toward our natural resources, corporate monopolies, and pressure politics. I am sure that there are many other revealing examples.

These issues speak to the so-called apolitical, laboratory scientist. My point is that scientists must press to understand the implications of their work. If one's research is seen to have no social or environmental implications, then it is possible that we are being shortsighted, or naive, or merely working on unimportant items. If implications were perceived, would they support current trends or would they suggest change. Once analyzed, what action does one take? Inaction most often supports current trends. Effective action to bring about meaningful change through democratic processes requires the utmost of skill and understanding. It is usually negotiated under a very different set of ground rules than scientists are used to in their laboratories. But to accept the responsibility to do research should also mean that one accepts the responsibility for communicating the impacts of one's research. This is the social responsibility of science and scientists. Because a scientist has more knowledge than the average layman, the scientist's responsibilities are therefore greater. We must raise our level of consciousness about these responsibilities so that we do our part to insure the continuity of the environment and human society.

NOTES

1. R. Cowen, "Is Science Out of Fashion?" *Christian Science Monitor* (June 4, 1975), p. 25.
2. D.H. Meadows et al., *The Limits to Growth* (New York: Universe Books, 1972).
3. Bruce Hannon, "Bottles, Cans Energy," *Environment* (March 1972).

4. ———, "System Energy and Recycling: A Study of the Beverage Industry," *CAC Document no. 23* (Urbana, Ill.: University of Illinois, Center for Advanced Computation, January 5, 1972).
5. U.S. Environmental Protection Agency, "Resource Recovery and Source Reduction," *Second Report to Congress* (Washington, D,C.: USGPO, EPA-SW-122, 1974), p. 83.
6. *Ibid.*
7. Edwin Lowry, Thomas Fenner, and Rosemary Lowry, *Disposing of Non-Returnables: A Guide to Minimum Deposit Legislation* (Stanford, California: Stanford Environmental Law Society, 1975), p. 8.
8. *Ibid.,* p. 9.
9. Hugh Folk, "Employment Effects of the Mandatory Deposit Regulation," *IIEQ Document 72-1* (Chicago: State of Ill., Institute for Environmental Quality, 1972), p. 27.
10. Rao B. Gondry, *Economic Analysis of Energy and Employment Effects of Deposit Regulations on Non-Returnable Beverage Containers in Michigan* (Lansing, Mich.: State of Mich. Dept. of Commerce, MPSC staff study 1975-3, October 1975); and Myron Ross, "Employment Effects of a Ban on Non-Returnable Beverage Containers in Michigan" (Kalamazoo, Mich.: Kalamazoo Nature Center of Environmental Education, April 1974).
11. N.Y. Senate Task Force on Critical Problems, *No Deposit, No Return: A Report on Beverage Containers* (Albany, N.Y.: New York State Senate, Task Force on Critical Problems, 1975).
12. U.S. Department of Commerce, Bureau of Domestic Commerce, *The Impact of National Beverage Container Legislation* (Staff Study: A-01-75, 1975), p. 3.
13. Edmund Faltermayer, "Metals: The Warning Signals Are Up," *Fortune* (October 1972), p. 110; *EPA Report to Congress on Resource Recovery* (February 22, 1976), p. 12.
14. Faltemayer, *op. cit.,* p. 110.
15. *Ibid.*
16. *U.S. EPA Report to Congress on Resource Recovery, op. cit.,* p. 10.
17. Peter Kakela, "Railroading Scrap," *Environment* (March 1975), pp. 27-33.
18. U.S. EPA, *Resources and Environmental Profile Analysis of Nine Beverage Container Alternatives* (Washington, D.C., U.S. Gov. Printing Office, 1974), p. 41.
19. *Rodale's Environment Action Bulletin* (Emmans Pa.: Rodale Press, 1975), p. 1.
20. William G. Shepherd, *Market Power and Economic Welfare* (New York: Random House, 1970), p. 152.
21. *Ibid.*
22. *Ibid.*
23. George Moneyhum, "Lobbists See Gains in Recycling Prospects," *Christian Science Monitor* (March 18, 1975), p. 3.
24. People's Lobby Against Non-Returnables, "The Oregon Leader Survey" (March 19, 1972).
25. Letter from Verdum Randolph, Associate Director, Illinois Department of Public Health.
26. Neil Seldman, "High Tech. Recycling Mainly Benefits Polluters," *Environment Action Bulletin* (February 7, 1976), p. 2.
27. Robert Metz, "Analysts Believe Competition to Grow Among Beer's Big 5," *Milwaukee Journal Balance* (June 16, 1975), p. 14.
28. Barry Commoner, "Testimony on Illinois Beverage Container Act; H.B. 1838," before the Environment, Energy, and Natural Resources Committee, Illinois House of Representatives (February 11, 1976).
29. Barry Commoner develops the notion of counter-ecological technology and its proliferation since WW II in *The Closing Circle: Nature, Man and Technology* (New York: Alfred A. Knopf, 1974).
30. Environments and People Students, Sangamon State University, Springfield, Illinois, December 1975, conducted a random survey of 100 people listed in the Springfield telephone directory and found that Springfield area residents tend to: buy beverage in throwaways (47% buy throwaways, 30% returnables, 22% buy half and half); realize beverage in returnables is cheaper (59%); support legislation to encourage returnables (69%); would return containers if all carried 5¢ deposit (81%). The greater public support for returnables than current purchasing practices suggests that freedom of choice is currently curtailed by stocking policy of retail outlets. Opinion Research Corporation, Princeton, New Jersey, in a February 9, 1975 survey for the Federal Energy Administration, found more than 73% support for mandatory deposits on all beverage containers in a nationwide public opinion poll.
31. William Shepherd, *Market Power and Economic Welfare* (N.Y.: Random House, 1970).

Environmental Change and Regional Limits to Growth

John J. Hidore
Indiana University

In 1972 the book entitled *Limits to Growth* (Meadows et al., 1972) created a furor by suggesting that there are limits to the economic growth on planet earth. The impact of the book was phenomenal. Some 2.5 million copies have been sold in several languages. The study was based upon a computer model of the world economic system considering three limiting factors; food supply, nonrenewable resources, and pollution. The conclusions were that there was an upper limit to world population numbers, an upper limit to economic growth, and that unless all growth were halted within two or three decades a complete collapse of the world socioeconomic system would result.

Limits to Growth stimulated limitless articles and several books which have attempted to refute the findings of the report. Partially as a response to the intensive criticism, a second volume (Mesarovic et al., 1974) sponsored by the Club of Rome reexamined the questions raised and assumptions made in the first volume. Considerable improvements were also made in the simulation model used. In the first volume the world was treated as a single unit with all parts of the world having the same levels of technology, pollution, and resource consumption. In the real world this is far from being the case with tremendous differences between the developed countries, the developing countries, and the underdeveloped countries. The rich and poor countries differ remarkably in population growth rates, per capita incomes, resource consumption, and level of technology. In the second report the world was broken down into ten subregions which gave a more representative model of the whole system. While some of the positions of the earlier work were indeed modified, the original conclusions still remain as possible events in the future.

More recently Herman Kahn (1976) provided a completely different picture of the future. He predicted some near-term problems and uncertainties for the human species but a favorable long-range picture. The thesis of physical limits to growth was rejected in his scenarios of the future, even for the poor countries of the world. The studies by the Club of Rome and by Kahn project the future anywhere from 50 to 150 years away. The reports seem to agree in some areas, including the belief that limits exist to population growth and to economic growth. There is no agreement as to whether there are physical limits to growth or when the limits will be reached. Since *Limits to Growth* went to press, the world population growth rate has begun to slow. Birth rates are falling rapidly in some areas and death rates have been very high in others.

It may well be that even conservative estimates of the population of the human species made five years ago will turn out to be high and that we will not again see a doubling of the earth's population, although this does not seem very likely.

In spite of the myriad of attacks on *Limits to Growth,* the thesis of the book persists. No one has been able to discredit the basic forecast. There seems to be too much evidence, and subsequently a widespread fear, that the prediction may be ultimately realized, even if not in the time projected. Coupled with this is a growing, and perhaps fully warranted, fear that mankind is setting into motion forces of a scale that we will be unable to control.

What is overlooked frequently in the discussions of changing consumer preferences, employment preferences, and population loads is that these concepts are irrelevant to the majority of the world's population. This jargon has meaning only in the rich countries of the world where such choices exist. The projections of a century or more in advance have no meaning where life depends on the rainfall of a single season. It is painfully clear also that the gap between the rich and poor countries is increasing and not decreasing. In much of Africa and Asia the future is one year, or five years, not 150 years.

Regardless of the dispute over the validity of the arguments presented in *Limits to Growth,* evidence exists to suggest that in some fairly large geographical areas population levels have reached and perhaps surpassed the limits which the system can sustain under present conditions. These areas are the extensive regions where subsistence economies still persist. Here the limits to growth are represented by the amount of food the environment can produce locally.

A major drought problem has existed in parts of Africa in recent years. Drought has actually been a problem in parts of Africa throughout historic times. Three years of successive drought may well have paved the way for the fall of the Khalifa in Sudan in 1896. Severe droughts existed in parts of West Africa from 1910 to 1914 and again in the 1940s. In recent years the hazard of drought has been increasing in frequency and areal extent. In Botswana some 60% of the population was reduced to starvation levels in 1965-1968. In 1973 world attention was temporarily focused on a widespread drought occurring in the Sahel of West Africa. An area from Ethiopia and Sudan on the east to the six West-African countries of Senegal, Mauritania, Upper Volta, Niger, Mali, and Chad was affected. In Chad between one-fourth and one-third of the entire population (approximately 3,000,000) had to migrate to more or less distant areas. While the total number of people crossing international borders as a result of the drought is not known it is estimated to be in the hundreds of thousands. The drought problem in the Sahel from 1968 to 1973 directly affected millions of people. The drought of 1973 to 1974 which occurred on the border of Kenya and Ethiopia was by far the worst in the memory of living persons and resulted in some 100,000 deaths.

The fundamental characteristics of the climate of these areas provide a basis for the problem but not the primary cause. In the area, precipitation is subject to very large fluctuations. The annual distribution is strongly periodic and is concentrated in a few weeks or months in the summer season.

The precipitation is also subject to cyclical fluctuations of various lengths and intensity. During the last 20,000 years, the climate of this region has probably changed several times from wetter periods to drier periods. No such change has been recorded during the last 2,000 years, but shorter fluctuations occur frequently. Dry and wet periods of several consecutive years have been recorded two or three times in this century. Precipitation and runoff were at least as low from 1907 to 1915 as during the recent period from 1968 to 1974. Other periods of reduced precipitation have also occurred in the last 100 years.

There is considerable literature suggesting the major factor is a long term decrease in rainfall. A major obstacle in analyzing this possibility is the lack of long-term records. Few recording stations for either precipitation or streamflow exist. However, no serious analysis of available data is known to show a falling trend of rainfall in the zone over the periods for which the records are available. No signs of general trend towards the drier or wetter climate could be found since the beginning of this century, either on the northern side of the Sahara or on the southern side. Similarly, attempts to document trends in rainfall in East Africa (Sansome, 1952) have not shown significant changes.

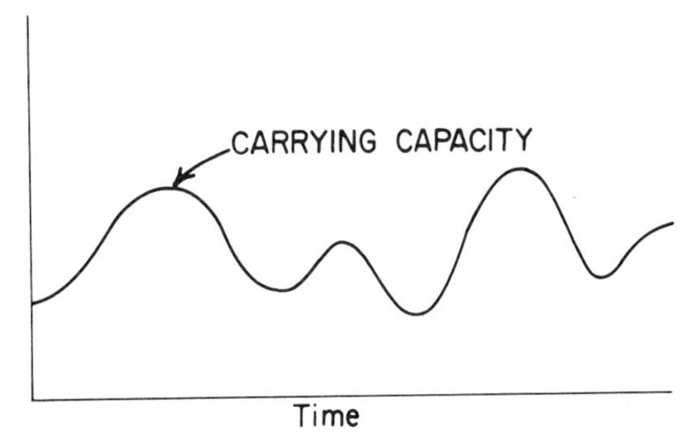

Figure 1. Precipitation and Carrying Capacity. The precipitation over much of sub-Saharan Africa is subject to marked cycles. Since most of the population engages in subsistance agriculture, the cyclical precipitation translates into cyclical carrying capacity.

ADAPTATION TO THE NATURAL SYSTEM

Much of the problem of environmental change is associated with human adaptation to the system. The environment is dominated by a seasonal rainfall regime which is longest in the South and short to nonexistent in the North. There is a continuum of vegetation zones, which increase in total biomass and variety of species as rainfall increases. On the desert margins, precipitation is sufficient only for the development of pasture. However, as the length of rainy season and amount of annual precipitation increase, conditions are reached where dry land agriculture is possible, at least in most years. Where dry land farming is marginal there is competition between the agriculturalists and pastoralists.

POPULATION INCREASES

There has been a phenomenal rise in the human population in the zone in the last few decades. Intertribal warfare and slave trading have been curtailed and public health efforts have made major strides toward reducing sleeping sickness, smallpox, cholera, and cerebrospinal meningitis. These changes have led to a marked reduction in the death rate, particularly among the children, while the birth rate has remained high. The rapid decline in mortality without a corresponding decrease in fertility rates has been one of the major factors responsible for the rapid upsurge in population. With rates of increase of 2.8% and 2.7%, Sudan and Niger are among the fastest growing countries of the world. In addition to the rapid natural increase the problem has been compounded by the migration of substantial numbers of nomads from areas which are in the worst condition to areas where conditions were less severe.

Parallel with the increase in human numbers has been an increase in animal populations. While it is difficult to determine accurately the numbers of animals on the range, various estimates agree on marked increases. The increase over the long run has been due to the control of animal diseases, the introduction of better breeds, and the sheer need to have additional animals to support the increasing number of people. In the nomadic societies where wealth is measured in terms of the number of animals an individual owns, there is continued effort to build the herds as large as possible. Also among some tribes such as the Masai of Kenya and Tanzania the building up of large herds is considered to be an insurance against losses from drought and disease.

Several innovations which have drastically increased the problem have been introduced in the decades since World War II. As a response to previous droughts, governments of the affected nations established programs to relieve the stress. Water was delivered to human settlements for drinking purposes, wells were dug and surface reservoirs constructed to store water for the animals. These measures did, of course, temporarily relieve the problem of water shortage. Whereas in previous climatic downturns there would be a dieback of animals and perhaps the human population as well, this loss was eliminated or greatly reduced.

In Kenya the government opened the game reserves to Masai herds during a recent drought. This saved many more animals than would otherwise have survived. These responses to the drought have tended to make the long-range problem much worse. When the rains increase in subsequent years there are larger herds to begin expansion than would otherwise be the case.

Lack of government action can be disastrous too as Haile Salassie discovered when the inadequate governmental response to the drought in Ethiopia apparently contributed to his overthrow in 1974.

The addition of water supplies has led to even greater pressure on the rangeland. In the space of a few years near wasteland has been produced for a radius of nearly fifty miles around each well.

Prior to colonial control the population growth took place slowly and was held in check by famine, warfare, and disease. The intensity of land use had probably reached the maximum sustainable under average climatic conditions. The food supply was thus marginal on a perennial basis. During the periods of below average rainfall overgrazing was commonplace and slow deterioration of vegetation and soils took place. Nomadic tribes moved away from an area when it became unproductive, eventually returning to it if the ecosystem recovered. The permanent migration of people out of some areas indicates that severe problems existed generations ago.

Under these circumstances almost any change in climate tends to have immediate effects on the carrying capacity of the system. During wet years, expansion and intensification of land use take place. These actions include grazing through enlarged herds, plowing and cultivation of new land, and wood collection around villages. The availability of additional pasture and water during wet years encourages nomads to increase the size of their herds. In this way the affected areas increase production and population growth and intensity of land use moves upward to the limits of the system under the more favorable environmental conditions. However, when an environmental downturn occurs disaster often results. The available pasture is reduced, water is insufficient and virtually all life suffers. It is precisely these kinds of drought problems which have affected the Sahel. The human and animal populations have grown during more favorable

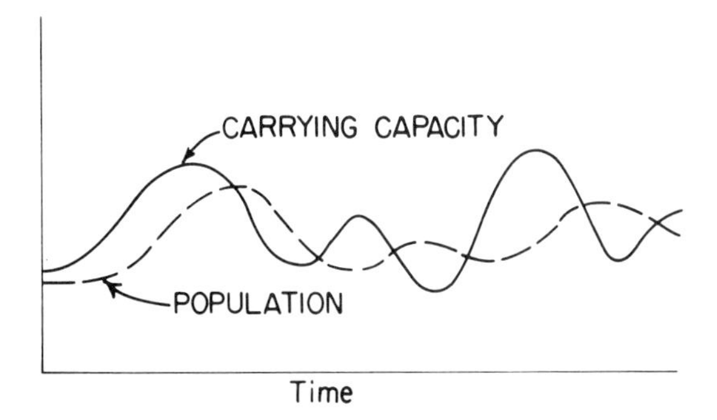

Figure 2. Population and Carrying Capacity in the Pre-colonial Area. In the pre-colonial era population size followed the precipitation cycles. When favorable periods occurred population expanded and when dry periods followed population die-back took place.

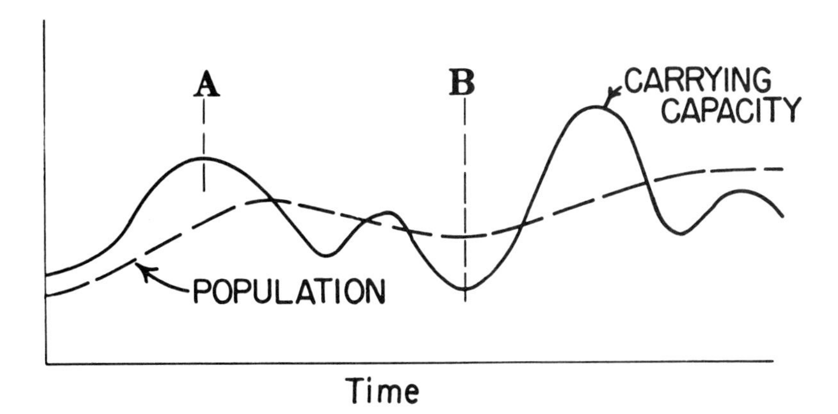

Figure 3. Population and Carrying Capacity Following Colonization. After colonization conditions changed to greatly increase population size. When the carrying capacity began to decline at point A, various forms of aid to the distressed area were instigated which kept down the death rate. Thus when environmental upturn takes place as at point B, a much larger carry-over population remains from which to continue growth. The end result is a population greater than the environment can support even under more favorable circumstances.

years to a level which cannot be sustained by the local ecosystem during periods of downturn. The first areas to suffer are the areas of subsistence economy, where population size is controlled by the ability of the local ecosystem to produce food. A drop in the production of food leads to a reduction in the human as well as the animal population.

ENVIRONMENTAL DEGRADATION

The ecosystems in much of the Sahelian zone will not support large numbers of people on a subsistence basis. The increases in human and animal populations, largely living on a subsistence basis, have created havoc with the local environment. Dry land cultivation practices are for the most part both primitive and exploitative. Cleared areas are farmed for several years during which the soil gradually deteriorates and loses fertility. When the soil is exhausted, it is generally abandoned with the assumption that it will rejuvenate with time. Because of the necessity of feeding increasing population numbers, the fallow period is shortened and there is insufficient time for the soil to recover. The introduction to commercial crops such as cotton and peanuts has served to accelerate the depletion of the soils.

The grasslands are very sensitive to overgrazing. The first effects are in terms of the quality of the pasture. The most nutritious species are consumed first with the least palatable species taking over. When the number of animals reaches a certain level, the pastures do not have time to recover due to frequent or persistent cropping. This may culminate in the disappearance of the groundcover, regardless of species. There will then be even more severe overgrazing in order to feed the ever-increasing herds. It is inevitable that large numbers of animals will die in drier than average years.

Overgrazing has long been a problem around villages and perennial water supplies, and the extent of overgrazing has increased rapidly with the increasing numbers of livestock and increasing concentration of herds around water supplies.

The end result of the increasing pressure on the land is a rapid expansion of the Sahara Desert southward into more humid areas. If this is not completely a man-made phenomenon, it certainly is being accelerated at a rapid pace. When the plant cover is destroyed beyond the minimum required for protection of the soil against erosion the process becomes irreversible. The collapse is precipitated because less water is retained, the soil dries out, and temperature increases. Bare soil appears after the fine soil particles have been removed by either wind or water. When large patches of bare ground are formed, the soil is exposed to wind erosion of even the larger sand-sized particles, and nutrients and humus are removed. The surface then becomes distinguished by denuded deflation pans. The blowing silt and sand further destroys the standing vegetation by stripping the remaining vegetation, or by burying it under dunes or slope-wash sediment. Degeneration into desert usually occurs in scattered patches of bare ground from a width of a few meters to several kilometers. It is not a desert expansion in the form of an even-advancing front of desert surface and hence is not easy to measure.

Charney and Stone (1975) added still another factor which adds to the self-destructing nature of the system. Using a general circulation model differing only in the prescribed surface albedo, they have shown that an increase in albedo resulting from a decrease in plant cover could cause a decrease in rainfall. Thus any tendency towards decreasing the vegetal cover would be reinforced by a decrease in rainfall, and could intensify or perpetuate a drought. It must be pointed out however that there has been no real world verification of decreased precipitation associated with the decrease in vegetal cover which has occurred.

Reports of the collapse of the grasslands to near desert or desert conditions exist in many areas. W.H. Pearsall (1957) has shown evidence that in the Serengeti combined pressure of cattle and wild game is in places reducing the range to desertlike conditions.

Severe overgrazing by domestic cattle in Karamoja in northeastern Uganda has left a hard compacted surface in many areas. When the rains occur there is little infiltration and plants associated with arid conditions are on the increase. The quality of grazing conditions continues to build up. Stebling (1953, p. 29) placed the area of Sudan threatened with immediate loss as lying between 13 °N and 15 °N, with the area further north conceded as lost.

In the drought of 1940-1941, an estimated 340,000 square kilometers in West Africa collapsed to desert and in the ten-year period 1964-1974 (Brabyn, 1975, p. 6) the Sahara had encroached as much as 150 kilometers into former nomad grazing lands. In the same area during 1972-1973 the ecological conditions deteriorated to such an extent north of 15 °N that hardly any livestock survived and agricultural production was nonexistent.

Millions of animals and thousands of people died as a result of the recent drought. Of the people who survived, and most did, a sizeable portion is now dependent on international aid while another portion moved into the lands of other people. The continuous migration of people and animals into already crowded areas is creating a complex of social and economic problems. In Africa, a conflict between sedentary and nomadic populations over such things as water rights and land rights probably has always existed. Throughout the Sahelian zone this conflict has been intensified in recent years by the extension of sedentary agriculture into the more arid zones, and since 1968 by the migration of nomads and their livestock to the south as the drought dried up northern grassland.

The complexity of the problem is illustrated by conditions in the western plains of Sudan. The northern part of these plains is occupied by camel-owning nomads while the southern parts are occupied by cattle-owning nomads. Traditionally, the nomadic movements have been synchronized in such a way that hardly any contact between these two distinct groups took place. But recently, the camel owners found it necessary to remain for longer periods in the southern areas and they have increased the pressure on the southern grazing lands. This has contributed to the deterioration or the elimination of pastures that were traditionally reserved for the cattle of the southern nomads. The camel owners also used to spend the winter season close to the heart of the Sahara desert in the vicinity of the Sudanese-Libyan borders in order to graze the so-called "gizzu" grass but for the last five years the "gizzu" pasture has ceased to exist and so the nomads remained further south. This has led to what seems to be an irreversible deterioration of the ecosystem. Also as a result of competition for pasture and water the frequency of tribal conflicts is increasing at a very fast rate. The situation is further aggravated by the eastward migration of the West African tribes especially the Tchadian tribes who are exerting further pressure on the environment.

CONCLUSIONS

On a worldwide basis the demand for food is increasing at a rate greater than the increase in production. This alarming situation is causing a worldwide concern and thus in the 1970s the cry for rapid economic development is heard throughout Africa, Asia, and Latin America. Part of this is the demand for expansion of agricultural production. Proposals for increasing production per unit of land and for bringing more land under cultivation exist in almost every country. At the same time the plea for rapid economic growth is heard, we hear more frequent rumblings of a worldwide shortage of agricultural products. Even though production increases, the shortages on a worldwide basis are increasing. Both the desire to expand the economy and the existing marginal surplus of food dictate that there will be great pressure placed on the earth's arable land to yield more food and fiber in the years to come.

The natural tendency will be to expand agriculture to the limits of the environment as they exist at the present, or as they exist during optimum conditions. However, because of the natural climatic changes which occur, such practices are fraught with hazards, and may inevitably cause great hardships for reasons discussed earlier. It seems vital that future development of regional environments take into consideration the natural environmental fluctuations. The limits to agricultural expansion must be established at a level sustainable during periods of climatic downturns whether these are in terms of moisture or temperature, depending on which of these factors is the most critical. There are two primary reasons why this must be done. The first is for obvious humanitarian reasons. Where the human species has found it acceptable to allow wild animal populations to be controlled by starvation during drought, it is not acceptable for the human species to allow the same thing to happen to its own kind. In the past and to a limited extent, at present, when starvation faced groups of people it was possible to migrate to better areas, or food supplies from other areas could be brought to the stricken region. These two alternatives are all but gone. In the first instance virtually all the most productive land in the world is already in use, and there is no new land to move to. In the second, while there have been massive food shipments to distressed areas in the decades since World War II, these are not likely to continue. These massive shipments have been possible due to the huge surpluses of grain that were produced. The world demand for food has increased to the point where the world's supply of surplus grain is of only a few months duration. Large surpluses simply will not be available to move to distressed areas.

The second reason why the intensity of economic development must be held to an equilibrium level is to prevent further deterioration of the environment. The present inadequate carrying capacities of the African tropical environments are under considerable strain and there is strong evidence to suggest that future environments may have even lower carrying capacities than those of the present day. Severe overgrazing or improper agricultural practices during time of environmental stress can disturb the ecosystem to the point that it may not be able to recover following the disturbance. It is this very process of overdevelopment, and destruction of environmental equilibrium, which has led to extensive abandonment of the grasslands of the world and to accelerated expansion of the deserts alluded to earlier.

The only long-range solution to the problem is to keep land use at an intensity level substantially below that which has traditionally been allowed. The levels of animal populations must be kept below the maximum carrying capacity represented by wetter-than-normal years. This is the only mechanism which will provide immediate relief from the drought hazard and at the same time give the pastures a chance to survive. The determination of these critical levels requires a thorough understanding of the regional environment as well as a close monitoring of the changing environments especially in these sensitive grassland environments.

REFERENCES

Brabyn, H. "Drama of 6000 Kilometers of African Sahel." *Unesco Courier* 28 (April 1975):4-9.

Bugnicourt, J. et al. "Sahel." *Unesco Courier* 28 (1975):10-32.

Charney, Jule; Stone, Peter H.; and Quirk, W.J. "Drought in the Sahara: A Biophysical Feedback Mechanism." *Science* 187 (1975):434-435.

Dow, T.E., Jr., "Famine in the Sahel: A Dilemma for United States Aid." *Current History* 68 (1975):197-201.

Dresch, Jean. "Reflections on the Future of the Semi-arid Regions." *African Environment.* London: International African Institute, 1975.

Dubois, Victor D. "The Drought in Niger: Part I." Also "Part II, The Overthrow of President Hamani Diori." Field Reports, American Universities Field Staff, Inc., vol. 15, 1974.

———. "Food Supply in Mali." *Field Reports* West African Series, no. 16. American Universities Field Staff, 1975.

El-Tom, Mahdi-Amin. Personal Communications, 1975.

Halwagy, R. "The Impact of Man on Semi-desert Vegetation in the Sudan." *Journal of Ecology* 50 (1962):263-275.

Lundholm, Bengt. "Environmental Monitoring Needs in Africa." *African Environment.* London: International African Institute, 1976.

Meadows, D.H.; Meadows, D.L; Randers, J.; and Behrens, W.W. *The Limits to Growth.* Washington, D.C.: Potomac Associates, 1972.

Mesarovic, M., and Destel, E. *Mankind at the Turning Point.* New York: Signet Books, 1974.

Miller, Norman N. "Journey in a Forgotten Land." Part I Food & Drought in the Kenya, Ethiopia Border, and Part II: Food and Drought: The Broader Picture. Field Reports, American Universities Field Staff, no. 19. Northeast Africa Series 6, 1974.

Ormeroo, W.E. "Ecological Effect of Control of African Trypanosomiasis." *Science* 191 (1976):815-821.

Prospero, J.M., and Carlson, T.N. "Vertical and Aerial Distribution of Saharan Dust Over the Western Equatorial North Atlantic Ocean. *Journal of Geophysical Research* 77 (1972):5255-5265.

Ripley, E.A. "Drought in the Sahara: Insufficient Biophysical Feedback?" *Science* 191 (1976):100-101.

Sansome, H.W. "The Trend of Rainfall in East Africa." East African Meteorological Department. Technical Memorandum #1, 1952.

Sheets, Hal. "Disaster in the Desert: Failures of International Relief in the West African Drought." *Humanitarian Policy Studies.* Washington D.C.: The Carnegie Endowment for International Peace, 1974.

Stebbing, E.P. "The Creeping Desert in the Sudan and Elsewhere in Africa." *Sudan Ltd.* McCorquodale and Co., 1953.

Sutton, L.J. "Temperature Trends in Egypt and Sudan." *Quarterly Journal of the Royal Meteorological Society* 62 (1936):120-122.

Temple, Paul H. "Lake Victoria Levels." *Proceedings of the East African Academy* 2 (1964):50-58.

Thomas, A.S. "The Vegetation of Some Hillsides in Uganda: Illustrations of Human Influence in Tropical Ecology." *Journal of Ecology* 33 (1945):153-172.

The Leaf Energy Budget Equation
A Review of the Literature

J.M. Davis
Indiana University

Gates (1968) presented one form of the complete energy balance equation for a leaf. This now familiar equation is given as:

$$Q_{ABS} = \epsilon\sigma T_1{}^4 + h_c(T_1 - T_a) + L(T)\,\frac{{}_s\rho_1(T_1) - \text{r.h.}{}_s\rho_a(T_a)}{r_1 + r_a}$$

where
Q_{ABS} = absorbed radiation (ergs cm^{-2} s^{-1})

h_c = convective heat transfer coefficient (ergs cm^{-2} s^{-1} K^{-1})

L = latent heat of vaporization of water (ergs g^{-1})

ϵ = leaf emissivity

σ = Stephan-Boltzmann constant (5.67×10^{-5} ergs cm^{-2} s^{-1} K^{-4})

T_1 = leaf temperature (K)

T_a = air temperature (K)

r.h. = relative humidity

r_1 = resistance to water vapor diffusion from the mesophyll cell walls through the stomate (s cm^{-1})

r_a = resistance to water vapor diffusion as a result of the leaf boundary layer (s cm^{-1})

${}_s\rho_1(T_1)$ = saturation water vapor density at leaf temperature in the intercellular air spaces of the leaf (g cm^{-3})

$\text{r.h.}{}_s\rho_a(T_1)$ = vapor density of the free air beyond the boundary layer of the leaf (g cm^{-3}).

Two of the most difficult terms to determine in this equation are the convective heat transfer coefficient and the stomatal resistance. Much recent literature has been directed towards these two terms and they will also be the primary focus of this review.

The first major portion of this paper deals with recent work proposing realistic alternate forms for the convective heat transfer coefficient. It often has been assumed that the convective heat transfer coefficient for a leaf can be modeled by using the Pohlhausen solution for laminar flow over a flat metal plate of uniform temperature. However, many problems are encountered when modeling a leaf, either as an individual item or in a canopy: the surface of a leaf is generally not isothermal (a direct result of the boundary-layer effect on the leaf), leaves have a distinct tendency to flutter in the wind (a property rarely displayed by a flat metal plate), leaves are rarely flat, and, in addition, the wind seldom blows exactly parallel to the leaf surface. Since the leaf is such a complex biological organ (Esau, 1965), the determination of its convective heat transfer coefficient requires a more sophisticated approach.

In recent years, stomatal mechanisms have generated much research interest, a large portion of which has been reviewed by Raschke (1975). The second main section of this review is directed toward literature concerned with the response of the stomate to environmental factors, particularly toward work beginning in the mid-sixties on the effect of exogenous environmental factors on stomatal resistance via the peristomatal mechanism. Much research has also been done on what Stalfelt in 1929 called the hydroactive and hydropassive processes. In this paper, these terms are used as Raschke (1975) used them. Hydropassive processes refer to turgor changes brought about by changes in the leaf water potential, the guard cell solute content remaining the same, and the word hydroactive describes the response to water stress when the guard cell solute content changes.

Other aspects of the stomate which will be considered include stomatal mechanisms analyzed in field research. Furthermore, recent years have witnessed a substantial increase in models of various aspects of the plant in its environment. Because of its role as a "control valve," the stomate has received wide attention in these models. Some of the means used to arrive at realistic values for stomatal resistance in these models will be reviewed.

Because of the vast amount of literature on these subjects, this review will necessarily be limited. Primary attention has been given throughout to the most recent literature in the field.

THE CONVECTIVE HEAT TRANSFER COEFFICIENT

The convective heat transfer from a leaf plays a vital role in the energy balance of the leaf. Kreith (1973) defines the rate of convective heat transfer to be:

$$q_c = \bar{h}_c A \Delta T \tag{1}$$

where q_c = rate of heat transfer by convection

A = heat transfer area

$\Delta T = (T_1 - T_a)$, where T_1 is the leaf temperature and T_a is the air temperature

$\bar{h}_c$ = average convective heat transfer coefficient

The term $\bar{h}_c$ depends on the geometry of the surface, the velocity of the moving fluid, the physical properties of the fluid and at times, the temperature differences. The convective heat transfer coefficient may be different from one point to the next on the surface. It is therefore possible to have a local convective heat transfer coefficient which is defined by:

$$dq_c = h_c dA(T_1 - T_a). \tag{2}$$

Heat transfer from one object can occur by either free or forced convection (Kreith, 1973; Kays, 1966). Because the amount of energy transferred by free convection is relatively small except at low wind speeds, only the forced convection case will be considered. Heat transfer engineers have determined that the convective heat transfer coefficient is a function of three nondimensional groups:

1. the Reynolds number, Re
2. the Nusselt number, Nu
3. the Prandtl number, Pr.

The general relationship for forced convective heat loss from spheres, planes, and cylinders, can be given by:

$$Nu = Pr^m Re^n \tag{3}$$

where and n and m are numerical constants. The Prandtl number, which specifies the relationship between the temperature and velocity distribution, has a numerical value near 0.72 for air.

Engineers (Kreith, 1973) have also shown that the relationship between the average convective heat transfer coefficient and the Nusselt number is given by:

$$\overline{h}_c = \overline{Nu}_L \frac{k}{L} \tag{4}$$

where k = thermal conductivity of the fluid
L = length dimension
$\overline{Nu}$ = average Nusselt number.

The local Nusselt number is given by

$$Nu_x = \frac{h_{cx}\, x}{k} = 0.332\, Re_x^{0.5} Pr^{0.33} \tag{5}$$

where h_{cx} = local heat transfer coefficient
x = a specific point on the object under investigation.

The average Nusselt number over a plate of length L is found to be:

$$\overline{Nu}_L = 0.664\, Pr^{0.33} Re_L^{0.5}. \tag{6}$$

Equation 6 is often referred to as the Pohlhausen solution, which holds for laminar flow over a flat plate with a uniform surface temperature. For a surface assumed to have a uniform heat flux, the equation is given by

$$\overline{Nu}_L = 0.680\, Pr^{0.33} Re_L^{0.5}. \tag{7}$$

For turbulent flow in a forced convection regime over a flat plate in parallel flow the equation in given by

$$\overline{Nu}_L = 0.036\, Re^{0.8} Pr^{0.33}. \tag{8}$$

The use of the Pohlhausen solution has generally been received with some skepticism for several reasons. First of all, is it safe to assume that the flow across the leaf is laminar at all times or at any time? Secondly, does a flat stationary plate in a wind tunnel provide a realistic model for a leaf in the environment?

Parkhurst et al. (1968) undertook the calculation of convection from broad leaves of arbitrary shape. Their basic assumption was that the equations for a flat plate in laminar flow were acceptable if the mean length in the flow direction were calculated and then weighted, based on the convection that would be expected from a rectangle of the length being averaged. This mean length would then be substituted into the rectangular plate equations. The results showed that for various shapes of leaves (electrically-heated copper models), convection does depend on $Re^{0.5}$. The electric heating wires were arranged to provide a uniform heat flux. However, due to lateral heat conduction the model was brought towards an isothermal state. Thus, data points of Nu versus Re_L were expected to lie between the uniform heat flux case and the uniform temperature case. This occurred in the majority of cases. The h_c value was estimated from an energy conservation relationship. It was shown that when the weighted-mean length was used to calculate Nu and Re, the relationship between them was close to the expected theoretical values.

The tilting of leaves in the wind was also examined since wind blowing over a leaf rarely remains parallel to the surface. Long narrow leaf convection was more affected than round leaf convection. It was also found that convection was increased when the model was turned about the shorter dimension and decreased when it was turned about the longer dimension. The results of simulated leaf flutter in the presence of other leaves indicated that flutter did not have a significant effect on convection for the case of a disk.

Pearman et al. (1972) set out to test whether the convective heat transfer coefficients which were obtained in natural winds over agricultural and nonagricultural surfaces were higher than those predicted from laminar heat transfer theory. Circular metal disks were used in the field work.

The authors employed three methods to obtain the heat transfer coefficients. One of these, the total energy balance method, yielded an estimate of the heat transfer coefficient which included both the mean heat transfer coefficient and an eddy covariance term.

The sensible heat flux density, H, is given by:

$$H = \overline{h}\ \overline{\Delta T} + \overline{h'\ \Delta T'} \tag{9}$$

where h = mean heat transfer coefficient
$\overline{\Delta T}$ = mean temperature difference
$\overline{h'\Delta T'}$ = covariance between the instantaneous deviations of h' and $\Delta T'$ from $\overline{h}$ and $\overline{\Delta T}$.

Their results showed that the covariance term was an insignificant transporter of convective heat under the range of conditions investigated. The effects of free convection were found to be insignificant for the data used in the study. The experimental results obtained were compared with the theoretical Nu for a circular disk of uniform temperature which is given by

$$\overline{Nu} = 0.74\, Pr^{0.33} Re^{0.5}. \tag{10}$$

The best-fit regression equation obtained in the Pearman et al. study was

$$\overline{Nu} = 1.08\, Pr^{0.33} Re^{0.5}. \tag{11}$$

This work demonstrated that the convective heat loss from a metal circular disk of uniform temperature in natural wind conditions was about 1.5 times greater than predicted by the laminar flow flat plate theoretical value. The authors attributed the increase to the turbulence in the natural wind. Thus, temperature differences between leaf and air may be less than the laminar theory predicts.

Monteith (1965) found that the leaf boundary-layer resistance was given by:

$$r_a = \frac{3.4\, L}{Re^{0.5}} \tag{12}$$

which when put into a form relating Nu to Re and Pr is given by:

$$\overline{Nu} = 1.55\, Pr^{0.33} Re^{0.5}. \tag{13}$$

Examination of the deviations of actual leaf convective heat transfer from the value indicated by the Pohlhausen solution was continued in papers by Parlange et al. (1971) and Parlange and Waggoner (1972).

In Parlange et al. (1971), the average boundary-layer resistance (the Pohlhausen case) was given as:

$$\overline{r_T} = \left[\frac{Pr^{0.67}}{0.664}\right] \left[\frac{1}{Re}\right]^{0.5} \tag{14}$$

The assumption of a uniform heat flux was felt to be more realistic than the assumption of a uniform temperature difference between plate and air upon which the Pohlhausen solution was based.

Theory predicts that if the variation in evaporation over the surface is small, the heat flux will be nearly constant and the leaf temperature will change with the square root of the distance from the leading edge. The average boundary-level resistance can be written to include the effect of turbulence (β) in the constant flux situation:

$$\overline{r_T} = \left[\frac{Pr^{0.67}}{0.680\beta}\right] \left[\frac{1}{Re}\right]^{0.5} \tag{15}$$

It is by coincidence that 0.664 and 0.680 are nearly the same, which as the authors pointed out, will yield nearly the same result for either the constant flux or the constant temperature approach.

The laboratory results revealed that for a still leaf in extremely turbulent air the value of β ranged from 2.4 ± 0.4 to 2.7 ± 0.4. Thus when compared to a laminar flow situation, the turbulence decreases the resistance by a nearly constant factor of 2.5. Leaf temperatures obtained in the laboratory were relatively close to those predicted by theory. Flapping leaves in extremely turbulent air were also investigated. It was found that the value was again near 2.5, indicating that there is little effect on heat transfer due to leaf flutter. Field experiments indicated the β value was again equal to 2.5, whether or not there was leaf flutter.

Schuepp (1972) used electrochemical methods in an electrolytic tunnel to investigate the forced convection mass and heat transfer from leaf models at Reynolds numbers $2 \times 10^3 < Re < 4 \times 10^4$. He compared his results with those obtained by using the transfer coefficients calculated from the laminar boundary-layer theory and found that, based on a similarity between convective heat and mass transfer, the following conclusions could be stated:

1. Convective transfer conditions from a single medium-sized leaf could not be adequately modeled by rigid plates. In the same way, clusters of plates do not adequately model leaves in a crop.
2. For plate models of leaves, the effect of flutter on the transfer coefficients depended on the eccentricity of the plate. Transfer coefficients were increased by a factor of 1.2 for disk and 1.8 for elliptical plates.
3. At Reynolds numbers $2 \times 10^3 < Re < 2 \times 10^4$ (possibly a transition zone from laminar to turbulent flow; see Monteith, 1965), convective heat and mass transfer coefficients were higher than the ones calculated on the basis of laminar flow by a factor of 1.4 ± 0.1 for plates which had the same surface area and outline as leaves.
4. Transfer coefficients of leaves in a crop may be greater by a factor approaching 2, which more than likely represents the upper limit.

It should be noted that the characteristic length concept of Parkhurst (1968) was used in the Schuepp (1972) study.

When comparing the Parlange et al. (1971) and the Schuepp (1972) results, it is interesting to note that in the fully turbulent flow regime of Parlange, the convective heat transfer enhancement factor was about 2.5, while in the transition zone investigated by Schuepp the enhancement factor was about 1.4. It seems likely that the 2.5, rather than the value of 2 suggested by Schuepp, represents an upper limit to the factor.

Clark and Wigley (1975) also investigated the validity of the application of engineering-derived heat transfer equations to heat loss from leaves. The local Nusselt number for constant flux heat transfer from a thin flat plate in laminar flow is:

$$Nu = 0.453\, Pr^{0.33} Re^{0.5}. \tag{16}$$

The authors in an earlier paper (Wigley and Clark, 1974) had reported deviations from the above equation when measuring the heat transfer from realistically-shaped models of leaves in turbulent and laminar flow. Wind tunnel results for parallel turbulent flow provided the following relationship for the local Nusselt number:

$$Nu = 0.045\, Pr^{0.33} Re^{0.84} \tag{17}$$

For laminar flow the turbulent intensity was about 1% to 2%, while within the working section of the wind tunnel, the turbulent intensity was between 30% and 40% for turbulent wake flow. Below local Reynolds numbers of 10^3, the data for both parallel laminar flow and parallel turbulent flow all fell along the same line, indicating a similarity in the transfer in both types of flow below a Reynolds number of 10^3. The equation which fitted the low Reynolds number flow is equation (17) above. In laminar parallel flow between Reynolds numbers of 10^3 and 2×10^4, the emprically fitted equation was

$$Nu = 0.22\,Pr^{0.33}Re^{0.6}. \qquad (18)$$

The results of the Wigley and Clark (1974) work indicated that deviations from the theoretical equation

$$Nu = 0.453\,Pr^{0.33}Re^{0.5} \qquad (19)$$

occurred for heat transfer from model leaves even in laminar flow.

Clark and Wigley (1975) calculated temperatures at a point (L) downwind on a dry flat surface in a 1 m s^{-1} wind, using three approaches to obtain the Nusselt number. These were:

1. Heat transfer from isothermal surfaces where the Nusselt number is expressed as

$$\overline{Nu} = 1.08\,Pr^{0.33}Re^{0.5} \qquad (20)$$

2. Local heat transfer from a constant flux surface where the Nusselt number is expressed as

$$Nu = 0.453\,Pr^{0.33}Re^{0.5}. \qquad (21)$$

3. Local heat transfer from a constant flux surface where Clark and Wigley (1975) determined that the Nusselt number could be expressed from field work as:

$$Nu = 0.10\,Pr^{0.33}Re^{0.75}. \qquad (22)$$

Substantial differences between the local temperatures in the constant flux case and the temperatures predicted for an isothermal surface were found.

The results stated in both Wigley and Clark (1974) and Clark and Wigley (1975) indicate that simple multiples of the Pohlhausen equation are not adequate to describe convective transfer in turbulent or natural conditions. Clark and Wigley (1975) stated that local deviations from the Pohlhausen equation are the result of the effects of leading and trailing edges. On large leaves, the effects will be minimal. However, the effects are important for grass and cereal leaves.

Further work on the effects of turbulence on heat transfer was carried out by Parkhurst and Pearman (1974), who tested whether or not convective heat transfer from a flat plate in wind tunnel flow would be enhanced by the action of periodic air flow partially simulating large turbulent intensities (of a single frequency) which are routinely found under natural conditions. The flat plate used was "semi-infinite," i.e., it crossed the wind tunnel completely, thus eliminating end effects. As the authors point out, caution is required in extrapolating the experimental results to real leaves.

For a plate of uniform temperature, the average convective heat transfer coefficient is

$$\overline{h_c} = 3.949 \times 10^3 \frac{(V)^{0.5}}{L}, \text{(ergs cm}^{-2}\text{ s}^{-1}\text{ K}^{-1}) \quad (23)$$

while for the constant heat flux case, the coefficient is

$$\overline{h_c} = 4.049 \times 10^3 \frac{(V)^{0.5}}{L}, \text{(ergs cm}^{-2}\text{ s}^{-1}\text{ K}^{-1}) \quad (24)$$

These differences are negligible in practice. The physical characteristics of the plate they used would place it between these two cases.

The results obtained for steady parallel flow indicated that empirical $\overline{h_c}$ values were about 13% greater than the values predicted by the above equations. The authors state that a possible reason for the discrepancy could have been an alignment error resulting in a slight deviation from parallel flow ($\alpha = 0$). As was increased from 0° in steady flow, the transfer coefficient increased to a maximum value at an angle of about 4° to 8°. The maximum value obtained was about 30% higher than would be expected for steady parallel flow. At higher angles, the coefficient decreased again. In the fluctuating parallel flow regime (velocity fluctuations of 10 and 20 Hz), the transfer coefficients remained about 14% above the values anticipated for steady parallel flow. The authors again assume some alignment error whereby $\alpha \neq 0$.

It was found that in the fluctuating nonparallel flow case, the heat transfer reaction to changes in α was similar to that found in steady nonparallel flow. However, the heat transfer values were higher in the fluctuating case. Maximum convection was found to occur when α was between 4° and 8°, regardless of the frequency. Convection at 20 Hz was greater than it was at 10 Hz for all angles. When was between 4° and 8°, the maximum increase was found to be 50%, relative to flux expected from laminar theoretical equations.

Aston et al. (1973) made a comparison of the heat transfer coefficients and resistances obtained using relationships given by Gates (1962), Raschke (1960), and Thom (1968). Raschke (1960) had presented an equation which showed the relationship between the convective heat transfer coefficient and the boundary-layer thermal resistance to heat transfer. The authors collected their data in a Flexiglas growth chamber in which air temperature, wind speed, humidity, and visible and thermal radiation were measured. From the data obtained in the experiment, the convective heat transfer coefficient was calculated using the relationships derived by Gates and Thom. Boundary-layer thermal resistances were then calculated using the Raschke equation and a similar equation derived by Thom. The results showed that the Gates and Thom methods of calculating the convective heat transfer coefficient yielded almost identical results, providing the area basis on which they depended included both leaf surfaces. Unfortunately, no experimental measurements of the convective heat transfer coefficient were made for comparison to the calculated values.

Perrier et al. (1973) made an important contribution to boundary-layer flow theory by comparing the wind tunnel flow characteristics over a soybean leaf with the flow characteristics over an artificial leaf (flat metal plate) at various velocities of the bulk air stream.

Free stream wind speeds of 39, 148, and 271 cm s^{-1} were used in the wind tunnel. At the low free stream velocities, there was little difference in the velocity profile shapes for the soybean and metal leaf. At 271 cm s^{-1}, the wind forces on the former initiated flutter, while the metal leaf showed no movement.

The authors pointed out that when fluttering began, the boundary layer was disrupted: its thickness almost doubled and it tended to separate and then reform. The turbulence level at which leaf flutter was initiated depended on such parameters as leaf geometry, size, curvature, roughness, and rigidity. As might be expected, Perrier et al., found that flow across the metal plate was always homogeneous—the turbulent eddies were statistically uniform in size. The soybean leaf could be described as homogenous only at the lower velocity, and only at the lowest velocity were the flow fields across the soybean leaf and metal leaf similar. When the free stream wind speed over the soybean leaf was 271 cm s^{-1}, the flow was turbulent, but it was not so at the lower wind speeds. The shape factor (δ^*/δ_m) was found to be the best indicator of laminar or turbulent flow, with δ^* being the displacement thickness and δ_m the momentum thickness. The major conclusion to be drawn from the paper is that the transition from laminar to turbulent flow occurs at a lower Reynolds number for a real leaf than for an artificial one.

Most of the experimental work cited in this paper indicates that under a variety of flow conditions the Nusselt number is proportional to the square root of the Reynolds number. In the laminar flow case, the relationship between the Reynolds and Nusselt numbers is given by equation (16) for a constant flux surface. In turbulent conditions, this coefficient has been found to increase to a value roughly between 1.5 and 2.5, thus suggesting the use of a multiplicative factor β to model turbulent flow. Wigley and Clark (1974) have questioned the entire concept of applying the β factor to the laminar flow equation to model turbulent flow. They found that in turbulent flow, the Reynolds number entered at the 0.84 power, which is close to the theoretical turbulent flow value of 0.8. A major aspect of the problem is based on whether the leaf boundary layer is laminar or turbulent. Parkhurst et al. (1968) have assumed that it is laminar, while Wigley and Clark (1974) have assumed it to be turbulent. The results of the latter raise several questions. At Reynolds numbers below 10^3 the similarity in transfer for both parallel laminar and turbulent flow might be expected, as the authors state, if the artificially-induced turbulence were being damped out. If it were assumed that the transfer in both flows is similar, one would expect the Reynolds number to enter at the 0.5 power rather than at the 0.84 power. It should also be noted that artificially-induced turbulence in a wind tunnel is not an adequate substitute for the complex situation existing in the natural air.

In most instances leaf flutter had little effect on the heat transfer coefficient. In nonparallel flow, heat transfer was found to be enhanced at incident angles between 4° and 8° but decreased at higher angles.

STOMATAL RESISTANCE

This section of the paper is concerned with the portion of the leaf energy equation which accounts for heat loss by transpiration. The equation is given by Gates (1968) as:

$$E = \frac{{}_s\rho_1 (T_1) - r.h.{}_s\rho_a (T_a)}{r_1 + r_a} \qquad (25)$$

where E = rate of transpiration (g cm^{-2} s^{-1})

${}_s\rho_1(T_1)$ = saturation vapor density at leaf temperature in the intercellular air spaces of the leaf

$r.h.{}_s\rho_a(T_a)$ = vapor density of the free air beyond the boundary layer of the leaf

r_1 = resistance to water vapor diffusion from the mesophyll cell walls through the stomate (leaf resistance)

r_a = boundary-layer resistance to water vapor diffusion.

The analogy between the above equation and the Ohm Law is readily apparent. The driving force in the above equation is provided by the vapor density gradient between the intercellular air spaces and the free air. Resistance to vapor movement is offered within the substomatal cavity and in the boundary layer over the surface of the leaf. The flux of water vapor ($g\ cm^{-2} s^{-1}$) is thus dependent on the concentration gradient and the resistance to flow. In Ohm's Law, the electrical current is dependent on the voltage (potential) gradient and resistance to current flow.

The boundary-layer resistance is given by Gates (1968) as:

$$(\text{CONSTANT}) \times \frac{D^{.35} W^{.20}}{V^{.55}} \qquad (26)$$

where D = leaf dimension in the direction of the wind (cm)

W = leaf dimension transverse to the wind (cm)

V = wind speed (cms^{-1}).

The constant term depends primarily on the W dimension. Values for the boundary-layer resistance range from about 0.1 s cm^{-1} to 1.0 cm^{-1}. The values for r_a are generally less than those for leaf resistance, r_l.

In the above equation for the rate of transpiration, the vapor density values are usually easily obtainable, except with halophytic plants, for which water potential must be considered (Taylor, 1971). It is also generally true that the boundary-layer resistance is about an order of magnitude less than the leaf resistance. Thus, an accurate determination of the leaf resistance is critical to an accurate estimation of the transpiration equation. There is abundant plant physiology literature on the stomate and how it functions. The most recent detailed review was done by Raschke (1975), while Meidner and Mansfield (1968) have devoted an entire book to the study of the various aspects of the stomate. Good discussions of stomatal processes can be found in most plant physiology books. For example, an excellent recent discussion is contained in Leopold and Kriedemann (1975). Stomatal resistance in general and in relation to plant water conditions is treated by Etherington (1975), Milthorpe and Moorby (1975), Hsiao (1974), Monteith (1973), Kramer (1969), and Slatyer (1967).

This section will review certain portions of the stomate-related literature which have not received as much attention as certain other areas. In particular, stomatal resistance will be examined from the following viewpoints:

1. How are realistic esimates of stomatal resistance obtained?
2. How do environmental factors affect stomatal resistance?

ELECTRICAL ANALOGUE OF STOMATAL RESISTANCE

Nobel (1974) and others decompose r_l into three component parts:

1. r_s, the stomatal resistance,

2. r_i, the intercellular air space resistance, and
3. r_c, the cuticular resistance.

Values for r_s usually range from about .5 to 5.0 s cm^{-1} to more than 20 s cm^{-1} for some tree species. r_c generally ranges from 30 s cm^{-1} to 200 s cm^{-1}, and r_a values are generally less than 1 s cm^{-1}.

Of these three, the stomatal resistance is the most important in controlling the rate of transpiration from the leaf and will therefore receive the primary focus of this review.

In electrical circuit theory, it is standard practice to "draw" the resistance network under consideration and to resolve the network by calculating a total resistance which takes into consideration the series and parallel combinations of all the resistance. Such a network can also be constructed for the plant water vapor network:

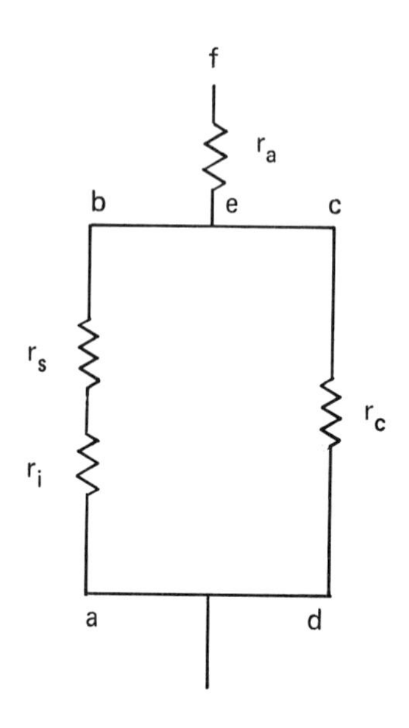

In constructing this diagram it has been assumed that for the transpirational network, the mesophyll resistance can be neglected (Slatyer, 1967).

For resistances in series, the total resistance is obtained by adding the value of all the individual resistances together. Thus, the resistance offered to water vapor flow in path ab is:

$$r_{ab} = r_s + r_i, \text{ while} \tag{27}$$

$$r_{cd} = r_c.$$

For resistance values in parallel, the reciprocal of the total resistance is equal to the sum of the reciprocal values of the component resistance, i.e.,

$$\frac{1}{r_t} = \frac{1}{r_1} + \frac{1}{r_2}, \text{ or } r_t = \frac{(r_1)(r_2)}{r_1 + r_2}. \tag{28}$$

Thus in our plant network the resistance for the block abcd would be the leaf resistance and would be given by

$$r_l = \frac{(r_i + r_s)(r_c)}{r_i + r_s + r_c} \quad . \tag{29}$$

Since it is generally true that

$$r_c >> r_i \text{, and } r_c >> r_s \text{,} \tag{30}$$

the above equation can be approximated by

$$r_l = r_i + r_s . \tag{31}$$

If the boundary-layer resistance is considered, then

$$r_{total} = r_l + r_a . \tag{32}$$

Following the development given by Nobel (1974), it can be shown that equation (32) is the leaf resistance for a hypostomatous leaf. For an amphistomatous leaf, another path (a parallel path in the resistance network sense) is open for the diffusion of water vapor from the leaf. The total resistance ($r_l + r_a$) in this case can be written as:

$$r_{total} = \frac{(r_l + r_a)_U (r_l + r_a)_L}{r_{lU} + r_{aU} + r_{lL} + r_{aL}} \tag{33}$$

where L = lower leaf surface, and
U = upper leaf surface.

This expression can be simplified by assuming that the boundary-layer resistances are the same on the top and bottom of the leaf. Further simplification could be obtained if water vapor could move equally well through the stomates at the top or bottom of the leaf—this implies equal resistance pathways. If resistance to the diffusion of water vapor were much greater through one side as opposed to the other, the resistance to diffusion through the low resistance path should adequately describe the resistance.

It should be noted that treating the plant resistance system in a way analogous to an electrical system does not eliminate the problem of obtaining values for the various resistance terms in the equation. Once estimates have been obtained for the various resistance components, the above equation permits the combination of components into one total leaf resistance value.

CALCULATION OF STOMATAL RESISTANCE

Nobel (1974), Meidner and Mansfield (1968), and Slavik (1974) give equations for the calculation of r_s based on the geometry of the stomata.

The equation given by Slavik (1974) is:

$$r_s = r_{tube} + r_{end} = \frac{4L}{\pi D d^2} + \frac{1}{2Dd} = \frac{1}{DA}\left(L + \frac{\pi d}{8}\right) \tag{34}$$

where r_{tube} = resistance of the stomatal tube itself.
r_{end} = resistance outside the ends of the tube
A = cross-sectional area of the diffusion pathway
d = diameter of stomatal tube
D = water vapor diffusion coefficient
L = length of diffusion path.

The end resistance is added to account for the fan-shaped diffusion flow lines that occur at the ends of the stomatal tubes. Diffusion from the ends of the tube was found to be proportional to the diameter of the tube and not to the cross-sectional area of the tube.

Some doubt remains about the validity of including the end effect in the calculation. Lee and Gates (1964) observed that there are two situations in which diffusion from the ends of a pore are proportional to the diameter or circumference of the pore:

1. when the thickness of the pore (L) is very small;
2. when the pore diameter to length ratio remains constant with a change in the pore radius.

The authors stated that neither condition is characteristic of leaf epidermis.

A comprehensive model of guard cell motion based on mechanical analysis has been devised by DeMichele and Sharpe (1973, 1974), who analyzed all the forces that act on the walls of the guard cell. The origin of the stomatal bending motion when guard and subsidiary cells are turgid was explained on the basis of moment analysis. Since guard cells come in many shapes and sizes, it was necessary for the authors to base their work on one particular form.

The conceptual framework for the model is based on the similarity between the guard cell and a rectangular beam. When the stomata open, the beam must deform in such a way that a pore is created between the two adjacent guard cells. As a test of the applicability of the model, the stomatal response to water stress was analyzed.

The authors derived an equation for stomatal resistance which depended upon stomatal dimensions, the diffusion coefficient for water vapor, the modulus of elasticity of the cell wall material, the moment of inertia of the guard cell about its neutral axis, and the resultant force acting upon the ventral wall. The last two equation components were obtained from other equations derived by the authors which again depended upon stomatal dimensions in the first case, and upon the guard and epidermal cell turgor pressures in addition to stomatal dimensions in the second case. Based on these equations the stomatal resistance could be calculated for various sizes and shapes of stomata under various water stress conditions.

A comparison of stomatal aperture and water potential using the model yielded results which were consistent with experimental results: a maximum stomatal aperture at a leaf water potential of -6.75 bars was found.

DeMichele and Sharpe point out that the length of the pore semicircumference has a great influence on the stomatal resistance because of its effects on pore area and the force that is needed to cause guard cell opening. The results showed that a nonlinear relationship exists between pore semicircumference and resistance for any given leaf water potential. Generally speaking, at a leaf water potential of -6.75 bars, the greater the length of the pore semicircumference, the lower the stomatal resistance. Guard cell size was found to have a major effect upon resistance for any given leaf water potential. At a water potential of −6.75 bars, the larger the guard cell, the lower the stomatal resistance.

MODELING STOMATAL RESISTANCE

In recent years there has been a significant increase in the number of models developed to describe plant physiological processes. Any model which contains a component to describe the water economy of the plant must come to grips with the problem of specifying the relationship between stomatal resistance and various plant and climatic factors. This section will discuss the methods that have been employed in various models to obtain realistic estimates of the stomatal resistance.

Shawcroft et al. (1973, 1974) presented a stomatal model based on field data that were collected for their soil-plant-atmosphere model (SPAM). A hyperbolic equation was fitted to the field data which gave the stomatal resistance functional relationship as

$$r_s = a_o + b_o/I \tag{35}$$

where r_s = stomatal resistance ($s\ cm^{-1}$)
I = light flux density in μ einstein $cm^{-2}s$
a_o, b_o = regression coefficients ($s\ cm^{-1}$, einstein cm^{-3}, respectively).

The a_o can be considered to be a minimum resistance value at high light intensity. This equation was considered to be the ideal no-stress case and showed that over a large range of light intensities up to full sunlight there is little change in the resistance. From the field data, it was observed that the minimum stomatal resistance increased as the degree of water stress increased. It should also be noted that as I becomes very small ($I \rightarrow 0$), the value of r_s becomes very large ($rs \rightarrow \infty$). To eliminate this problem, the authors introduced I_o, which corresponds to a maximum finite resistance, r_c (cuticular resistance) or to a resistance which occurs when the stomata are closed. To account for water stress and the condition when $I \rightarrow 0$, the following equation was proposed:

$$r_s = a + b_o/(I + I_o) \tag{36}$$

where a = f (water stress), and as $I \rightarrow 0$, $r_s \rightarrow r_c$.

The limitation to the above formulation was stated by the authors. It is an empirical relationship needed to describe a complex biological process and is not to be taken as an exact relationship. Subsequent testing of the stomatal resistance model revealed some of its weaknesses, and the authors concluded that further work was needed to produce a more realistic model.

Miller (1973) developed a model which calculates the energy budget and rates of transpiration, respiration, and gross and net photosynthesis for sunlit and shaded leaves. The model was based (with certain modifications) on the heat transfer for a single leaf as presented by Gates (1962). In the model, leaf resistance to water loss (r_l), is a function of leaf water deficit (WD) and absorbed solar radiation (S_{abs}). The model equations are:

$$r_l = r_{lmin} - WD/(10.75 - 40.0\,WD), \text{ and} \tag{37}$$

$$r_l = r_{lmin} - 0.002/(0.001 + S_{abs}). \tag{38}$$

The WD term was calculated as the difference between water uptake by the leaf and water loss by the leaf over a period of time, added to the previous water deficit. Model validation was carried out using data on the microclimate, stand structure, leaf temperatures and resistances collected in a low uniform red mangrove forest located in Florida south of Miami. Leaf resistances were important components in calculating the evapotranspiration rates. The model estimated the transpirational loss from the canopy to be about 0.12 cm per day. Calculated evapotranspiration rates (using the heat balance—Bowen ratio method) for the canopy were 0.15 cm per day.

Tenhunen and Gates (1975) used regression analysis to develop an equation to relate leaf diffusion resistance to light and temperatures. The basic data for the model were collected in the field at Douglas Lake in Michigan, where a diffusion porometer was used to obtain leaf resistance for broad leaves.

The form of the model was given by

$$R = a\left(\frac{1}{L}\right) + bT + c \qquad (39)$$

where R = leaf diffusion resistance from the cell walls to the leaf surface ($s\ cm^{-1}$)

a = empirical constant ($s\ cal\ cm^{-3}\ min^{-1}$)

L = incident shortwave radiant flux density ($cal\ cm^{-2}\ min.^{-1}$)

b = empirical constant ($s\ cm^{-1}\ 1C^{-1}$)

T = leaf temperature (C)

c = empirical constant ($s\ cm^{-1}$).

The authors suggest that physiologically both independent variables have their effects by means of the changes in the CO_2 concentration in the intercellular air spaces (IAS). Light has the effect of reducing the concentration of CO_2 in the IAS by increasing net photosynthesis according to a Michaelis-Menten type of response. IAS CO_2 concentration can either be increased or decreased by increasing temperature according to the effect of temperature on photosynthesis and respiration.

It was noted that a better form of the model might be:

$$RL = a + bTL + cL. \qquad (40)$$

Regression equations were derived for common milkweed (*Asclepias syriaca*) and red oak (*Quercus rubra*) leaves. The explained variance for the pooled regressions of resistance on temperature and reciprocal light was 0.6. Temperature accounted for about 5% of this value while light alone accounted for about 55% to 60% of the variation. The correlation coefficient for the light and temperature data was about 0.37. When the soil was dried from its original state (13 g water/100 g dry soil to 9 g water/100 g dry soil), the regression coefficients predicted for milkweed were significantly different and the response to temperature changed substantially.

Leaf resistance was found to decrease with increasing values of shortwave radiation intensity up to about 0.3 cal cm^{-2} $min.^{-1}$ for both the wet and dry plots of *Asclepias* at leaf temperatures of 15, 20, and 25C. Above the 0.3 level, leaf resistance was constant. It should be noted that in the wet plot, the temperature effect on the leaf resistance-radiation relationship was weak, while in the dry plot, at a given radiation value, the resistance decreased significantly as the temperature decreased. For *Quercus,* the leaf resistance also decreased with increasing shortwave radiation. The leaf resistance-radiation relationship was found to be strongly temperature dependent. One of the main problems with the expression is that it omits the influence of plant and soil water potential. Directly.

In a paper by Ripley and Saugier (1975), fluxes of sensible heat, water vapor, and carbon dioxide were calculated above a natural grassland using measured profiles of wind speed, temperature, humidity, and CO_2 concentration. Radiation and soil heat flux readings were also made. These data were related to soil, plant, and atmospheric factors, and several models were constructed. In the water model the following equations, based on field measurements, were used to calculate stomatal resistance:

$$r_s = 1.2/G^{(0.20-0.02\,\psi_p)} \qquad \text{for } \psi_p > -15 \text{ bars} \tag{41}$$

$$r_s = 1.2\,\text{EXP}\,(-0.14\,(\psi_p + 15))\;/G^{(0.20-0.02\,\psi_p)} \qquad \text{for } \psi_p < -15 \text{ bars} \tag{42}$$

where G = global radiation (ly min^{-1})

ψ_p = plant water potential (bars)

r_s = stomatal resistance (s cm^{-1}).

When $\psi_p = 15$, the second equation becomes the same as the first one. From equation (42), it is apparent that when ψ_p drops below -15 bars, the value of r_s increases rapidly. The incorporation of the water potential term in the radiation term allows for a more gradual response to light under water stress. Testing of the model indicated good overall agreement between model predictions and observed data.

Two of the most comprehensive models of stomatal mechanisms are those of Penning de Vries (1972) and Cowan (1972). Penning de Vries presented a model designed to simulate the rate of transpiration from a nongrowing leaf for a day under changing environmental conditions. A major portion of the study was devoted to modeling stomatal aperture. Cuticular resistance, which was fixed at 20 s cm^{-1}, and stomatal resistance were treated as parallel resistances. Stomata aperture changes are known to be the result of deformation in the guard cell walls caused by guard and subsidiary cell volume changes. Considering these deformation effects as additive, Penning de Vries obtained the stomatal aperture by summing the aperture due to guard cell volume and to subsidiary cell volume. One of the main working hypotheses was that the effects of those factors which caused changes in the stomatal aperture (leaf CO_2 concentration, leaf water potential, and light) were additive. It was also assumed that the subsidiary cells made proportional changes in the relative stomatal aperture based on their turgor. In relation to guard cells, it was assumed that the guard cell pressure potential consisted of three componenets which depended on light intensity, CO_2 concentration, and relative water content. The sum of the three pressure potentials is the total pressure potential of the guard cells. Not considering the subsidiary cells, a linear relationship was assumed between stomatal aperture and pressure potential. Light has a small direct influence. Its main influence is via photosynthesis by lowering leaf CO_2 concentration. The author also assumed that in the steady state regime a fraction of the relative stomatal aperture could be accounted for by leaf CO_2 concentration. Considering that the model was tested with turnip, for which many parameters and functions were estimated, the agreement between measured and calculated rates of transpiration can be said to be good. The main differences between measured and simulated values occurred at low light intensities.

Cowan (1972) constructed a model for oscillations (see Barr, 1971) in stomatal conductance. His primary concern was with water-based stomatal oscillations with periods of 20 to 100 minutes. The Cowan model, designed to account for the observed occurrence of steady oscillations in stomatal conductance in cotton, analyzed the propagation within the plant of a change in turgor brought about by a change in the evaporation rate. In order to investigate the behavior of the model, Cowan allowed four components to be perturbed: the potential rate of transpiration; root resistance to water uptake; the water potential in reservior which the roots draw on; and, the guard cell osmotic potential. The model was found to be successful in simulating the properties of sustained oscillations in stomatal conductance, transpiration rate, and water flux which are observed in real plants.

FIELD STUDIES OF STOMATAL RESISTANCE

In a series of papers (Brown and Rosenberg, 1970a and 1970b, and Rosenberg and Brown, 1973), Rosenberg and Brown carried out field studies on parameters relating to stomatal resistance.

In the first 1970 paper, the effects of windbreaks, consisting of two rows of corn, on stomatal resistance was studied for sugar beets (*Beta vulgaris.*) The mean stomatal resistance (r_S) for the exposed plot, which was calculated from measurements of stomata aperture and density, was found to be greater than the r_S for the sheltered plot, while stomatal resistance for plants in both plots increased with decreasing soil water potential (ψ_S). Midday stomatal closure occurred, with the degree of closure being more pronounced in the exposed plot. On several days, cyclical oscillations in r_S with a period of about two hours were observed. The amplitude of the cycles was more pronounced when ψ_S was less than -0.45 bars. No evidence was found that wind speed, CO_2 concentration, air temperature, or vapor pressure oscillated in any way which might be connected with the r_S oscillations.

The conclusion drawn from the study was that micrometeorological conditions, not physiological changes, were responsible for the variations in r_S from the open to the sheltered plot. It was also found that as ψ_S becomes limiting, the r_S values in the sheltered plot have lower values than those found in the open plot.

In the second 1970 paper, the influences of leaf age, illumination, and leaf surface on sugar beet r_S were reported. Results showed that r_S for leaves similarly exposed was independent of physiological age except for the most immature leaves, and as the leaves expand, the stomatal size increases at nearly the same rate as the density decreases. The two effects seem to be mutually compensating so that r_S is left unchanged with age. The largest r_S values were found on older leaves that were in the lower part of the canopy. This was attributed to microclimatological differences, such as reduced illuminations, rather than to physiological age changes.

In a related study, Jordan et al. (1975) examined the role of leaf age and position on a cotton plant in the stomatal response to soil water deficit. The plants were grown in controlled environments during a period of soil moisture stress and the evaporative demand was maintained at a constant level. The authors found that as soil water stress increased, stomatal resistance also increased, beginning on the lower leaves. It was noted that adaxial stomata were more sensitive to decreasing leaf water potential than were abaxial surfaces. Stomatal closure (and thus an increase in stomatal resistance) proceeded from the oldest leaves to the youngest as the water stress became more severe. The stomatal resistance of each leaf was found to be related to its own water potential, which was modified by leaf age and the radiation environment during development.

In Rosenberg and Brown's 1973 paper, the energy balance equation for a leaf was adapted for use on a whole crop by using a crop resistance (r_{cr}) in place of the r_s value for the single leaf. Model results indicated the expected outcome that the latent heat flux was inversely proportional to r_{cr}. It was also found that the dependence of evaporative heat loss on r_{cr} was greatest when r_a (air resistance), e_a (vapor pressure), and net radiation were low. The relationship between evaporative heat loss and r_s was less sensitive to air temperature than to other micrometeorological parameters. The findings also indicated that the stomatal control of LE (latent heat loss) was greatest under cloudy conditions, when the radiation intensity was low (early morning or late afternoon), and when the leaf was shaded rather than sunlit. As wind speed increased, the control of LE by the stomates tended to increase in importance.

Turner studied the stomatal behavior in maize, sorghum, and tobacco (Turner, 1973 and 1974, and Turner and Begg, 1973). In Turner (1973), the effects of illumination and leaf resistance (calculated from adaxial and abaxial stomatal resistance values) on transpiration were examined. An increase in leaf resistance due to rapid decline of illumination with depth into the canopy was observed in all three crops at midday. The upper canopy leaves were well illuminated at this time, but by late afternoon, their stomatal resistance had increased in all three species. By early evening, the resistance was equally large in all leaves. The vertical profile of illumination in the canopy by late afternoon had the same general shape as it had at midday. However, the difference in illumination between the upper and lower leaves was smaller.

In contrast to Brown and Rosenberg (1970b), Turner showed that senescence in older yellower leaves of maize was responsible in part for the resistance increase in the lower canopy. In sorghum, there were not as many yellow leaves but, r_s was greater in the lowermost leaves at corresponding illumination. If the older leaves were ignored, there was a hyperbolic relationship between r_s and illumination in maize and sorghum. Under constant environmental conditions, sorghum and tobacco had minimum r_s values at maximum illumination (8900 and 7600 ft-c, respectively). In both species, the minimum r_s was doubled at 4000 ft-c. Abaxial stomata which are normally exposed to bright light were found to close at 2000 ft-c for both tobacco and sorghum, while those normally exposed to dim light did not close in either species until lower illumination levels were reached. Turner's main conclusion was that in well-watered field crops, light is the primary factor in determining stomatal resistance. Essentially the same conclusion was reached by Turner and Begg (1973).

Turner (1974) found that at low soil water potential, the development of low leaf water potential, and turgor pressure, not illumination, had the main effect on the diurnal changes in stomatal resistance. It was observed that stomatal resistances were high in the leaves of all species at sunrise. The resistance decreased in the upper leaves as the illumination increased, then increased again early in the morning as a result of increased water stress in the plants.

As ψ_l (leaf water potential) values became lower, there was a marked increase in stomatal resistance at a critical value of ψ_l in the three species tested (maize, sorghum, and tobacco). This critical value was discovered to vary from −13 bars in tobacco to −20 bars in sorghum, while it was −17 bars in maize.

The stomatal aperture, and thus stomatal resistance, is dependent on the difference in turgor between guard and subsidiary cells. Turner found that leaf turgor could decrease over a considerable range of values without having any effect on stomatal resistance in any of the three species. This closure of the stomate over a rather narrow range of turgor values suggested that this mechanism is an adaptive feature of the plant that permits high rates of CO_2 exchange, an advantage under field conditions where the diurnal range in turgor can be great.

Beadle et al. (1973), working with corn and sorghum in a growth chamber, reported a marked increase in corn diffusion resistance (stomatal resistance plus boundary-layer resistance with the boundary-layer resistance generally constant and at least an order of magnitude smaller than the stomatal resistance), occurred at leaf potentials between −8 and −11 bars. For sorghum, substantial increases in diffusion resistance occurred between −10 and −14 bars.

Teare and Kanemasu (1972) also investigated the variation of stomatal resistance (adaxial and abaxial surfaces in parallel) and leaf water potential at various heights in a soybean and sorghum crop. Their results, similar to Turner's (1974), indicated a signficant increase in stomatal resistance when leaf water potential exceeded −15 bars. As expected, the hourly trends in stomatal resistance and water potential of the upper leaves of soybean and sorghum canopies paralleled each other. Leaf position and water potential were also examined. At noon, it was found that the upper leaves had a lower water potential than the lower leaves. Thus in the morning the water potential gradient was strong from lower to upper leaves for both crops, while in the afternoon, the gradients remained in the sorghum canopy but disappeared in the soybean canopy. This indicates that stomatal resistance increased in upper leaves of soybeans but not in sorghum, resulting in a decrease in the CO_2 diffusion rate in the former but not in the latter. The significance of this difference is clearly demonstrated by the fact that sorghum was found to produce nearly three times as much dry matter as soybean did.

Brun et al. (1973) examined three methods of estimating transpiration resistance in a sorghum crop, which he applied to the Monteith (1965) evapotranspiration model, the results of which were compared with lysimeter data. The three procedures were:

1. taking the harmonic average of the r_S of all leaves on a plant;
2. using layers within the crop canopy and weighting each layer's resistance by its leaf area index; again all leaves were taken into account;
3. taking the harmonic average of the r_S of only the upper three leaves of the plant.

Field results showed that the first method provided a better estimate of evapotranspiration than either method two, which in addition required information on the leaf area of each canopy layer, or method three, which underestimated the transpiration resistance and thus resulted in a 12% to 18% overestimation of evapotranspiration.

Gardner's (1973) approach to stomatal resistance involved relating plant leaf water potential (ψ_l) to soil water potential (ψ_S) and to the transpiration rate, and relating r_S to leaf water potential (ψ_l).

A linear relationship between stomatal conductance (C_S) and turgor potential was shown for both the abaxial and adaxial leaf surfaces. Since the turgor potential is influenced by the transpiration rate, a direct coupling between C_S and transpiration is implied. When ψ_S decreases below -1 bar, a direct connection is established between C_S and ψ_S.

An examination of the relationship between C_S and transpiration rate at different values of ψ_S shows that it is a nonlinear one, best approximated by a hyperbola. In the range of ψ_S values between 0 and -1 bar, C_S is more strongly influenced by the transpiration rate than by ψ_S. When ψ_S decreases below -1 bar, it becomes more important in determining C_S than the transpiration rate, which is influenced by atmospheric factors.

Gardner makes several interesting points from an agroclimatological perspective:

1. The influence of the external water regime on plant growth can be quantitatively estimated after determination of the relationship between C and turgor potential and between transpiration and turgor potential.
2. The stomates provide the link between the growth rate and the transpiration rate.
3. In evaluating the water factor in agroclimatology, the C_S appears to be the most useful single measurement.
4. If the $\psi_l - C$ connection can be determined, it will be possible to predict the effect of environmental factors upon plant growth when water is limiting.

While hundreds of papers have been written concerning these four points, precise relationships have not been, and may never be, established.

In a similar vein, Brady et al. (1975) attempted to establish a relationship between stomatal resistance, measured on both surfaces of soybean leaves, and soil water potential (ψ_S). The results indicated that r_S increased with decreasing ψ_S for both the adaxial and abaxial leaf surfaces. The adaxial surface r_S was found to be more sensitive to ψ_S than the abaxial surface. Because of the difference in stomatal densities of the two sides, the adaxial surface has a higher resistance than does the abaxial surface. For both surfaces, the relationship between r_S and ψ_S was nonlinear. It was pointed out that when attempting to relate field measurements of r_S to ψ_S, other important factors are plant growth stage, location of measurements, evaporative demand, wet leaves, and recent irrigation.

The authors used least squares to fit second order polynomials to the $r_S - \psi_S$ data for the two leaf surfaces. r_S measurements on the adaxial surface of the upper leaves indicated that the physiological stage of growth had little effect on r_S. It was found that r_S might not recover immediately after a plot is irrigated. As a result, in some cases r_S will reflect plant water status, but might not reflect ψ_S.

Findings for evaporative demand indicated that for adaxial r_S of soybeans irrigated at 40% depletion, r_S increased with potential evapotranspiration (PET). In as much as even on the days of highest PET, r_S values of the irrigated plants did not increase above 4 s cm^{-1}, r_S greater than this value result from low ψ_S. Thus r_S measurements can be used to determine when soil water is limiting in soybeans, eliminating the need for determining root depth and extraction patterns.

EXOGENOUS EFFECTS ON STOMATAL RESISTANCE

Lange et al. (1971) studied the effects of humidity on stomatal responses with the use of epidermal strips of fern. The inner surface of each strip was left in contact with air spaces of uniformly high humidity, while the outer strip surface was treated with air of varying degrees of humidity. The results showed that treatment of the outer side of the epidermis with dry air led to a rapid closing of the stomata while moist air caused opening.

The authors postulated that stomatal behavior was the result of peristomatal transpiration. The guard cells were then able to function as humidity sensors which sensed water potential differences between the inside and the outside of the leaf. The stomatal aperture of each stomate was then controlled by its transpiration condition. The authors noted the importance of this mechanism in a plant's water economy. The plants seemed to have the ability, through an increase in stomatal resistance, to reduce their transpirational water loss during periods of decreasing humidity in the ambient air without changing the water status of the whole leaf. This direct cutical water loss of the guard and subsidiary cells permits a much more sensitive reaction by the stomate in comparison to the hydropassive mechanism which operates by changes in whole leaf water status.

Raschke (1970) found that in *Zea mays,* the stomata responsed to humidity changes in the ambient air. *Zea mays* is different in structure from the epidermal strips of fern and herb species used by Lange et al. (1971).

Previously, work on the reaction of stomata to air humidity, and temperature was done by Drake et al. (1970), whose objective was to ascertain whether the energy partitioning between transpiration and convection is controlled by changes in the transpiration resistance, with these resistance changes resulting from responses to changes in the physical environment of the leaf. The work was carried out under controlled conditions using single attached leaves of *Xanthium strumarium L.* in a wind tunnel. Results indicated that with increasing temperature, there was a decrease in leaf resistance caused by increased stomatal apertures. Leaf resistance was found to be higher in dry air than in moist air under constant air temperature conditions.

Schulze et al. (1972), using a field laboratory equipped with temperature and humidity controlled curvettes, studied the stomatal responses to changes in humidity for plants growing in the Negev Desert under varying soil water conditions. As they point out, Drake et al. (1970) did not attempt to determine if humidity had a direct effect on the stomata (peristomatal mechanism) or whether the effect were an indirect one acting on the total leaf water status. It was this aspect that Schulze et al. (1972) wished to investigate.

The field results on intact plants confirmed the results obtained with the epidermal strips used by Lange et al. (1971), i.e., plant stomata were found to open at high air humidity and close at low air humidity. It was noted that at a higher water vapor deficit in the atmosphere, the stomata resistance increased although the water content to the leaves was increasing as a result of the reduced transpiration rate. These findings caused the authors to rule out a reaction by means of the water potential in the leaf tissue. In addition, the humidity-induced response in the stomata was maintained under intense drought conditions in the soil. It was discovered that the humidity-controlled response intensified when soil water became limited in availability.

Schulze et al. (1973) used the field apparatus mentioned in connection with Schulze et al. (1972) to study stomatal responses to changes in temperature at increased water stress. The water vapor concentration difference between the leaf and the surrounding atmosphere was held constant at all temperatures so that stomatal aperture changes could be attributed only to the effects of temperature.

When the difference in water vapor concentration between the mesophyll and the atmosphere was held constant, the authors found that at low moisture stress plant stomata opened with temperature increases in the range of 25C to 40C. It was reported that with increased water stress, this temperature-induced stomata response was reversed, i.e., the stomata tended to close with increasing temperature.

Several climatic benefits result from the stomatal action. Under desert conditions, stomatal opening in response to increased temperatures (when plants are not under stress conditions) results in an increase in transpirational cooling. In desert plants at low plant water potentials, temperature-induced stomatal closing would be beneficial because it would cut down on water loss. The reduction in water loss at this critical time would be more important to the plant than transpirational cooling. The experiments of Schulze et al. (1973) indicated that for some species, the temperature-controlled stomatal response was operating independently of the effect of humidity in the stomatal opening. Thus, the relationship between the impulse of the stomata to open in response to increased temperature and the impulse to close in response to increased differences in water vapor concentration must be determined.

Schulze et al. (1974) investigated this relationship by attemping to show to what degree the daily course of stomatal resistance can be simulated and explained by the response to temperature and air humidity. The data were obtained in experiments carried out in a Negev Desert field laboratory under constant conditions with only one variable factor. A constant boundary-layer resistance was maintained in the chamber. Experimental work was carried out on two days: a moist, cool day and a hot, dry day.

The results showed that on the dry, hot day, both leaf temperature and vapor pressure difference reached maximum values approximately between 1100 and 1400 hour, and both declined after 1400 hour. Observed resistance values reached a peak around 1600 hour and then began to decline. The transpiration rate climbed to a steady value at about 0900 hour, remained near that level until about 1300 hour, and then began to decline. The water vapor concentration difference acted to cause stomatal closure. The plant was not under high water stress. Therefore, the high afternoon temperatures promoted stomatal opening. According to a simulation model designed by the authors, about a third of the potential humidity-controlled stomatal closure was counterbalanced by the stomatal response to the temperature increase. The decline in the observed resistance value in the afternoon indicated that the stomata reopened.

As expected, on the cool, moist day, the leaf temperature and the water vapor concentration difference remained lower than on the hot, dry day. The maximum temperature on the latter was near 45C, while on the former, it was near 37C. The maximum water vapor gradient on the cool, moist day was about 32 mg H_2O 1^{-1} which was about 23 mg H_2O 1^{-1} less than the maximum value on the hot, dry day. It was estimated that on the moist day the humidity-controlled stomatal closure should increase the diffusion resistance by about 9 s cm^{-1} as compared to an expected (based on the Schulze et al. simulation model) increase of 21 s cm^{-1} on the hot, dry day. The counterbalancing effect of the temperature on the moist day reduced the estimated change in stomatal resistance to 5 s cm^{-1}. In general, the model provided an adequate simulation of the course of stomatal resistance. Low transpiration rates were found to occur on dry days while transpiration rates on moist days were higher. Thus, if a moist day followed a dry day, there was still enough reserve water in the plant and soil to permit a substantial water loss.

The daily transpiration function is determined by the climatic control of stomatal response. As discussed before, it would be expected that when the diffusion resistance is high, the transpiration rate is low and vice versa. Thus, on moist days transpiration is greater, and the total amount of water loss was found to be considerably smaller on a dry day than on a moist one. As the authors point out, the stomata are closed in spite of an improved situation in the water supply, and they open in moist air in spite of an increased loss of water. It was therefore concluded that the observed midday depression of net photosynthesis and transpiration is based mainly on stomatal closure induced by air temperature and air humidity. For the apricot tree used by Schulze et al., the humidity-controlled response was found to dominate. This was observed to be a mechanism to prevent excessive water loss.

In the first of two companion papers, Schulze et al. (1975a) explored the relation between stomatal control systems regulated by the temperature and humidity of the apricot and plant water stress and intercellular air space CO_2 concentration. Further evidence of the independence of stomatal resistance and leaf water status was provided by a plot of relative water content (RWC) and water vapor concentration difference between leaf and air. The graph clearly showed that the RWC decreased with increasing water vapor concentration difference down to a value near 15 mg H_2O l^{-1}. However, as the concentration difference continued to increase, the RWC increased.

It was indicated that CO_2 can change stomatal resistance through changes in CO_2 concentration in the intercellular air spaces. Reduction in stomatal resistance can logically be associated with an intercellular decrease in the CO_2 concentration while an increase in stomatal resistance can result from an increase in CO_2 concentration in the intercellular spaces. Emphasis was placed upon the importance of knowing if the control system which functions on the basis of CO_2 were modified or somehow overridden by other forces when the midday photosynthesis depression is being considered. The results showed that in general, the effects of the plant internal CO_2 concentration mechanism were either overruled or modified by the external climatic factors of air temperature and humidity gradient.

In the third paper in the series, Schulze et al. (1975b) analyzed temperature—and humdity-induced stomatal response and photosynthetic activity by the use of a simulation procedure based on water use efficiency.

Hall and Kaufmann (1975a) also investigated exogenous effects on stomatal resistance. The results for sesame showed that leaf resistance increased as the humidity gradient increased, with the strongest response occurring at lower temperatures. This increase was found to regulate the upper limits to the transpiration rate. However, the interaction with ambient temperature allowed a greater transpiration rate at higher temperatures. Similar results were obtained by Schulze et al. (1974, 1975a, 1975b).

Tests were also carried out at 20C and 30C with progressive increases in the humidity gradient for sesame, sunflower, and sugar beet. Results showed that at the higher temperature, the humidity gradient had little influence on the stomatal resistance for sunflower and sugar beets, but the sesame response was similar to that described above. At 20C, all three exhibited increased leaf resistance in response to humidity gradient increases; however, sunflower was still less responsive than sesame. Hall and Kaufmann's principal conclusion coincides with that of the Schulze et al. papers, i.e., the effects of the humdity gradient on transpiration indicate that the stomata, rather than the incipient drying on the walls inside the leaf, are the controlling agent. It is possible, according to Hall and Kaufmann, that the stomatal responses are not due primarily to changes in the water status of the leaf as a whole but rather, to the "peristomatal transpiration" principle.

The authors point out that if the humidity gradient were not controlled, any stomatal response could occur. If the humidity gradients were permitted to increase with increasing temperature, stomata would tend to close at high temperatures. When the humidity gradient is kept constant, the stomatal resistance will either remain constant or decrease with increasing temperature. They suggest that it is likely that extremely high temperatures will cause stomatal closure regardless of the humidity gradient because of the occurrence of metabolic lesions.

Hall and Kaufmann (1975a) make reference to an earlier work by Raschke and Kuhl (1969), in which an investigation of stomatal responses to changes in atmospheric humidity and water supply was made in CO_2-free air by exposing leaf sections of *Zea mays* to CO_2-free air. This limitation was imposed to prevent interference by the CO_2 in the guard cell mechanism. The results indicated that the stomates did not close in response to changes from moist to dry air. The air was both passed over the leaf and forced through the intercellular spaces. However, it was found that the stomatal opening became narrow (increased resistance) when the water potential in the liquid supplying the leaf was lower. Raschke and Kuhl concluded that the guard cells are tightly coupled to the leaf water supply system and are indirectly related to atmospheric conditions by a negative feedback of transpiration on the water potential in the total leaf. The Hall and Kaufmann (1975a) findings showed that low CO_2 concentrations decreased but did not eliminate resistance response to humidity. They suggest that totally CO_2-free air may have eliminated the humidity-induced stomatal response observed by Raschke and Kuhl.

In another paper, Hall and Kaufmann (1975b) reported on the interaction between the effects of humidity gradients, ambient temperature, and CO_2 concentrations on stomatal resistance. The effects of humidity gradients on r_m' (mesophyll resistance) to CO_2 were also reported. With relatively constant values of CO_2 concentration (near ambient levels), increases in leaf resistance to water vapor and small decreases in mesophyll resistance to CO_2 were recorded as the air became drier. When the CO_2 concentration was reduced to 54 μ l l^{-1}, it was found that as the humidity difference increased, the resistance again increased, but not as much as had been observed at the higher CO_2 concentrations. The CO_2 concentration was found to have a greater effect on resistance in dry air than in humid air. Under small humidity differences, leaf resistance decreased less than 0.2 s cm^{-1} when the CO_2 concentration was decreased by 170 μ l l^{-1}. A subsequent increase in CO_2 concentration also resulted in small changes in leaf resistance. At large humidity differences, a one-step decrease in CO_2 concentration from 337 to 64 μ l l^{-1} resulted in a resistance decrease of about 1 s cm^1. The process was reversible. Thus, decreasing the CO_2 concentration had a greater effect in lowering the leaf resistance in dry air than in moist air. The dependence of resistance on temperature (at 20C and 34C) at constant CO_2 concentration near ambient levels was also investigated. At the lower temperature, increases in the humidity difference (from 5 to 12 μ g cm^{-3}) resulted in a 2 s cm^{-1} increase in resistance. At the higher temperature, no change in resistance was noted when the humidity difference was increased from about 7 to 27 μ g cm^{-3}. These observations agree with those found by Schulze et al. (1974), where the humidity-induced stomatal closure in dry air was modified at higher temperatures.

Lange et al. (1975), working in a control chamber in the desert, showed that external factors were important in stomatal control. His results were similar to those of other papers just reviewed.

In a related study, Drake and Salisbury (1972) researched the after effects of low and high temperature pretreatment on leaf resistance transpiration and temperature, using attached leaves from greenhouse-grown *Xanthium strumarium L.* plants. The experimental work took place in a wind tunnel, where after 72 hours of low temperature treatment (10C day, 5C night), leaf resistance was increased compared to the control plants. Plants which were pretreated with high temperatures (40C day, 35C night) showed little change from the control plants.

The consequences of the increased leaf resistance for the low temperature treated plants were found to be elevated leaf temperature and a reduction in transpiration. The authors discovered that in the air temperature range of 20C to 40C, the low temperature pretreated leaf was between 6C and 10C higher than the control or high temperature pretreated plants. They speculated (and other papers reviewed in this section have answered some of these questions) that temperature pretreatment after effects on leaf resistance were caused by changes in stomatal aperture and not by changes in the leaf surface waxes.

Evidence to support the results reported in Schulze et al. (1974) has been found by many experimenters. Slavik (1973) reported on transpiration resistance in maize leaves which were grown in humid and dry air. During the 12-hour day, the dry condition relative humidity was 40 ± 5% while the wet condition humidity was 85 ± 5%. The daytime temperature in the phytotron chamber was 28C. At night, the relative humidity was 40 ± 5% and 100% with a temperature of 20C. Measurements made under dry conditions indicated that in plants grown in humid air the transpiration rate was higher and the transpiration resistance was lower than in plants grown in dry air. However, it was noted that the transpiration rate in the humid air plants decreased very rapidly after the transfer to dry air. Slavik attributed this rapid change to increasing water deficit and hence hydroactive closure of the stomata. However, the peristomatal mechanism should be considered as a possible explanation, especially because the humid air plant response to the dry air occurred so quickly.

O'Leary (975) studied the influence of the environment on total water consumption by whole plants. To determine the effects of relative humidity, he placed 20 bean plants (*Phaseolus vulgaris L.*) into each of three growth chambers which were maintained at 30% to 35%, 70% to 75%, and 95% to 100% relative humidity, respectively. The chambers were programmed to give 12 hours day, 1 hour dusk, 10 hours night, 1 hour dawn, with air temperature maintained at 24C, 21C, 18C, and 21C, respectively.

Results showed that transpiration rates declined with increasing relative humidity whether calculated from the short-term experiment or from the daily water consumption. O'Leary plotted vapor pressure gradient from leaf to outside air against transpiration rate and found that the points for the high and medium humidities fell on a line that passed through zero as predicted by diffusion theory. However, the point for low humidity did not conform to the theoretical diffusion theory value. If the transpiration were in direct proportion to the driving gradient, the transpiration rate for the plants at low humidity would be about 3.5 times the measure value. The author suggested that the plants were exerting control over the transpiration rate by reducing the stomatal aperture and increasing stomatal resistance. Again the peristomatal mechanism discussed by Schulze et al. (1974) seems to be in operation. In addition, measurements revealed that the total diffusion resistance was higher at the lower humidity. Thus, the transpiration rate did not increase in direct proportion to the increase in the gradient of vapor pressure.

The importance of the work done by Schulze et al. (1974) to the stomatal oscillation research is also worth considering. A complete discussion of the work that has been done on stomatal oscillations is beyond the scope of this review. The reader is referred to the following papers for detailed information on the subject: Barrs and Klepper (1968), Barrs (1968), and Barrs (1971). Additional material can also be found in Kriedemann (1971) and Cowan and Troughton (1971). Of note is the observation by Barrs and Klepper (1968) that the generally accepted explanation for cycling is based on the facts that the stomates open in the light, transpiration increases, leaf water content falls, and the associated water stress brings about stomatal closure. The leaf then regains its turgor as water flows into it in response to the steepened water potential gradient. The water potential in the leaf rises again, allowing the stomate to open, and the cycling begins anew.

Recent work by Farquhar and Cowan (1974) supports this explanation. They point out that the observed oscillations are caused by the properties of a loop in which evaporation rates affects, through physiological processes, the stomatal aperture which in turn affects the evaporation rate.

The evaporation rate per unit leaf area is:

$$E = \frac{\Delta\chi\, g_1 g_b}{g_1 + g_b} \qquad (43)$$

where $\Delta\chi$ = difference in absolute humidity across the epidermis and external boundary layer of the leaf
g_1 = leaf conductance
g_b = boundary-layer conductance.

The authors have defined "environmental gain" as

$$G = \frac{\partial E}{\partial g_1} = \frac{\Delta\chi\, {g_b}^2}{(g_1 + g_b)^2} \qquad (44)$$

In an experiment the ambient humidity was decreased, causing an increase in $\Delta\chi$; thus the evaporation rate increased. The initial plant response was to open the stomates and thus to add to the effects of reduced humidity. Subsequently, the stomata closed and the evaporation rate was reduced. This sequence of events is often referred to as hydropassive stomatal movements. The leaf conductance initially increased and was then followed by a series of damped oscillations until conductance became steady at a new lower level.

A further reduction in ambient humidity rendered the system unstable: the initial fluctuations in conductance were followed by successively larger fluctuations. It was found that under these conditions of instability, the oscillations tended to become unsteady and to be maintained for long periods of time.

The authors demonstrate that the hydropassive opening of the stomata constitutes a positive feedback action since it results in an increase in the rate of transpiration. From the gain equation, the amount of feedback can be seen to depend on the effect of leaf conductance on rate of evaporation. The environmental gain (G) is seen to increase as the ambient humidity decreases.

The authors' data show clearly that the leaf conductance mean value decreased following each reduction in the ambient humidity. The decrease in conductance was such that there was no significant increase in mean evaporation rate, even though there was a greater difference in humidity across the leaf epidermis. The indication is thus that the gain of the internal transfer function was extremely high; however, Farquhar and Cowan mention the element of open loop control of leaf conductance, i.e., the stomata are responding directly to changes in the ambient humidity through the peristomatal mechanism. The strength of this effect in the oscillation mechanism needs to be given further consideration.

REFERENCES

Aston, A.R.; Millington, R.J.; and Peters, D.B. "The Energy Balance of Leaves." In R.O. Slatyer (ed.) *Plant Response to Climatic Factors.* Paris: UNESCO, 1973.

Barrs, H.D. "Effects of Cyclic Variations in Gas Exchange Under Constant Environment Conditions on the Ratio of Transpiration to Net Photosynthesis." *Physiologia Plantarum* 21 (1968):918-929.

———. "Cyclic Variations in Stomatal Aperture, Transpiration, and Leaf Water Potential Under Constant Environmental Conditions." *Ann. Rev. Plant Physiol.* 22 (1971):228-286.

Barrs, H.D., and Klepper, B. "Cyclic Variations in Plant Properties Under Constant Environmental Conditions." *Physiologia Plantarum* 21 (1968):711-738.

Beadle, C.L.; Stevenson, K.R.; Neumann, H.H.; Thurtell, G.W.; and King, K.M. "Diffusive Resistance, Transpiration, and Photosynthesis in Single Leaves of Corn and Sorghum in Relation to Leaf Water Potential." *Canadian Journal of Plant Sciences* 53 (1973):537-544.

Brady, R.A.; Goltz, S.M.; Powers, W.L.; and Kanemasu, E.T. "Relation of Soil Water Potential to Stomatal Resistance in Soybean." *Agronomy Journal* 67 (1975):97-99.

Brown, K.W., and Rosenberg, N.J. "Effect of Windbreaks and Soil Water Potential on Stomatal Diffusion Resistances and Photosynthetic Rate of Sugar Beets (*Beta vulgaris*), *Agronomy Journal* 62 (1970):4-8.

———. "Influence of Leaf Age, Illumination, and Upper and Lower Surface Differences on Stomatal Resistance of Sugar Beet (*Beta vulgaris*) leaves." *Agronomy Journal* 62 (1970):20-24.

Brun, L.J.; Kanemasu, E.T.; and Powers, W.L. "Estimating Transpiration Resistance." *Agronomy Journal* 65 (1973):326-328.

Clark, J.S., and Wigley, G. "Heat and Mass Transfer from Real and Model Leaves." In D.A. deVries and N.H. Afgan (eds.) *Heat and Mass Transfer in the Biosphere. I. Transfer Processes in the Plant Environment.* New York: Wiley, 1975.

Cowan, I.R., and Troughton, J.H. "The Relative Role of Stomata in Transpiration and Assimilation." *Planta* 97 (1971):325-336.

———. "Oscillations in Stomatal Conductance and Plant Functioning Associated with Stomatal Conductance: Observations and a Model." *Planta* 106 (1972):185-219.

DeMichele, D.W., and Sharpe, P.J.H. "An Analysis of the Mechanics of Guard Cell Motion." *Journal of Theoretical Biology* 41 (1973):77-96.

———. "A Parametric Analysis of the Anatomy and Physiology of the Stomata." *Agricultural Meteorology* 14 (1974):229-241.

Drake, B.G.; Raschke, K.; and Salisbury, F.B. "Temperature and Transpiration Resistance of *Xanthium* Leaves as Affected by Air Temperature, Humidity, and Wind Speed." *Plant Physiology* 46 (1970):324-330.

Drake, B.G., and Salisbury, F.B. "Aftereffects of Low and High Temperature Pretreatment on Leaf Resistance, Transpiration, and Leaf Temperature in *Xanthium*." *Plant Physiology* 50 (1972):572-575.

Esau, K. *Plant Anatomy*. New York: John Wiley, 1965.

Etherington, J.R. *Environment and Plant Ecology*. New York: John Wiley, 1975.

Farquhar, G.D., and Cowan, I.R. "Oscillations in Stomatal Conductance: the Influence of Environmental Gain." *Plant Physiology* 54 (1974):769-772.

Gardner, W.R. "Internal Water Status and Plant Response in Relation to the External Water Regime." In R.O. Slatyer (ed.) *Plant Response to Climatic Factors*. Paris: UNESCO, 1973.

Gates, D.M. *Energy Exchange in the Biosphere*. New York: Harper and Row, 1962.

———. "Transpiration and Leaf Temperature." *Annual Review of Plant Physiology* 19 (1968):211-238.

Hall, A.E., and Kaufmann, M.R. "Regulation of Water Transport in the Soil-Plant-Atmosphere Continuum." In D.M. Gates, and R.B. Schmerl (eds.) *Perspectives of Biophysical Ecology*. New York: Springer-Verlag, 1975.

———. "Stomatal Response to Environment" with *Sesamum Indicum L. Plant Physiology* 55 (1975):445-459.

Hsiao, T.C., and Acevedo, E. "Plant Responses to Water Deficits, Water-use Efficiency, and Drought Resistance." *Agricultural Meteorology* 14 (1974):59-84.

Jordan, W.R.; Brown, K.W.; and Thomas, J.C. "Leaf Age as a Determinant in Stomatal Control of Water Loss from Cotton During Water Stress." *Plant Physiology* 56 (1975):595-599.

Kays, W.M. *Convective Heat and Mass Transfer*. New York: McGraw-Hill, 1966.

Kramer, P.J. *Plant and Soil Water Relationship: A Modern Synthesis*. New York: McGraw-Hill, 1969.

Kreith, F. *Principles of Heat Transfer*. New York: Intext.

Kriedmann, P.E. "Photosynthesis and Transpiration as a Function of Gaseous Diffusive Resistance in Orange Leaves." *Physiologia Plantarum* 24 (1971):218-227.

Lange, O.L.; Losch, R.; Schulze, E.D.; and Kappen, L. "Responses of Stomata to Changes in Humidity." *Planta* 100 (1971):76-86.

Lange, O.L.; Schulze, E.D.; Kappen, L.; Buschbom, U.; and Evenari, M. Photosynthesis of Desert Plants as Influenced by Internal and External Factors." In D.M. Gates and P.B. Schmarl (eds.) *Perspectives of Biophysical Ecology*. New York: Springer-Verlag, 1975.

Lee, R., and Gates, D.M. "Diffusion Resistance in Leaves as Related to Their Stomatal Anatomy and Micro-structure." *American Journal of Botany* 51 (1964):963-975.

Leopold, A.C. and Kriedemann, P.E. *Plant Growth and Development*. New York: McGraw-Hill, 1975.

Linacre, E.T. "Further Studies of the Heat Transfer from a Leaf." *Plant Physiology* 42 (1967):651-658.

Meidner, H., and Mansfield, T.A. *Physiology of Stomata*. New York: McGraw-Hill, 1968.

Miller, P.C. "A Model of Temperature, Transpiration Rates, and Photosynthesis of Sunlit and Shaded Leaves in Vegetation Canopies." In R.O. Slatyer (ed.) *Plant Response to Climatic Factors*. Paris: UNESCO, 1973.

Milthorpe, F.L., and Moorby, J. *An Introduction to Crop Physiology*. London: Cambridge University Press, 1974.

Monteith, J.L. "Evaporation and Environment." *The State and Movement of Water in Living Organisms*. 19th Symposia of the Society for Experimental Biology. New York: Academic Press, 1965.

———. *Principles of Environmental Physics*. London: Edward Arnold, 1973.

Nobel, P.S. *Introduction to Biophysical Plant Physiology*. San Francisco: W.H. Freeman, 1974.

O'Leary, J.W. "Environmental Influence on Total Water Consumption by Whole Plants." In D.M. Gates and R.B. Schmerl (eds.) *In Perspectives of Biophysical Ecology*. New York: Springer-Verlag, 1975.

Parkhurst, D.F.; Duncan, P.R.; Gates, D.M.; and Kreith, F. "Windtunnel Modelling of Convection of Heat Between Air and Broad Leaves of Plants." *Agricultural Meteorology* 5 (1968):33-47.

Parkhurst, D.F., and Pearman, G.I. "Convective Heat Transfer from a Semi-infinite Flat Plate to Periodic Flow at Various Angles of Incidence." *Agricultural Meteorology* 13 (1974):383-393.

Parlange, J.Y.; Waggoner, P.E.; and Heichel, G.H. "Boundary Layer Resistance and Temperature Distribution on Still and Flapping Leaves. I: Theory and Laboratory Experiments." *Plant Physiology* 48 (1971):437-442.

Pearman, G.I.; Weaver, H.L.; and Tanner, C.B. "Boundary Layer Heat Transfer Coefficients Under Field Conditions." *Agricultural Meteorology* 10 (1972):83-92.

Penning de Vries, F.W.T. "A Model for Simulating Transpiration of Leaves with Special Attention to Stomatal Functioning." *Journal of Applied Ecology* 9 (1972):57-71.

Perrier, E.R.; Aston, A.; and Arkin, G.F. "Wind Flow Characteristics on a Soybean Leaf Compared with a Leaf Model." *Physiologia Plantarum* 28 (1973):106-112.

Raschke, K. "Heat Transfer Between the Plant and the Environment." *Annual Review of Plant Physiology* 11 (1960):111-126.

———. "Stomatal Responses to Pressure Changes and Interruptions in the Water Supply of Detached Leaves of *Zea mays L.*" *Plant Physiology* 45 (1970):415-423.

———. "Stomatal Action." *Annual Review of Plant Physiology* 26 (1975):309-340.

Raschke, K., and Kuhl, U. "Stomatal Responses to Changes in Atmospheric Humidity and Water Supply: Experiments with Leaf Sections of *Zea mays* in CO_2-free Air." *Planta* 87 (1969):36-48.

Ripley E., and Saugier, B. "Energy and Mass Exchange of a Native Grassland in Saskatchewan." In D.A. deVries and N.H. Afgan (eds.) *Heat and Mass Transfer in the Biosphere. I. Transfer Processes in the Plant Environment.* New York: Wiley, 1975.

Rosenberg, N.J., and Brown, K.W. "Measured and Modelled Effects of Microclimate Modification on Evapotranspiration by Irrigated Crops in a Region of Strong Sensible Heat Advection." In R.O. Slatyer (ed.) *Plant Response to Climatic Factors.* Paris: UNESCO, 1973.

Schuepp, P.H. "Studies of Forced-convection Heat and Mass Transfer of Fluttering Realistic Leaf Models." *Boundary-Layer Meteorology* 2 (1972):263-274.

Schulze, E.D.; Lange, O.L.; Buschbom, U.; Kappen, L.; and Evenari, M. "Stomatal Responses to Changes in Humidity in Plants Growing in the Desert." *Planta* 108 (1972):259-270.

———. "Stomatal Responses to Changes in Temperature at Increasing Water Stress." *Planta* 110 (1973):29-42.

———. "The Role of Air Humidity and Leaf Temperature in Controlling Stomatal Resistance of *Prunus Armeniaca* L. Under Desert Conditions. A Simulation of the Daily Course of Stomatal Resistance." *Oecologia* 17 (1974):159-170.

———. "The Role of Air Humidity and Leaf Temperature in Controlling Stomatal Resistance of *Prunus Armeniaca* L. Under Desert Conditions. II. The Significance of Leaf Water Status and Internal Carbon Dioxide Concentration." *Oecologia* 18 (1975):219-233.

———. "The Role of Air Humidity and Temperature in Controlling Stomatal Resistance of *Prunus Armeniaca* L. Under Desert Conditions. III. The Effect on Water Use Efficiency." *Oecologia* 19 (1975):303-314.

Shawcroft, R.W.; Lemon, E.R.; and Stewart, D.W. "Estimation of Internal Crop Water Status from Meteorological and Plant Parameters." In R.O. Slatyer (ed.) *Plant Response to Climatic Factors.* Paris: UNESCO, 1973.

Shawcroft, R.W.; Lemon, E.R.; Allen, L.H., Jr.; Stewart, D.W.; and Hensen, S.E. "The Soil-Plant-Atmosphere Model and Some of Its Predictions." *Agricultural Meteorology* 14 (1974):287-307.

Slatyer, R.O. *Plant-water Relationships.* New York: Academic Press, 1967.

Slavik, B. "Transpiration Resistance in Leaves of Maize Grown in Humid and Dry Air." Plant Response to Climatic Factors (ed. R.O. Slatyer), Paris: UNESCO, 1973.

———. *Methods of Studying Plant Water Relations.* New York: Springer-Verlag, 1974.

Taylor, S.E. "Ecological Implications of Leaf Morphology Considered from the Standpoint of Energy Relations and Productivity." PhD Thesis, Washington University, St. Louis, Missouri, 1971.

Teare, I.D., and Kanemasu, E.T. "Stomatal-diffusion Resistance and Water Potential of Soybean and Sorghum Leaves." *New Phytology* 71 (1972):805-810.

Tenhunen, J.D., and Gates, D.M. "Light Intensity and Leaf Temperature as Determining Factors in Diffusion Resistance" (D.M. Gates and R.B. Schmerl, eds.). *Perspectives of Biophysical Ecology.* New York: Springer-Verlag, 1975.

Thom, A.S. "The Exchange of Momentum, Mass and Heat Between an Artificial Leaf and the Airflow in a Windtunnel." *Quarterly Journal of the Royal Meteorological Society* 94 (1968):44-55.

Turner, N.C. "Illumination and Stomatal Resistance to Transpiration in Three Field Crops." In R.O. Slatyer (ed.) *Plant Response to Climatic Factors.* Paris: UNESCO, 1973.

———. "Stomatal Behavior and Water Status of Maize, Sorghum, and Tobacco Under Field Conditions. II. At Low Soil Water Potential." *Plant Physiology* 53 (1974):360-365.

Turner, N.C., and Begg, J.E. "Stomatal Behavior and Water Status of Maize, Sorghum and Tobacco Under Field Conditions. I. At High Soil Water Potential." *Plant Physiology* 51 (1973):31-36.

Wigley, G., and Clark, J.A. "Heat Transport Coefficients for Constant Energy Flux Models of Broad Leaves." *Boundary-Layer Meteorology* 7 (1974):139-150.

Variable-sized Point Symbols in Thematic Mapping

R. Kingsbury
Indiana University

Variable-sized point symbols are commonly used in thematic mapping to show quantitative aspects of data applying to specific points or areas. One of their uses is to make comparisons among urban centers or individual communities that collectively form urban centers. Very often, they function as data summaries for statistical areas such as countries, states, counties, or census tracts. By segmenting these symbols, it is possible to indicate subdivision of the data or to show changes through time. These symbols generally are drawn in a proportional or graduated fashion according to the quantities they represent.

Quantitative point symbols may be one, two, or three dimensional in form. One-dimensional symbols usually are seen as bars or columns where the lengths represent amounts and the widths are arbitrary. Two-dimensional symbols may be circles, squares, triangles, or other shapes where the areas represent amounts. Three-dimensional symbols include spheres and cubes where the volumes represent amounts.

While one-dimensional symbols are the easiest of the three to construct and interpret, they are not usually suited to displaying data where the range of statistical values is considerable. In order to portray such data effectively, a scale must be devised so that bars or columns representing small quantities often will become extremely long and be necessarily placed far outside the location or area to which they apply. If not labeled very carefully, this might bring considerable graphic confusion. Even when it does not, it often results in drastically lowering the readability of the visual composition.

By contrast, three-dimensional symbols are well suited to displaying data where the ranges are great. However, research has indicated that users consistently read three-dimensional symbols as though they were two-dimensional and, thus, they are unsuited to effective data transmission.[1] Indeed, Dickinson, after administering tests using a number of different point symbols, stated that "Proportional spheres proved quite hopeless . . ." because of the high user error in evaluation.[2]

POPULARITY OF TWO-DIMENSIONAL SYMBOLS

The two-dimensional symbol is the most useful and the most commonly employed of the variable-sized point symbols. While some cartographers have urged use of the square instead of the circle,[3] the circle is usually favored over the square, triangle, or other forms. This may result from its ease of construction, its grace in appearance, and its compactness, which allows it to fit neatly within the boundaries of statistical subdivisions. In his original work on the graduated circle as a quantitative map symbol, Flannery reported that 60% of all map users preferred circles over triangles, squares, and rectangles.[4] Frequent use of this cartographic technique in journals, atlases, and other published works suggests that map makers believe that graduated circles make extremely effective thematic maps in that circles are easily interpreted by the map user. It is this last assumption that is investigated here. How effective a method of communication is graduated circle symbolization on thematic maps?

One good indication of the popularity of these maps today may be seen by examining current research literature. During the five-year period, 1971-1976, approximately 370 different quantitative distribution maps appeared in the *Annals* of the *Association of American Geographers*. Of these, 80 maps or about 22% of the total used graduated circles as their main form of symbolization. Clearly, many map makers consider graduated circles a very useful means of communication.

Unfortunately, many maps employing graduated circles are examples of poor cartographic design. A common fault is the inclusion of area patterns, linework, other point symbols, and lettering "background noise" as it has been called. Sometimes, individual proportional circles are labeled by place name or statistical subdivision. Supplementary information such as political boundaries, rivers, cities, and other information may be drawn and named. Occasionally, there is an excessive overlap of circles. In these cases, the map has been overloaded with information and the reader is lost in a maze of detail. In this case, the data transmission ability of the graduated circles has been lost or at least impaired seriously.

USE OF SIZE CLASS CIRCLES

Graduated circle maps generally are fashioned following one of two different systems. The circles may be drawn in size classes or they may be drawn on a continuously proportional basis. By the former method, a small selected group of distinctly sized circles is used to represent specific ranges of data. By the second method, continuously graduated circles are drawn to represent continuously changing data. Under the latter system, if 100 different statistical quantities were being represented, there would be in theory 100 different circle sizes on the map.

Some studies have suggested that size class circles are easier to evaluate than continuously graduated circles, and, hence, maps using the former are the more readable of the two.[5] This is so where the map contains a very limited number of different circle sizes and each circle size is clear and distinct in itself. Dickinson observed that size class maps are especially suitable for statistical data where ". . . the map reader needed a certain amount of guidance with information sorted out into different degrees of importance."[6]

In spite of the seemingly simple, forthright visual form of many size class circle maps, user evaluation may be neither uniform nor necessarily accurate. The responses received in preliminary testing of a sample map suggested that many users did not limit their appraisal to the legend information but offered impressions, inferences, and conclusions, often false, that the map display was never designed to transmit. Internal map cues as well as external information contributed to their analysis of the visual product. Many users appeared to be dissatisfied with the legend information and attempted to differentiate between symbols of the same size and class, even though the map provided no real data for this. Other users failed to read the legend completely or were unable to understand categories of information as presented in the legend.

Size class gradational circles are often seen in atlas reference maps. On many of the maps in one popular classroom atlas, five categories of city population sizes are used including the open-ended "1,000,000 and over" which may prove confusing to many readers. In this same atlas, one category makes cities of 50,000 to 500,000 have the same symbol, which certainly would be misleading to many users. It would appear that the facts of urban population size or other statistical information being displayed are often hidden in size categories, and, thus, such maps are only of limited value as a form of communications.

VISUAL PERCEPTION OF CONTINUOUSLY GRADUATED CIRCLES

A useful method of design involves using continuously graduated circles. Usually such maps present circles which have been drawn so that the area of each is in proportion to the statistical quantity being depicted. For example, a circle representing 4,000 will contain exactly twice the area of one representing 2,000.

Truth in statistical representation should result from continuously graduated circles. Unfortunately, truth in representation and truth in user perception are two quite different matters. Research has indicated that maps using continuously graduated circles are subject to user perception error.[7]

A relatively simple map of the western United States (Figure 1) was devised to check the results of existing research findings and to discover the extent of user perception error. The map

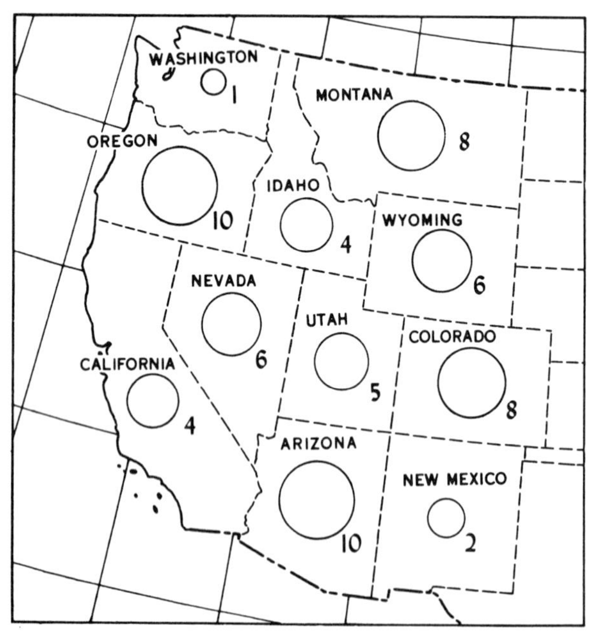

Figure 1. Test Map.

contained open graduated circles, one per state, which had been drawn with exact ratio of area increases. Thus, Montana was eight times larger than Washington, Arizona was five times larger than New Mexico, and so forth.

The sample map was used in association with questions which sought to have users examine and evaluate circle sizes throughout the map. In each case, they were requested to estimate how many times larger one circle was than another. These ten questions were asked:

1. Oregon is ______ times larger than New Mexico. (5)
2. Montana is ______ times larger than New Mexico. (4)
3. Colorado is ______ times larger than Washington. (8)
4. California is ______ times larger than Washington. (4)
5. Nevada is ______ times larger than New Mexico. (3)
6. Arizona is ______ times larger than Washington. (10)
7. Utah is ______ times larger than Washington. (5)
8. Arizona is ______ times larger than Idaho. (2 1/2)
9. Wyoming is ______ times larger than Washington. (6)
10. Oregon is ______ times larger than Washington. (10)

Correct answers are shown in parentheses. The range of differences between circles compared varied between 1 and 10 which is typical of some such maps but not typical of others. Thus, current literature contains graduated circle maps which show less range than this while at the same time it also contains maps which show far greater range.

This map of open circles and its questions were given to 103 freshman college students in a beginning geography course. These students made 1,030 separate comparisons of circle sizes. Some 71.0% of the comparisons underestimated the size of large circles compared to small circles. Only 18.3% of the comparisons were correct and 10.7% were overestimates. See Figure 2a.

An additional series of tests using the same circle sizes but varying the shading of the circle or its background were given to other groups of students. Variations included circles with line shading; circles with dot screens; heavy open circles; solid black circles; open circles on a 60% dot screen background; and open circles with a heavy "background noise" of line work simulating transport routes. The results from using these new maps were not substantially different from the results of the open circle test. The cumulative results of all test maps in this Series A are seen by Figure 2b. Some 258 respondents made 2,566 separate comparisons of circle sizes. Some 66.8% of the comparisons underestimated the size of large circles compared to small circles. Only 22.7% of the comparisons were correct and 10.5 % were overestimates.

Similar results were obtained by Dickinson in a series of tests given to ascertain user perception of the values of various forms of thematic map symbolization.[8] Among other forms, he tried circles, squares, and spheres arranged in simulated map presentations. The circle test, which contained ten different graduated circles ranging in proportion from 3 to 37 in size was administered to 44 university students. In a first exercise, he provided the value of the smallest circle and asked respondents to estimate on this basis the value of each of the other nine circles. The number of correct answers was only 2.6% and 39.1% of the estimates showed an error of

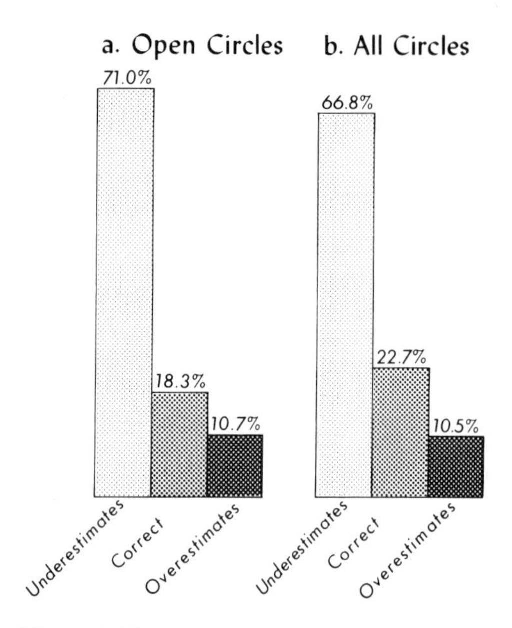

Figure 2. Test A—Circles Proportional to Area.

40% or more. In a second exercise, he provided the value of both the smallest and the largest circles and asked respondents to estimate on the basis of the value of each of the other eight circles. This additional cue served to improve the answers considerably. The number of correct estimates was raised to 4.9% and estimates with error of 40% or more was decreased to only 7.4%.

The Dickinson tests were rather severe in that they asked respondents to give whole number answers which in the second series were between 12 and 148. The correct answers were 28, 40, 44, 52, 76, 100, 112, and 124, and anything else was incorrect. When the results were tabulated so that the number correct or to within 10% of the correct answer were considered satisfactory, the correct answer rate increased substantially to 46.7%. Still, the high percent of error in the use of such maps led Dickinson to conclude that his respondents ". . . experienced very real difficulty in estimating and differentiating quantities, two processes which they are frequently required to do when using statistical maps."[9]

CORRECTING CIRCLE SIZES TO AID EVALUATION

The results of these studies confirmed the work of similar or closely related research reported by Williams, Flannery, Ekman and Junge, Crawford, and others.[10] All agreed that underestimation of the size and, hence, the value of large circles relative to the value of small circles on such maps appeared to be a fact.

To compensate for this underestimation, statistical procedures were employed to decide upon corrective factors which could be used to increase the size of large circles in comparison to small ones. Meihoefer discusses the process:

> . . . Thus by adapting some method of scaling the sizes of circles that allows for this consistent bias in visual perception it should be possible to ensure accuracy in quantitative estimation.
>
> . . . The amount of distortion can be described by referring to an exponent of the degree of difference between one statistically accurate circle and its perceptually appropriate modified counterpart.[11]

The exponents derived by various studies included .86 and .87 by Flannery and .90 and .91 by Crawford; others included .76, .80, .82, and .86.[12] While these results suggested relatively close agreement, there were differences which resulted from the individual formulation of tests including the range of circle values used, the inclusion of various stimuli, and the varying number and type of respondents. Some tests used the circle in a map context while others did not. Since no tests were exactly the same, the results understandably were slightly different.

Without passing any judgment on the relative merits of these various studies, the work of Flannery clearly has had by far the widest exposure in American cartography. This is because the major textbook in cartography used in college courses initially reported Flannery's research and conclusions in 1960 and expanded on them in a revised edition in 1969. The latter provided a table of radius index values and a full explanation on its use to correct graduated circle sizes based upon Flannery's research.[13]

Practically then, it is the Flannery correction that is readily available to those that design and draw graduated circle maps. From a realistic view, it is probably the *only* correction system which most such designers and draftsmen are likely to employ at the present time.

In order to examine the Flannery correction system in some detail and under a variety of map conditions, a series of test maps were prepared. In its simplest form, open circles were used on a white background. In Figure 3, the original open circle map (Test A) reported earlier is seen to the left. The new increased sized open circle map (Test B) using the Flannery correction is seen to the right.

This new test map and its several versions all used a series of circles which varied proportionally in size from 1 to 10, just as had the original map series. This represented only a modest range of circle sizes compared to the work of some researchers. Thus, Flannery initially used tests where the circle size differences ranged primarily from 1 to 50. On the other hand, circles in the Dickinson test ranged only from 1 to 12.3.[14] Because of the relatively small differences in circles it was expected that results of these new tests might show very accurate user results of perception of differences in circle sizes.

The open circle map with Flannery specified circle sizes was given to 200 college students. The exact same comparison questions were used as in the original test maps. In addition, supplemental maps varying the shading of the circle or its background, also paralleling the original tests, were given to 263 other students. Samples of these supplemental maps used in Test B are seen in Figure 4. Results from use of these various maps were about the same with one exception which will be mentioned later.

The test results of using both the open circle map alone and all the circle maps are shown by Figure 5a and b. There were little differences found. In all the Flannery circle map tests, 463 respondents made 4,489 separate comparisons of circle sizes. Some 40.8% underestimated the

Test A — "Normal" Circles

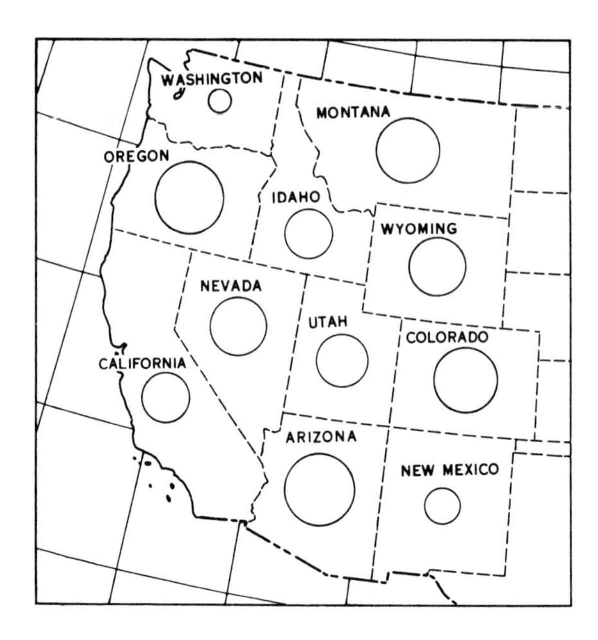

Test B — Flannery Correction

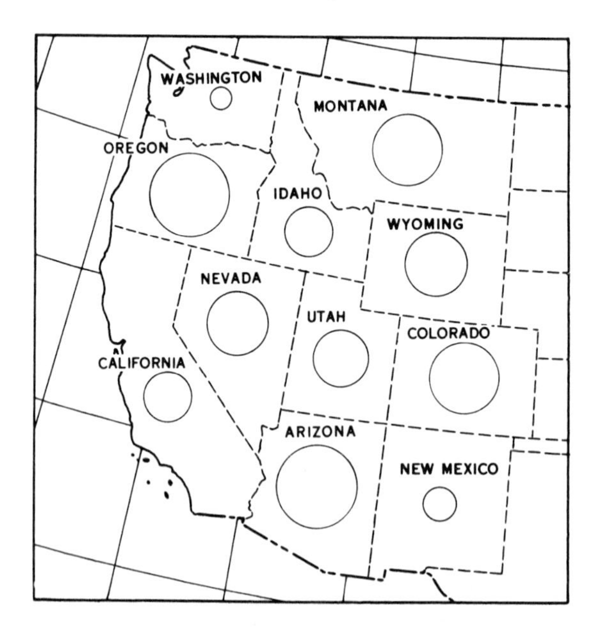

Figure 3.

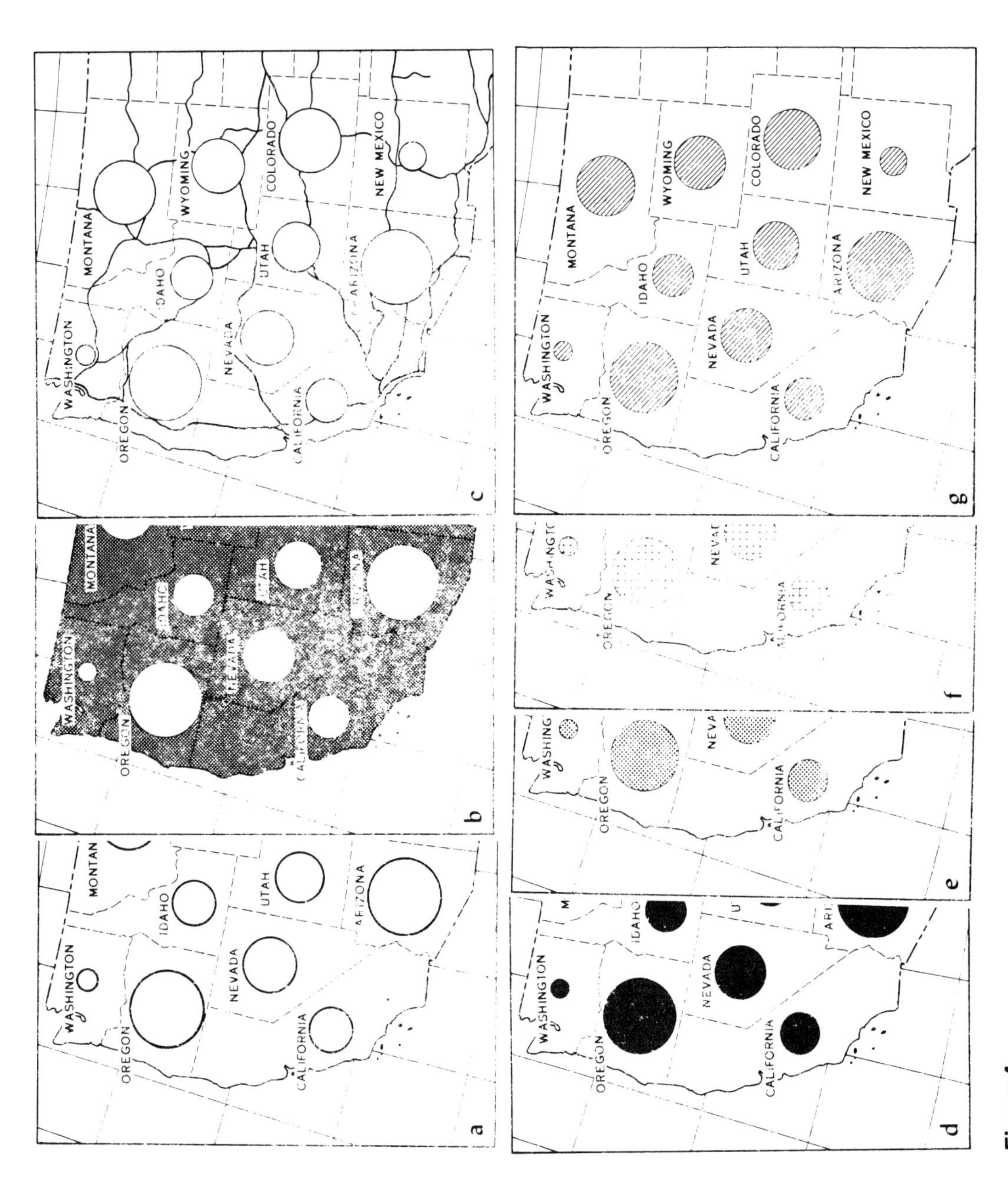
a
WASHINGTON
OREGON
IDAHO
MONTAN
NEVADA
UTAH
CALIFORNIA
ARIZONA
b
WASHINGTON
OREGON
MONTANA
NEVADA
UTAH
c
WASHINGTON
OREGON
MONTANA
IDAHO
NEVADA
UTAH
WYOMING
COLORADO
CALIFORNIA
ARIZONA
NEW MEXICO
d
WASHINGTON
OREGON
NEVADA
CALIFORNIA
e
OREGON
CALIFORNIA
f
g
WASHINGTON
OREGON
MONTANA
IDAHO
NEVADA
UTAH
WYOMING
COLORADO
CALIFORNIA
ARIZONA
NEW MEXICO

Figure 4.

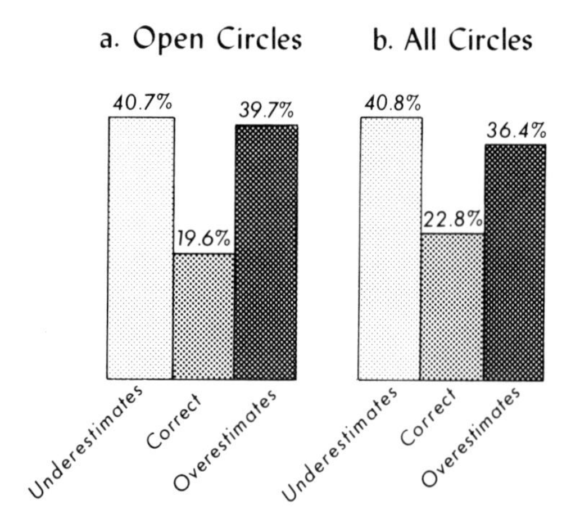

Figure 5. Test B—Flannery Correction of Circle Sizes.

size of large circles compared to small circles, 22.8 had correct answers, and 36.4% overestimated them. The increased circle size used on the Flannery maps resulted in respondents giving on the average higher estimates of value than in the original series. Thus, the Flannery series contained far fewer underestimates and more overestimates.

According to Flannery, ". . . in any series of tests of estimates of the value or relative size of the symbol, there should be approximately as many overestimates as underestimates but with a concentration of estimates near the correct value.[15] These results showed that overestimates and underestimates were about the same in the Flannery series. Unfortunately, the concentration of estimates of correct value was little different between series. Thus, these results alone failed to prove conclusively that the Flannery circles were an improvement over the old method of constructing them.

But Flannery only said that there should be a concentration of estimates "near" the correct value. If one circle is exactly four times the size of another but a map user perceives it as three times that size, the user error is only 25%. The impression being received is close to fact and perhaps this may be adequate for many kinds of information presented. Shown in Figure 6 are the Test B results when the answers were tabulated to 25% of the correct answer. So many underestimates and overestimates were within 25% of the correct answers that the correct answer rate was raised to 49.1%. This same type of tabulation was made with the Test A circle results but the correct answer rate was raised to only 31.7%. Clearly, on the basis of this comparison, user perception of graduated circles made following the Flannery correction appeared to be higher than those following the old method.

As mentioned earlier, the introduction of various maps changing the shading of the circles of their background did not affect user perception markedly. The single exception proved to be sample maps with open circles on a dark shaded background (Figure 4b). In both Test A and Test B, students on the average estimated circle sizes more accurately on these than on any other maps.

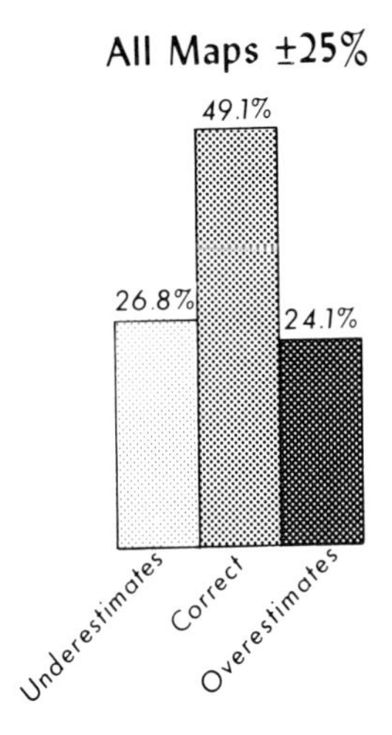

Figure 6. Test B—(Flannery Correction).

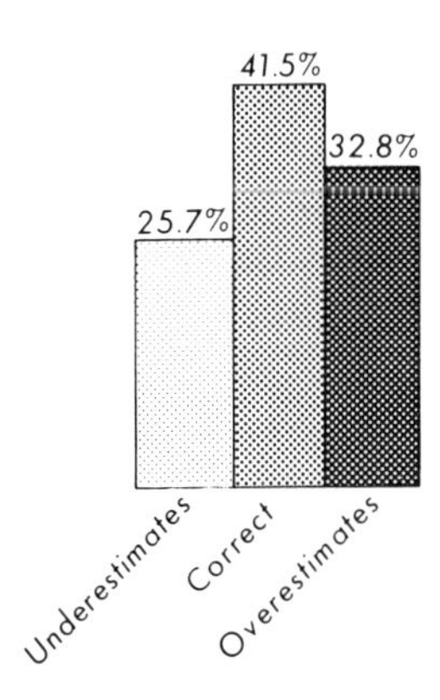

Figure 7. Test B—(Flannery Correction).

The greatest improvement was recorded by the Test B answers. As shown in Figure 7, when open circles were used with a dark shaded background, correct answers increased from 22.8% to 41.5% and the underestimates and overestimates correspondingly decreased. Apparently this dark background assisted users in some fashion in perceiving circle size differences. However, no conclusion are offered as this test was tried by only 65 respondents who made about 650 size comparisons—a rather small sample.

THE VALUE OF LEGENDS

Even when recorded to 25%, the percent of correct answers remained disappointingly low. It is the contention of Meihoefer that no modified scale of circle sizes, either the one developed by Flannery or any others, can solve this problem. He states "The actual values that a circle represents cannot be perceived by most persons on the basis of visual inspection and comparisons of circle sizes."[16]

These experiments by Meihoefer involved mainly testing several versions of a map of West Germany which contained 35 proportional circles drawn in seven distinct size classes. He found that a legend illustrating the various circle sizes depicted on the map aided the reader enormously in perceiving correct circle sizes.

This and other work seemed to suggest that map users are unable or at least unaccustomed to evaluating quantitative symbols without the visual crutch or cue of a legend. After all, if users need a legend for understanding size class circle maps as Meihoefer demonstrated, then they probably need one even more for understanding continuously graduated circle maps. All the sample maps tried so far had required respondents to make their estimates entirely on the basis of comparing various circles within the map neatlines. It was decided that a legend showing the meaning of circle values, even when given in map data figures, should be of assistance to users in evaluating relative circle sizes.

A variety of legends may be prepared for graduated circle maps. Some possibilities are shown by the seven samples in Figure 8. Samples D and E, shown drawn horizontally, are also often seen in a vertical arrangement. While all these forms have been used at times, probably the most commonly seen is that shown by Sample B. On a review of all maps in the *Annals of the Association of American Geographers,* 1971-1976, which used continuously graduated circles and also contained a legend, over three-fourths of them were of the Sample B type.

Robinson pointed out that a legend similar to Sample D ". . . is probably easier than a nested set *(viz., Sample B)* for the map reader to use," and later adds that, ". . . side-by-side array is more efficient for the reader."[17] Studies by Clarke indicated that legends similar to Samples F and G were much preferable for visual evaluation by readers.[18]

Given the choice of these seven forms of legend, Samples A, B, and C have the advantage of taking up relatively little space on the map. Thus, their presence is more likely to complement the map display than detract from it. By contrast, samples F and G consume considerable amounts of map space, and with some ranges of statistical data, can actually appear as prominent as the map itself. Thus, they may offer more interference than help in the readability of the graphic display.

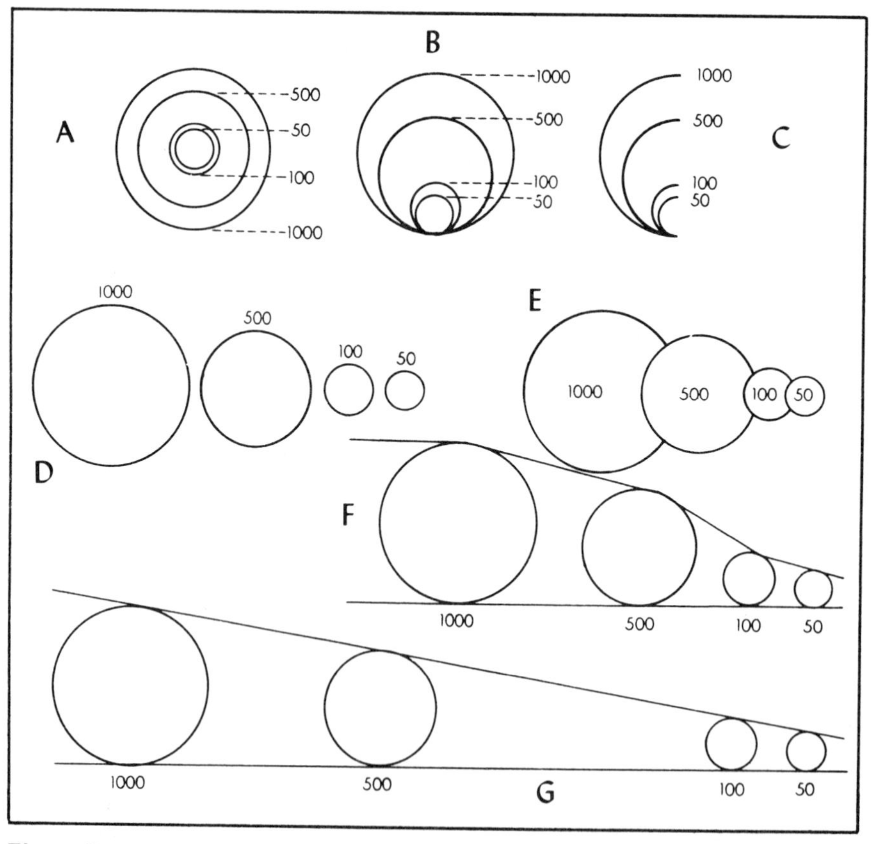

Figure 8.

Sample B appears to be an attractive, visually appealing legend, and that may explain its persistent use. It is certainly more attractive than Sample A and if not more attractive, probably more meaningful to users than the half circles in Sample C. Samples F and G may appear unattractive to many users while Samples D and E, the midpoints between the extremes, are reasonably pleasing and may be more functional than Sample B.

In order to examine the importance of a legend in graduated circle map evaluation as well as to discover if different types of legends affected this evaluation, two legends were added to the bottom of different Test B open circle maps. Sample B and D legends were chosen. Sample B was selected because of its common usage. Sample D was selected because it appeared to be clearer and possibly more informative than B. In both uses, several representative circles with simulated data values were given and examples of both the smallest and largest circles found on the map proper were included in the legends. Both legends used the same number and same size of circles. Only the visual arrangement of these circles was different.

These two new versions of the Test B map series using the same questions as before were administered to 60 new respondents with 30 respondents examining each form of legend. A tabulation of results of those using the map with the Sample B or common usage legend showed a marked increase in accuracy of size perception over the previous legendless map. While the percent of underestimates and overestimates continued to be about equal, the percent of correct answers increased from 19.6% to 31.1%. When the results were tabulated to 25% of the correct answers, the correct answer rate was raised substantially to 57.8%.

Results of those using the Sample D or expanded form of legend were markedly superior to those using the legendless or Sample B legend. While, as before, the percent of underestimates and overestimates continued to be about the same, the percent of correct answers increased to 45.8%, which is well over two times the initial legendless test and markedly above the Sample B legend test. When the results were tabulated to 25% of the correct answers, the correct answer rate was raised dramatically to 77.5%.

These new tests seemed to indicate the importance of a prominent, clear legend for accurate visual evaluation of graduated circles. In addition, although the sample of respondents was rather small, the results suggested that the expanded legend form of Sample D was superior to the conventionally used legend form of Sample B. These results are displayed in Figure 9.

OBSERVATIONS ON USER PERCEPTION

One observation on circle perception that can be offered is that, regardless of the test map used, a high percent of respondents were consistent in that they either underestimated most of the comparisons, overestimated most of them (and often overestimated them excessively), or, in the case of some respondents, answered most correctly or nearly so. In short, there are enormous differences in perception among individual map users but there is consistency in that a large group of users are underestimaters and another large group are overestimaters. The Flannery change in relative circle size improved the average estimate but it did not eliminate this grouping of underestimaters or overestimaters.

The proximity of the circles being compared was important too. Thus, the percent of correct or nearly correct answers when comparing Oregon with Washington was very much higher than when comparing Arizona with Washington, even though the circles for Oregon and Arizona were exactly the same size.

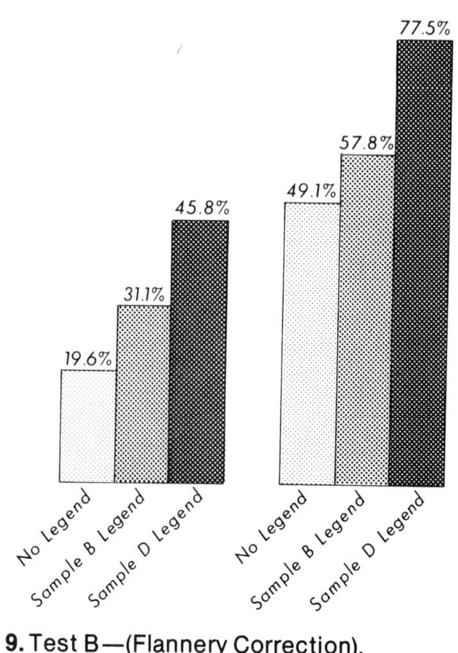

Figure 9. Test B—(Flannery Correction).

Some tests requiring users to combine circle values were attempted. As a sample, users were asked how many times larger was Arizona than Washington and Idaho combined. Five such questions were tried with a small sample of respondents, about half using the Series A maps and half using the Series B maps. The answers as a whole were extremely poor and suggested that map users can not combine the areas of graduated circles with any ease. Still, those using the Series B or Flannery correction maps had decidedly better estimates than those using the Series A maps. In spite of publication of the Flannery research over 15 years ago, there appears to be no widespread acceptance of his circle size correction system. This is suggested by examining current maps in the *Annals* and other professional journals. Careful measuring indicates relatively few such graduated circle maps are employing the Flannery correction. Many appear to be using proportional area circles as used in the Test A maps. Others are using graduated circles that simply defy understanding of the system being used, if any. Flannery himself observed this lack of acceptance of his technique and suggested: "Perhaps the system is suspect."[19] It seems more likely that this research is being disregarded through ignorance of it as most graduated circle maps probably are being produced by draftsmen with insufficient training in thematic cartography who give little or no thought to accurate user perception of symbolization.

In summary, the continuously graduated proportional circle map, in spite of its seemingly simple and graceful appearance, presents a complicated form of symbolization not readily understandable by many map users. There is a high degree of inaccuracy in user perception of circle values. The Flannery correction of circle size improved user perception of circle values and improved it considerably when results were tabulated to 25% of the correct answers. Results of the Flannery circle tests were improved even further when open circles against a shaded background were used and when a large prominent legend was provided the user. Clearly, additional examination of this problem, including testing of various types of legends, is warranted before any final conclusions can be reached on the ideal method of using graduated circles as a means of visual communications.

NOTES

1. G. Ekman, R. Lindman, and W. William-Olsson, "A Psychophysical Study of Cartographic Symbols," *Reports from the Psychological Laboratory, The University of Stockholm,* no. 91 (1961), p. 12. T.W. Birch, *Maps—Topographic and Statistical* (Oxford: Clarendon Press, 1964), p. 207.
2. G.C. Dickinson, *Statistical Mapping and the Presentation of Statistics* (London: Arnold, 1963), pp. 87-88.
3. Birch, p. 205; Dickinson, p. 87.
4. As reported in: J.J. Flannery, "The Relative Effectiveness of Some Common Graduated Point Symbols in the Presentation of Quantitative Data," *The Canadian Cartographer* 8 (1971):97.
5. Hans-Joachum Meihoefer, "The Visual Perception of the Circle in Thematic Maps—Experimental Results," *The Canadian Cartographer* 10 (1973):83.
6. Dickinson, p. 113.
7. Flannery, pp. 96-109. J.J. Flannery, *The Graduated Circle: A Description, Analysis, and Evaluation of a Quantitative Map Symbol,* unpublished Ph.D. dissertation (Department of Geography, University of Wisconsin, 1956). P.V. Crawford, "Perception of Grey-Tone Symbols," *Annals, Association of American Geographers* 61 (1971):721-735. R.L. Williams, *Statistical Symbols for Maps: Their Design and Relative Values* O.N.R. Project Report NRO 88-006, NONR 609 (3), Yale University, 1956. J.R. Mackay, "Experiments with Some Symbols and Map Projections," *Annals, Association of American Geographers* 44 (1954):225-226. G. Ekman, and K. Junge, "Psychophysical Relations in Visual Perception of Length, Area, and Volume," *Scandinavian Journal of Psychology* (1961), pp. 1-10.
8. Dickinson, pp. 85-89.
9. ———, p. 88.
10. See footnote 7.
11. Meihoefer, p. 65.
12. Nineteen circle size tests are compared and summarized in: B.D. Dent, "Communication Aspects of Value-by-Area Cartograms," *The American Cartographer* 2 (1975):158.
13. A.H. Robinson, *Elements of Cartography* (New York: John Wiley, 1969), pp. 125-126, 368-369.
14. Flannery, 1971, p. 98; Dickinson, p. 86.
15. Flannery, p. 98.
16. Meihoefer, p. 68.
17. Robinson, 1969, pp. 126-127.
18. John I. Clarke, "Statistical Map Reading," *Geography* 44 (1959):96-104.
19. Flannery, 1971, p. 100.

The Teaching of College Geography

Introduction

Accompanying the impressive gains in geographic research in recent years has been an exponential growth in the number of students taking courses in geography in colleges and universities in the United States. This expansion of enrollments and the growing concern of college students, geography faculty, and administrators for effective teaching have led to an increasing emphasis on the examination of the goals, methods, and outcomes of geographic education. The attention paid to these problems by diverse committees of the Association of American Geographers and the National Council for Geographic Education attests to the importance assigned to pedagogical improvements in geography.

The first three papers in this section reflect the experience of geographers who have attempted to enhance the quality of their teaching by introducing new approaches to their courses. Judging by student evaluations and other sources, these pedagogical innovations have succeeded admirably in attaining their objectives.

The transformation of introductory courses in physical and human geography into an audio-tutorial format is the theme of the papers by Robert Kingsbury and Alan Backler. Both authors cite their indebtedness in this respect to Samuel Postlethewait of Purdue University who has been the dominant figure in the development of the audio-tutorial instructional approach in general. Backler also was able to use the cumulative experience of Kingsbury, who introduced this mode of instruction to introductory courses at Indiana University in the late 1960s. Over 2,000 students a year on the Bloomington campus attend audio-tutorial courses in physical and human geography. They seem to welcome the opportunity to proceed at their own pace in mastering the course materials presented simultaneously by slides and tapes in our audio-tutorial laboratories and to be able to obtain assistance from instructors in the laboratories in answering any questions which might arise. These courses also have carefully defined objectives, expected performance levels, and thorough procedures of evaluation. In addition to evaluations from students at the time they took his course, Backler participated in telephone interviews of former students, grouped by final-grade levels, and gained some important insights in the reasons for student successes and failures in this course.

Another approach designed to enhance the effectiveness of geography courses is the classroom use of game simulations which assign students various decision-making roles with respect to a hypothetical version of a real-world problem with a locational component. John Jakubs provides an enlightening analysis of the pedagogical uses and limitations of gaming-simulation methods by examining two games which he devised and used in courses. One of these simulation games deals with the conflicts associated with the location of undesirable public facilities and has been quite successful. The other game treated some of the difficulties resulting from residential and political decentralization in large cities and it encountered a wide range of problems. On the basis of his experience with these games, Jakubs cites some caveats which should be taken into account for gaming simulation to fulfill adequately its important pedagogical functions.

The concluding paper by Gary Manson focuses on the general nature and problems of introductory courses in geography. Most of this paper is devoted to a survey of the diverse views concerning four fundamental elements involved in introductory courses: purpose, content, modes of learning, and evaluation. On the basis of this review, he combines these views into a conceptual model comprised of four continua, each of which corresponds to one of the elements cited above. He then positions introductory courses in geography on each continuum according to his perception of their current orientation. Finally, Manson derives some ideal-type combinations of these continuous variables on the basis of the purposes of introductory courses.

Advantages of the Audio-tutorial Learning System

R. Kingsbury
Indiana University

Audio-tutorial instruction in the Department of Geography at Indiana University was introduced ten years ago after we learned of the innovative work of Dr. Samuel Postlethewait of Purdue University in this field. At the time, we had become discouraged over the use of graphic materials in certain courses. These materials encompassed a wide variety of maps, diagrams, photographs, and models which came in many different forms, including slides, filmstrips, motion pictures, overhead transparencies, wall displays, printed sheets, and single copy items suitable for desk use only. The difficulties of handling such a mix in a conventional classroom had become increasingly apparent. Because the Postlethewait technique offered a potentially viable way of incorporating these materials effectively into the learning situation, we decided to experiment with his system.

We converted a sophomore level course in Maps and Aerial Photographs first and, then, parts of two other courses. After comparative testing and evaluation by outside experts seemed to prove the system worked in these courses, the initial audio-tutorial version of our beginning physical geography course was written. Since then a version of our introductory human geography course has been converted to the system also.

Most of the comments in this paper apply to procedures and experiences in the freshmen level physical geography course, "Physical Systems of the Environment." This course was first offered in the late 1960s and has been rewritten several times. It is now given every semester and most summers. Currently, the audio-tutorial version is taken by about 1,200 students yearly at the Bloomington campus and two of our regional campuses. Enrollment in the present semester at Bloomington is about 450 students.

The outsider often thinks that the audio-tutorial system consists of lectures placed on tape. This is not the case. Lectures delivered to a large group of students in a classroom or auditorium tend to be boring and ineffective when heard on tape by a single student in a carrel. Further, in the carrel situation the instructor can introduce many techniques and materials not appropriate to the classroom. Thus, course conversion requires the writing of a completely new presentation for each audio-tutorial lesson. Development of such new presentations is a slow process and the typical lesson in our physical geography course has taken an average of 50-55 hours to complete.

Each audio-tutorial lesson represents a carefully sequenced program of study. To this end, many different approaches are used in a coordinated series of activities where the learner is motivated to become involved in participating in the learning process whenever possible. In each lesson, the unit of work is discussed in detail with major emphasis placed upon a series of learning or performance objectives. The student is asked numerous questions and requested to complete a variety of tasks that involve working with: color filmstrips; aerial photographs and stereoscopes; maps, graphs, and diagrams; readings from the textbook and other sources; relief and other models; and other materials.

After experimentation with audio-tutorial the past ten years, a number of advantages are evident and some of those are discussed below. Included also are some of the results of our last comprehensive student course evaluation.

STUDENT INVOLVEMENT

The student changes from a passive receiver to an active participant in the learning process. Postlethewait has called this an emphasis on student learning rather than on instructor teaching. Essentially the tape acts as a private tutor to involve the student in his own learning. Whereas formerly only one person in the classroom answered a question, now every student considers the question and answers it. To this end, all lessons require the student to examine maps, graphs, and diagrams, interpret and often plot data on them, and draw conclusions based upon their work. A variety of other techniques using readings and equipment is employed to engage the student in active pursuit of learning. In a written evaluation, 88% of the students believed that this involvement had improved substantially their own level of learning.

STUDENT RESPONSIBILITY

The student is given responsibility for his own learning. He must decide himself when he will use the Learning Center where the independent study carrels are located and for how long a period. Since he is using his own time, he is likely to use this time effectively. The best learning is done through concentration and most students seem to come to the Learning Center to concentrate. In this connection, our evaluation indicated that 92% of the students thought the complexity and length of course assignments were reasonable.

STUDENT ATTENTION

When using the tape lesson, the student occupies an individual, private carrel with his own tape player, filmstrip viewer, and other materials. Headphones are used always. Tapes demand attention. Students are not distracted by other students, by the mannerisms of instructors, and by extraneous noises. When a student tires of working, he turns off the player, walks around, perhaps obtains a soft drink, and otherwise relaxes before continuing the lesson. In evaluations, 78% of the students believed their level of concentration was higher when using a taped presentation than when listening to an auditorium lecture.

FLEXIBILITY OF SCHEDULING

The student can use the lesson at the time of his own choosing. Our Learning Center is open 38-40 hours a week including morning, afternoon, and evening hours. Students prefer to use the lesson at varying times depending upon their own maximum efficiency periods, their class schedules, their outside work hours, and other factors. Some students come once a week while others divide the lesson into two or more visits a week. Some prefer daylight hours and other prefer evening hours. These differences can be accommodated so that most students can use the materials at the time best suited to them. In addition, the system easily serves the student who misses a lesson because of illness or other reasons or the student who desires to review a previous lesson. Previous lessons are available in the Learning Center and in special carrels in the nearby Geography and Map Library. In evaluations, 94% of the students cited this scheduling flexibility factor as a major attraction of the course.

ACCOMMODATION OF INDIVIDUAL DIFFERENCES

The system accommodates individual differences among students in mastering the subject matter. Students can adopt their study pace to their ability to assimilate new materials. Some students spend an hour with the weekly lesson, while others spend two or three or even more. For some students, repetition is necessary, but for others, it is boring. Thus, students who fail to understand materials the first time reverse the tape and repeat the appropriate sections. Better students omit sections of the tape lesson with which they are already familiar. In this way, the time of better students is used more effectively and their interests are not dulled by needless repetition. In evaluations, 85% of the students agreed that the taped material was presented at a pace suitable for maximum comprehension.

INDIVIDUAL ATTENTION

A qualified associate instructor is on duty at all times during the open hours of the Learning Center. When a problem arises, students will seek help from this instructor. To encourage this, early in the course the student is requested to turn off the tape player and discuss with the instructor a specific lesson objective. Last year, in addition to the associate instructors, we used undergraduate tutors. These were paid undergraduate students who had taken the course previously and were particularly qualified in working with fellow students. This undergraduate peer instruction proved both popular and effective but unfortunately the funds for its continuance were not available this year. In evaluations, 90% of the students thought the associate instructors of the Learning Center were well prepared in the course subject matter and 77% said they were helpful in problems relating to the lessons. In addition, 64% indicated they consulted regularly with the undergraduate tutors.

RELATIONSHIP BETWEEN SENIOR INSTRUCTOR AND STUDENTS

The carrel situation allows a simulated one-to-one relationship between senior instructor and student. In writing and recording each lesson, the aim has been to further this relationship by a conversational type of presentation rather than a formal lecture. With the instructor's time freed from preparation and formal lecturing, he can become involved more actively with individual students and with small groups of students. Such involvement has included: working scheduled hours each week in the Learning Center; helping with the quiz sessions; visiting discussion classes; having his own discussion sections; holding review sessions; and offering oral quizzes to students who prefer that means of testing. These kinds of activities aid in making the senior instructor both well known and readily available to students in the course. In evaluations, 74% of the students appeared satisfied with opportunities for personal contact with the senior instructor.

COURSE ORGANIZATION

Each lesson is far better organized when taped than when given as a lecture. We use 60- minute tape cassettes, including one per lesson and one per week. Each lesson requires the use of the course textbook and a specially prepared study guide which has been published commercially. Coordination of ideas and materials to produce a meaningful, effective lesson in a specific amount of time requires more than just an outline before the tape is made. It requires a word-for-word script, several rehearsals, and numerous script revisions and retaping. If this were done properly, it would result in a carefully assembled and coordinated presentation. In evaluations, 94% of the students indicated they thought the taped presentations were well prepared and organized.

LESSON GOALS

The student is made fully aware of the expected level of achievement in each lesson by a series of six to ten learning or performance objectives. These are mentioned at the beginning of each lesson in the course study guide and discussed in detail by the taped presentation. With the objectives of the lesson clearly stated, the student can prepare in a purposeful way for the testing that follows. A self-test board in the Learning Center helps in this preparation as do review questions which are given at the end of each lesson in the study guide. In evaluations, 92% of the students said they knew what was expected of them in the course, 91% indicated they had no doubt about the learning objectives or what was needed to meet them, and 94% indicated the tests constituted a fair examination of these objectives.

INTEGRATION OF MATERIALS

The carrel situation permits use of any appropriate mix of graphic and textual materials as desired for a particular lesson. Maps, diagrams, and photographs are used extensively and many of them have been placed on specially made color filmstrips. Other graphics in the form of atlases, sheet maps, aerial photographs, and relief models are found in the Learning Center. The student brings both the textbook and the study guide to each carrel lesson. Discussion of textbook maps, graphs, and readings is part of every lesson. The tapes even contain deliberate pauses for the student to open his textbook to particular pages under discussion. Maps and diagrams which are not found in the textbook and which are unsuited to color filmstrip display are included in the study guide and these are also used in each taped lesson. In evaluations, 83% of the students felt this integration of materials was well handled.

SUPPLEMENTARY MATERIALS

The opportunity to examine topics in more depth is provided to the rapid learner as well as other students who are willing to exert the additional effort and spend the additional time.

Exercises and reports known as "quests," which involve using supplementary materials furnished in the Learning Center and in the Geography and Map Library, are available for each lesson. Since the student may submit these for grading, all quests are changed each semester. Under our adaptation of the mastery learning system, completion of a certain minimum number of quests is mandatory for a "B" or "A" final course grade. Thus, these supplementary materials are used each week by a considerabl number of the students enrolled in the course. In evaluations, 78% of the students considered the quests useful in expanding their knowledge of the course subject matter and 87% approved the mastery learning system as used in the course.

EQUIPMENT AND MATERIAL COSTS

While establishing and operating an audio-tutorial facility does require expenditure for initial purchase and maintenance of equipment, this equipment need not be either elaborate or very costly. Thus, our Learning Center functions very well using inexpensive portable tape players and filmstrip viewers. The cost of equipment is under $100 per carrel. In addition, certain previous costs are lowered by the use of the audio-tutorial system. This is because the use of some materials and equipment is spread out over a longer period each week than in the usual classroom or laboratory situation. As an example, we needed 44 lens stereoscopes and 44 copies of each stereogram formerly to operate two laboratory class sections simultaneously, but now one-fourth of this amount is adequate inasmuch as only small numbers of students in the Learning Center are using the equipment at the same time. For the same reason, smaller numbers of atlases, topographic maps, weather recording instruments, relief models, and other equipment are required.

WEEKLY DISCUSSION CLASSES

A weekly discussion class attended by all students makes an extremely important contribution to the success of the audio-tutorial program. The discussion classes are limited to about 22 students per section and are led by qualified associate instructors. The weekly session is important in terms of contact among students and interaction between students and instructors. Review questions in the study guide generally are used by instructors as a basis for examining the lesson. The discussion class becomes one of review, reinforcement, and correction preparatory to the testing session. During one week, a role-playing activity is discussed on the taped lesson and used in the class. A bonus of the discussion class is feedback to the instructors on the success or failure of various features of the instructional program. In evaluations, 75% of the students expressed approval over operation of the discussion classes.

USE OF THE COURSE ON OTHER CAMPUSES

The system has contributed to the improvement of instruction at off-campus centers. This introductory physical geography course, utilizing the same audio-tutorial tapes and materials, is now in use at two of our regional campuses. Library carrels at the regional campuses duplicate most of the Learning Center facilities at Bloomington. Instructors and other staff personnel have been tutored in the techniques of operating the course and its discussion classes.

In summary, the audio-tutorial system has many advantages. Some of those given are peculiar to the system, whereas others are an outgrowth of our use of the system and might be applied equally to conventional courses. In our comprehensive evaluation, 88% of the students expressed approval of the system by indicating they would take another audio-tutorial course if it were offered in a subject of interest.

Converting a course from lecture to audio-tutorial requires a very considerable amount of time and thought. Once converted, the course has become "public" in that it is open to examination and criticism by anybody. While this may have some drawbacks, the advantages far exceed the disadvantages. The instructor now can view his course in its entirety and weak or ineffective features are more easily discovered and corrected. Student and staff evaluations help enormously in recognizing where other changes should be effected. Experience has indicated that the process of change begins almost immediately upon conversion and never ends. Thus, each year major revisions of tapes and graphic and written materials are made in an attempt to improve the course.

The audio-tutorial learning system is *not* presented here as the ultimate answer. Nevertheless, for some instructors and for some courses the system appears to work well and is an effective alternative to conventional classroom teaching.

REFERENCE

Postlethewait, S.N.; Novak, J.; and Murray, H.T., Jr. *The Audio-tutorial Approach to Learning.* Minneapolis: Burgess, 2nd ed., 1969.

The Audio-tutorial Approach in the Teaching of Human Geography

Alan L. Backler
Indiana University

The description of an introductory audio-tutorial course in human geography, which was developed at Indiana University, Bloomington, should provide additional insights into the nature of this approach. The course uses a self-paced, mastery-oriented approach to learning and has been in operation for six semesters, with an average enrollment of 240 students per semester.

BACKGROUND

When teaching a mass introductory course at a large state university one encounters students with diverse backgrounds, abilities, and interests: freshmen, graduate students, foreign students, rural students, urban students, majors, business students, elementary education students. One is always wondering where to "pitch" one's lectures. Many of the students are confused by the abstract, while certain students become bored with the concrete. Then there is the question of pacing. Are too many or too few examples being used? With these frustrations as motivation, I began to search for alternate teaching-learning strategies. Discussions with Robert Kingsbury convinced me that the audio-tutorial approach addressed itself to my major concerns. And the present version of the course, which will be described here, represents the second revision of the program which was instituted in the spring semester, 1973.

PHILOSOPHY UNDERLYING COURSE

Most students (perhaps over 90%) can master what we have to teach them in a particular course, regardless of their aptitude for that course. Course mastery assumes, however, that the appropriate learning procedures are used. Educational psychologists are just beginning to understand which educational procedures do in fact lead to mastery. Many of them argue that by incorporating into a course such procedures as specifically stated objectives, self-pacing, immediate feedback on progress, and active student responding, student mastery can be approached. The audio-tutorial technique is a learning strategy which incorporates the above mentioned procedures, among others, and is therefore an attempt to approximate the mastery model.

COURSE ORGANIZATION

Introductory Phase

During the first eleven weeks of the course, students are introduced to a variety of human geographical concepts, conceptual frameworks, and skills for manipulating geographically interesting data. The introductory phase is divided into eleven lessons, each lasting one week. The lessons are structured so that the students are not only introduced to the concepts, skills, and conceptual frameworks used by geographers but are given the opportunity to test their understanding and ability to make use of them.

Application Phase

In the last four weeks of the course, students are given the opportunity to apply what was learned in the Introductory Phase to a "real world" societal issue. Students are given raw materials concerning the issue and are asked to analyze it, making use of the appropriate geographical concepts, conceptual frameworks, and skills previously learned. The purpose of this phase of the course is to enable students to demonstrate to themselves that they are capable of analyzing societal issues from a geographic point of view as a result of what they have learned in the course.

COURSE COMPONENTS—INTRODUCTORY PHASE

General Assembly Session

The only time the entire class meets as a group is during the general assembly sessions. During these sessions, films used to introduce course units are shown, examinations are administered, and examinations are reviewed. No lectures are presented.

Audio-tutorial Presentation

The content of the course is taught through audio-tutorial presentations. Each weekly audio-tutorial presentation makes use of the student guide, a filmstrip specifically designed for the course, and a tape. The tape acts as a tutor guiding the student through the maze of readings, diagrams, pictures, tables, music, and maps contained on the student guide and filmstrip.

A number of special features of the audio-tutorial presentations are worth mentioning.

Performance Objectives: For each lesson, specific performance objectives are written into the student guide just after a statement outlining the lesson for the students. These behaviorally specific statements reflect exactly what we expect students to be able to do when the lesson is completed. Students are encouraged to examine the performance objectives before starting the audio-tutorial presentation and to use them to structure their work. The content and sequencing of material contained in the lessons is carefully selected to assist students in the task of mastering the objectives. Furthermore, the exams are constructed so as to reflect these objectives as exactly as possible.

Active Response Exercises: Periodically during the audio-tutorial presentations students are asked to develop definitions, construct graphs, plot data, and analyze readings. Each of these activities is tied to a specific performance objective designed to give students practice in

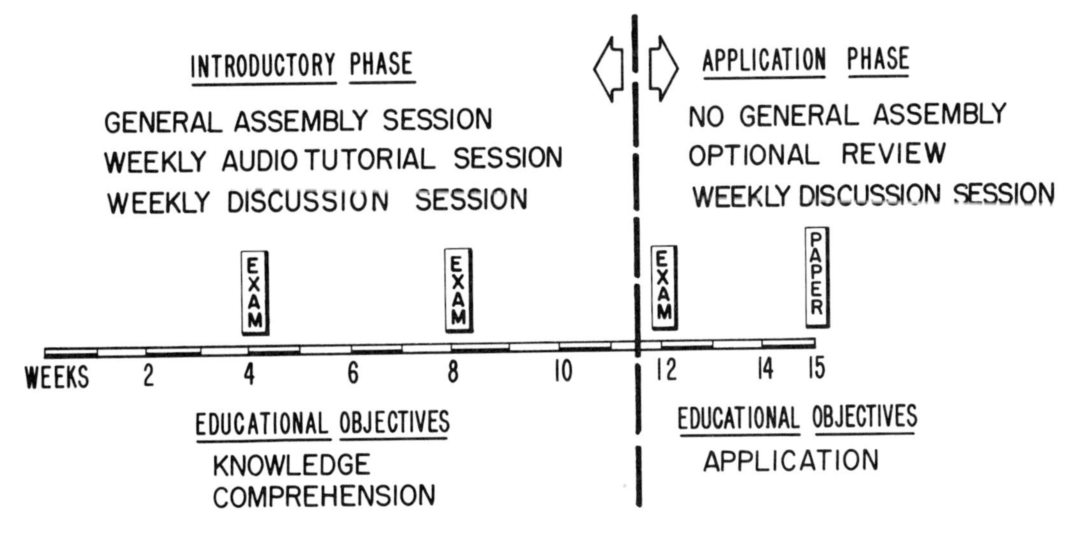

Figure 1. Course Design, Human Geography G-110.

performing the objectives and to encourage them to become actively involved in the learning process. Students also are encouraged to discuss the results of these exercises with the learning center instructor, thereby getting immediate diagnostic feedback on their progress through the lesson.

Practice tests: At the end of each lesson students can take a diagnostic practice test covering the lesson. Each practice test contains the performance objectives for a particular lesson and associated test questions similar to those which appear on the actual examination. By taking the practice tests students see how specific performance objectives are translated into the test questions and determine whether they are able to perform the objectives for a particular lesson. Learning center instructors are available to make corrective suggestions on the basis of practice test results.

Discussion Session

Each week the students meet in small groups for a discussion session which is associated with that week's audio-tutorial presentation. The purpose of the discussion session is to provide students with an opportunity to practice performing the weekly objectives using material not contained in the audio-tutorial presentation.

COURSE COMPONENTS—APPLICATION PHASE

During the application phase of the course, students work individually on their final projects. A taped presentation, student guide, and resource material packet are used to introduce this phase. The student guide contains a copy of the evaluation which will be used to grade final projects. Thus, the student knows beforehand exactly what criteria are used in evaluating his work. Weekly newsletters are used to answer general procedural questions asked by students. During the discussion sessions (there are no audio-tutorial presentations or general assembly sessions in this phase) student progress is monitored. Students are given specific assignments to complete before attending discussion sessions which each week more closely approximate the final project.

COURSE STAFF MEMBERS

In a course of this sort the instructor plays a somewhat nontraditional role. Most of his (or her) work is done as part of the preparation of the course while his assistants play a major role in the actual presentation of the course to students.

From the student's point of view, the most important members of the staff in this course are the associate instructors who are graduate students in geography. The A.I.'s handle the discussion sections and manage the audio-tutorial learning center. They administer, evaluate, and provide constructive feedback on the practice tests and on the active response exercises. They are generally available to help students with course problems.

The course instructor is responsible for: the selection of all materials used in the course; the organization, sequencing, and mode of presenting this material; the construction of practice tests, examinations, and the final project; the training and monitoring of associate instructors in the course; the administration and monitoring of the various instructional evaluation instruments; and the final evaluation of each student's progress. It is also the instructor's duty to handle general assembly sessions and to act as a clearing house for requests and complaints. The instructor of course, arbitrates in any cases of disagreement between students and associate instructors. Students are given a statement describing the features of the course on the first day of class, including information on grading which will be discussed in a later section of this paper.

CONTENT AND STRATEGIES

Before introducing the audio-tutorial form of G-110, I taught the same course with a lecture format for six semesters. My experience indicated that almost 100% of the students enrolled had never taken a human geography course at the college level, and had not taken geography since their world studies class in grade seven. Freshmen at Indiana University (about 50% of the enrollment in G-110) do not have strong mathematical backgrounds. A vast majority of the course enrollees (approximately 85%) only take one human geography course—to fulfill a social studies requirement in education or a social science requirement in Business and Arts and Sciences.

With these student characteristics in mind, the content of the course is non-mathematically oriented and assumes no previous work in geography. The course exposes students to all of human geography and provides them with a "bag of geographic tools" which can be applied elsewhere. In the introductory phase of the course, five general "research themes" are developed—man-environment relationships, location theory, spatial interaction, spatial diffusion, and regionalization. The selection of these particular themes was stimulated by *Geography*, Edward Taaffe, ed. (Prentice-Hall, 1970). Within each theme, associated concepts, skills, and conceptual frameworks (called "organizers") are developed. A list of the concepts, skills, and organizers used in the course at the present time is included in this report. (See Table 1)

Table 1
Concepts, Organizers, Skills

Theme	Concepts	Organizers	Skills
Man-land Relationships Lessons 1, 2, and 3	1. Human Geography 2. Physical Environment 3. Environmental Determinism 4. Natural Resource 5. Environmental Perception 6. Man, The Active Agent 7. Technology 8. Man's Attitude Towards Nature	9. Man-environment Organizer	10. Scattergramming
Location Theory Lessons 4 & 5	11. Site 12. Situation 13. Central Place City 14. Transportation City 15. Specialized Function City	16. Concentric Zone Organizer 17. Sector Organizer 18. Multiple Nuclei Organizer	
Spatial Interaction Lessons 6 & 7	19. Complementarity 20. Intervening Opportunities 21. Transferability 22. Accessibility 23. Time Space Convergence 24. Spatial Interaction		25. Graphing
Spatial Diffusion Lessons 8 & 9	26. Expansion Diffusion 27. Relocation Diffusion 28. Hierarchial Diffusion 29. Contagious Diffusion 30. Physical Barriers 31. Cultural Barriers 32. Voluntary Migration 33. Intervening Obstacles 34. Forced Migration 35. Spatial Diffusion	36. Voluntary Migration Organizer	37. Isopleth Mapping
Regionalization Lessons 10 & 11	38. Formal Regionalization 39. Formal Region 40. Functional Region 41. Nodal Region 42. Administrative Region 43. Gerrymandering 44. Stacked) 45. Cracked) Gerry 46. Packed) Mandering 47. Silent)		48. Choropleth Mapping

A variety of strategies and materials is used to develop the concepts, skills, and organizers. For example:

1. Scattergramming is taught using a narrated filmstrip sequence developed specifically for this course. After proceeding through the sequence, students are asked to create a scattergram and discuss it with the learning center instructor. The scattergram is introduced in Lesson 2—as a technique for establishing the existence of a relationship between physical environmental variables and human behavioral variables.
2. In Lesson 3 students draw maps of Bloomington from memory including on them those activities which they consider significant. These maps are then compared to other student maps drawn in response to the same instructions. The contrasts among the maps are used to illustrate the notion that people can perceive the same environment quite differently.
3. In Lesson 8 students listen to the song "The Streets of Laredo," which they identify as a traditional American cowboy song. They then hear another song "The Unfortunate Rake" which happens to be an ancient Anglo-Irish ballad which folklorists argue was diffused to the new world where it was adopted and adapted to local conditions as "The Streets of Laredo." The songs, then, are used to introduce some basic diffusion processes.

At the present time, no textbook is used in the course. Instead, material has been selected from a variety of courses and is presented as part of the weekly student guide.

In the Application Phase of the course, students use the concepts, skills, and organizers learned, their "bag of geographic tools," in the analysis of a societal issue. The issues presented are similar to ones which they would confront as concerned citizens. For example, in the fall semester, 1975, the students considered world hunger. They were given global data dealing with variety of social, economic, and population factors and articles discussing hunger taken from journals such as UNESCO *Courier*. They were asked to organize and make sense of these materials using the concepts, skills, and organizers learned.

EVALUATION

Evalution of Students

During the Introductory Phase of the course, three objective exams are administered, each covering 3-4 lessons. As noted above, specific performance objectives are written for each lesson. Students can see how these objectives are translated into exam questions and whether they are able to perform the objectives by taking the practice tests each week.

Exam questions are written to translate performance objectives as exactly as possible. Since the performance objectives are consciously written to reflect at least the "Knowledge" and "Comprehension" levels of Bloom's Cognitive Taxonomy, the exam questions must also reflect these levels. A further conscious effort is made to write exam questions reflecting *all* performance objectives in order to evaluate student mastery of all aspects of the Introductory Phase.

The pool of exam questions is improved each semester through the use of an item analysis program available from the Indiana University Bureau of Educational Studies and Testing. The program allows each question to be examined in terms of the proportion of students answering correctly and the correlation between student performance on that item and performance on the whole exam.

While exams are traditionally used to evaluate students, they can also be used to evaluate the course. The Division of Development and Special Projects developed a computer program which allows us to examine each students' response to each exam question individually and to arrange these data in a variety of ways. Using this program a number of course dimensions can be examined. For example, we can investigate if students who do poorly on the exam were able to answer "Knowledge" level questions but no "Comprehension" level questions. This has implications in the area of sequencing. If questions related to active response exercises were answered incorrectly by students who make use of the A.I.'s—this would have implications in the area of staff training. If students consistently answered questions related to specific lessons incorrectly, this would have relevance for the area of course design. We are just beginning to make use of the data generated by this program.

Letter grades for the exams are determined by the proportion of course material mastered rather than position on a curve. Clearly the adoption of a mastery philosophy necessitates the rejection of the notion that student grades will aproximate a normal curve. What we get instead is a very skewed exam grade distribution.

There is a small significant proportion of students who do not do well on the exams. In the fall semester 1974-1975, we began to examine these students. Our first strategy was to ask them why they had done poorly on the exams. Almost invariably some unique phenomena, such as deaths in the family, and thefts of notes, were used by the students to explain exam performance. Clearly another tactic was needed to investigate the problem of low achievement. In the spring semester, 1974-1975, graduate students in a seminar dealing with learning differences were each assigned to work with four to six, D and F students in G110. The graduate students (all in Instructional Systems Technology) examined the low achievers' study strategies as they related to the course and to their work in general. They also gathered data about their academic programs and backgrounds.

They concluded that the D and F students differed from other students along a variety of dimensions: (1) They had lower SAT scores, (2) they valued the outcomes of education less, (3) they spent less time at their studies, (4) they did not see how the course fit into their larger academic program, (5) they used ineffective planning and study strategies. The graduate students were able to help certain students alter some of these dimensions with fairly positive results.

In the Application Phase, students are evaluated on the basis of a final project which takes the form of a 10-15 page paper. As was mentioned previously, the students are given a copy of the evaluation form which is used to grade their papers when the final project is distributed to them. Each discussion leader grades his (or her) students' papers. To assure consistency in grading, all discussion leaders grade several papers in common and then compare and discuss results before going on to grade their assigned papers. During the Application Phase, the students are given specific assignments to complete before attending discussion sessions which each week more closely approximate the final project and focus on a particular aspect of the project evaluation form. This phase of the course is designed to test students' ability to operate at the "Application" level of Bloom's Taxonomy.

Evaluation of the Instruction

Beginning with the first revision of the course, made in the summer of 1973, a number of evaluation instruments have been periodically used. The instruments were specifically developed for the course with the aid of the Division of Development and Special Projects.

The first evaluation instrument which will be discussed was concerned with assessing the students' reactions to individual audio-tutorial presentations. In the Introductory Phase, we wanted to develop a feedback mechanism which could be used to make specific revisions in the development of materials and exercises. The instrument developed consists of Likert-type items with five response choices ranging from strongly agree to strongly disagree. Students complete the forms immediately upon finishing each lesson. The instrument was administered during three semesters.

Each weekly evaluation contained ten questions on a variety of topics. Some topics were included in each questionnaire, while others were included in only one or two of the weekly evaluations. The complete list of topics is as follows:

1. Student repetition of portions of the audio-tutorial presentations.
2. Helpfulness of the student guide.
3. Value of student guide introductory statement in orienting the student to the lesson.
4. Relationship between performance objectives and lesson content.
5. Student self-evaluation of his/her ability to perform objectives.
6. Tape voice quality satisfaction.
7. Helpfulness of learning center instructor.
8. Effectiveness of particular instructional techniques.

The data collected in the fall semester 1973-1974 covering Lessons 1-6 indicated that the content of the lessons was well organized, the objectives were clearly stated, the instructional material was clearly related to the objectives, and the instructional techniques were effective.

These data allowed us to identify weaknesses in the course, as well as strengths. For example, while the learning center instructors were expected to give students feedback, encouragement, and assistance in the active response exercises, the weekly evaluations indicated that fully a third of the respondents were undecided as to whether or not the learning center instructor was helpful. This information contributed to the establishment of a more rigorous associate instructor teaching program.

In a course of this sort it is argued that the student who makes use of the learning procedures included in each lesson will master the material as measured by exam grade. An evaluation instrument was developed and administered in the fall semester 1973-1974 to determine whether this argument was valid. Fourteen learning procedures available to the students in this course were identified as possibly contributing to differences in exam results (see Table 2). The relationship between these procedures and exam grades was examined while controlling for several aptitude and status variables.

Table 2

G-110 Learning Procedures

Read Objectives
Read Outline
Take Notes
Perform Activities
Discuss Work with A.I.

A.I. Was G-110 A.I.
Repeat Tape
Skip Sections
Complete Lesson Before Discussion
Advance Preparation for Discussion

Attendance at Discussion Group
Participate in Discussion Group
Practice Objectives in Discussion Group
Time Spent in Lab

The data were collected from 36 of the 156 students in the course. Twelve students were randomly selected from each of the high, medium, and low scoring thirds on the first exam. They were interviewed by telephone.

The most significant finding from the telephone interviews was that almost all the students with high and medium scores on the first test read the objectives for each of the first three lessons before turning on the tape, while less than half of the students with low scores did so. Most of the low-score students read the objectives only occasionally. The higher scorers were also slightly more likely than the low scorers to read the outline of the lesson before beginning to discuss their work with the A.I. in the learning center, and to advance preparation for the discussion group meetings. There were no noticeable differences among score groups in any of the remaining variables included in the interview. Most respondents indicated that they were making use of each of these other strategies.

One dimension of the analysis involved an examination of the relationships among time spent in learning center, SAT score, and test score. It was found that students who spent a lot of time in the learning center did well on the exam regardless of SAT scores. While this finding suggests that, in this course, effort is more important than ability, further tests of this relationship are needed before the results are broadcast widely.

One use to which these findings have been put is in the area of student advising. We can suggest corrective strategies which are based on empirical information to students who are not doing well.

A follow-up interview schedule was developed to assess the students' reactions to several aspects of the course. The instrument was administered six weeks after the fall semester, 1973. We were interested in determining whether students understood the final project in the course and its purpose, and the relationship between the Introductory and Application Phases of the course. Their general reaction to the audio-tutorial approach also was sought.

The sample of students interviewed was selected in a manner similar to that used in the Achievement Correlates Evaluation (it was also of similar size). However, in addition to stratifying the sample on the basis of exam score (in this case, total exam score) the sample was also stratified on the basis of whether they had been previously interviewed.

The results of this interview indicated that over 80% of the students recalled what the final assignment was and its purpose. Over 90% of the students interviewed recalled that there was an instructional relationship between the Introductory and Application Phases of the course, and they were able to describe what that relationship involved. Over two-thirds of the students expressed a favorable attitude towards the way in which the course was organized.

With respect to the audio-tutorial approach itself, over 90% of the students expressed a favorable attitude towards it. An even greater percentage (97%) indicated that the audio-tutorial approach provided an effective and efficient means of learning the material. However, when asked if the course were enjoyable, 47% said yes, while 43% were indifferent, and 10% said no. Even so, over 90% said that they would not have preferred another method of presenting the lessons. Since these data were generated, efforts have been made to increase the contact between student and A.I. Hopefully, this will have an impact on an affective as well as cognitive level.

Other information on the effectiveness and efficiency of the audio-tutorial approach was gathered in quite a different fashion. In the fall semester 1974-1975, a presentation describing the G-110 audio-tutorial course was made to a group of advisors, who counsel students considered high academic risks. As a direct consequence of the presentation the course was deluged by these students in the spring semester. Interestingly the grade distribution that semester was not significantly different from previous semesters' distributions. The course continues to get a large number of referrals from these advisors with similar results.

The follow-up interview measured the impact of certain dimensions of the course over a short period of time (six weeks after the semester). We are in the early stages of developing an evaluation instrument which will measure the long range impact of having taken the course. We plan to interview students who took the course two years ago. Their retention of course material will be measured. We hope, also, to measure the extent to which they make use of what was learned in the course in their activities.

Some of the instruments which are presently used for evaluation in this course will be used for a slightly different purpose in the spring semester 1975-1976. For the first time G-110 is being taught using the audio-tutorial approach on the Indiana University regional campus in New Albany. The major problem involved in implementing the course there, so far, is staffing. We have been forced to use undergraduates as learning center instructors. We will use the lesson- by-lesson evaluation instrument to monitor learning center instructor behavior and be able to make training changes as needed during the semester. The Achievement Correlates Evaluation instrument will be administered in New Albany to monitor student behavior. Comparisons will also be made with similar data gathered for Bloomington.

Gaming-simulation in Geography: Problems and Issues in the Design of Classroom Games

John F. Jakubs
Indiana University

INTRODUCTION

Although games, or synonymously, "gaming-simulations" or "simulation games" are not new to geographic education, interest seems to have flourished during recent years. This is indicated by: recent publications (Kibel, 1972; Taylor & Walford, 1972; Walford, 1969); activities of the High School Geography Project; the continuance of game-related activities at meetings of the National Council for Geographic Education; and, the emergence of a similar interest within the Association of American Geographers. In addition, a national professional organization and a major journal devoted to social science educational gaming have been established. This burgeoning interest provides the rationale for the comments in this paper as does the increasing role of self-designed games or the adaptation of existing games. There are some pitfalls to beware of and a sharing of experiences might be useful in this context. I would like to discuss some problems in this area by drawing examples from two efforts at game design. One of these was successful with respect to the finished product and the other succeeded in providing insight into game construction from an otherwise ill-fated effort.

GAMES

Educational games are taken to mean classroom activities consisting of a set of players, a set of alternative actions available during the course of play, a set of outcomes dependent upon the actions taken by one or more of the players, a termination rule, and possibly, chance rewards or payoffs. Taylor and Walford (1972, p. 17) characterize a gaming simulation as follows:

> "1. Players take on roles which are representative of the real world, and then make decisions in response to their assessment of the setting in which they find themselves. 2. They experience simulated consequences which relate to their decisions and their general performance. 3. They 'monitor' the results of their actions, and are brought to reflect upon the relationship between their own decisions and the resultant consequences."

One or two good games included within a course can provide an alternative mode of learning which some, but not all, students will find more rewarding than a lecture/discussion format. Further, some concepts or processes, especially those of high abstraction or complexity, can be understood more easily by experiencing the process than by having it described. Games have been characterized as tools to encourage active rather than passive learning. Promoting the use of educational games is not a purpose of these comments, however. Most advantages of using simulation games in the classroom depend on the particular game itself. As is true of lectures, discussions, films, and field trips, games range widely in value as learning devices, and much depends upon the administration of the activity. Games must be ''delivered,'' and particular skills are necessary to do this.

The design of a game is critical to ease of delivery and administration. The two efforts at game design discussed below should make this apparent.

ISSUES IN THE USE AND DESIGN OF GAMES

Port Sivad (Jakubs et al., 1971), is a role playing simulation concerned with the location of noxious, or locally undesirable, public facilities. Specifically, the objective is to locate a sewage treatment plant as quickly as possible in a hypothetical urban area. Players have both public images and objectives and private agendas which in some cases are in conflict with each other. Port Sivad addresses the following issues: (1) inequities from community to community in the allocation of noxious public facilities; (2) the effects of citizen organization on public decision-making processes; (3) the potential influence of monetary power upon these processes; (4) civil disobedience as an instrument of power; (5) the difficulty which time constaints cause for decision makers; and (6) the general nature of group interaction and its effect upon locational decisions.

Large classes present game designers with special problems. Port Sivad was designed from the outset for classes ranging from 15 to 80 students. It is not easy to construct a game which encourages active participation by as many as 80 students yet does not require excessive classroom time. The earlier versions addressed this problem by designating only 15 students as players. These people were assigned roles and were given game rules, including background information pertaining to the sewage treatment plant issue and the reputations or public images of the other players as well as their own personal role descriptions. The remaining students were to act as the public, to follow the action of the game, to vote should referenda be sponsored, and following a discussion at the termination of play, to assess the quality of play by each of the 15 players.

This approach failed. Unwittingly, we had simulated the ''real'' world too closely. Apathy among the electorate was a severe problem. The best way to involve students in a game is to give them some power to affect the outcome and some motive for doing so. The structure of the game was altered to achieve this goal. Six of the fifteen roles had been those of private citizens and these became leaders of citizen groups. The fifteen players were now titled ''major game participants.'' The remaining students became ''minor game participants'' and were divided equally among the six citizen groups. Each group member was given a role, with goals, motives, and complete background information specified. Because all group members were to be treated similarly, this required the creation of only six additional roles. The addition of minor game participants did not affect the balance of the game, because the allocation was to result in equal numbers in each group. Actions from citizens who were not members of any of these particular groups would be represented probabilistically (through throwing of dice) when called for.

This approach has worked well. Minor game participants affect the course of action through their group leader. This occurs in the one way flow of suggestions from members to leader and certain actions of group leaders are predicated upon the approval of members.

Another problem of motivation existed in earlier versions of Port Sivad. As mentioned earlier, the (apathetic) general population, without roles, was to assess the quality of play of each of the fifteen participants. These fifteen were equipped with verbal descriptions of their values and beliefs, goals, and motivations. These verbal passages were interpreted rather loosely by players on certain occasions. We are modest enough to agree that the fault may have rested with the descriptions themselves. In any case, some players were following their own moral codes instead of those with which they had been provided. To alleviate this problem, a system of scoring was devised. Each participant, major and minor, was provided with an individualized pay-off table, with points allocated for all possible game outcomes and for various occurrences during the course of play. These values were made to correspond as closely as possible to the verbal descriptions, which were retained. Players then were instructed to accumulate as many points as possible, a goal which could be accomplished best by closely following their assigned roles. This amendment succeeded in clarifying the roles and demands of role-playing. Further, it alleviated the problem of evaluation which surfaced with the deletion of the (apathetic) general population described above.

Simulation games are necessarily simplified versions of reality whereby aspects of reality not essential to the achievement of game purposes are deleted or held constant. In Port Sivad, we did not want players to spend excessive amounts of time searching for and evaluating alternatives in the hopes of finding a noncontroversial site. We wanted the decision makers to realize that their actions would be unsatisfactory to some residents of Port Sivad. Thus, we constructed the scenario at the outset to include only three possible locations for the sewage treatment plant. We explained that an earlier study, carried out by private, nonlocal consultants at considerable public expense, concluded that the physical characteristics of the region identified these three as by far the best choices. The possibility of compensatory payments in the form of "salutary" public facilities (parks, playgrounds, or whatever was desired by the community) on three expenditure levels formed twelve possible outcomes, three different compensation levels and a no-compensation solution for each of the three possible locations. In addition, "no solution" was a possibility, inasmuch as players labor under a time constraint for decision making. Thirteen possible solutions have seemed to be a reasonable number which is large enough for considerable debate and small enough to keep frustration levels low. These potential outcomes are specified clearly, and players spend their time on serious discussion of each instead of searching for an ideal location which would be completely satisfactory to all parties.

Port Sivad suffered from a number of defects in earlier versions. These were corrected and the simulation has been used successfully many times since, both by the authors of the game and by others. The scenario is not an abstract one. Students are immersed easily into the problem of negotiating a solution to a different situation and are made aware of some of the potential levers in arriving at satisfactory locational decisions.

"Debriefing" is an essential part of any educational simulation game and for Port Sivad this is an easy task. It consists first of having all major game participants explain their roles and how they attempted to achieve their goals. They then generalize the issues in Port Sivad to the general problem of locating noxious public facilities in urban areas. Thus, we discuss the influence which an organized citizenry can have upon elected decision makers. Often this occurs simply because of the existence of an organized group. For instance, one of the three potential locations in Port Sivad has no organized citizen opposition, and it is a rare occasion when the players in Port Sivad do not attempt to locate the plant there from the start. Typically the area selected as the site will have been offered some compensation, and this use of side-payments or compensatory mechanisms serves as a springboard for a discussion of equity in public facility location in general. Likewise, the actual play of the game naturally raises issues of civil disobedience and the dependence of political action upon private monied interests.

Some games are not as easy to "debrief," however. "Succeed" is a game which was constructed to simulate the process of residential and political decentralization in larger cities and the central city-suburban fiscal disparities problem which results (Jakubs, 1973). The game is plagued with many problems. One of these is the debriefing problem. Succeed is very abstract. It is so abstract, in fact, that most players do not know the general problem being simulated during play. They are allocated markers on a game board, financial resources, and income levels which vary from player to player, and are instructed to attempt to maximize wealth. One way to accomplish this is through the movement of markers and the construction of "zones," which represent autonomous political entities. The debriefing task is a formidable one. If the fiscal disparities issue were not discussed in class prior to playing Succeed, the entire problem must be addressed and Succeed would be useful only as a model to present the topic in a simplified manner. The debriefing typically has developed into a lecture. On the other hand, if the fiscal disparities problem is treated prior to Succeed, then the game is redundant. This is the most fundamental flaw of Succeed, and of many other games. There must be substantive issues of satisfactory number and complexity to warrant the classroom time spent playing the game. Simulation games often involve relatively indirect ways to impart knowledge or to aid students in conceptualizing a process. Classroom time, a scarce resource, must be allocated wisely.

There is a serious danger in employing simulation games carelessly and I wish to conclude these comments by treating it briefly through a return to Port Sivad. The scenarios set forth in simulation games are necessarily artificial in their simplicity but they represent the game designer's view of the world and can be artificial in this context as well. Students seem to be much more willing to accept uncritically the assumptions involved in a scenario than they are when these assumptions are unmasked and explained in a more straightforward manner. For instance, in Port Sivad we designed the balance of power to reflect what we considered to be the most common case, one in which the more affluent communities are the most powerful. When this point is raised in a discussion it is quite obviously an opinion. The presentation of supporting evidence might result in its acceptance as a logical conclusion but such evidence clearly would be required. It is rare, however, when the assumptions underlying the scenario in Port Sivad are questioned, yet there is no such supporting evidence. During the debriefing this issue should be raised to make all players aware of the opinion on which the game is based, and to provide some of the evidence which is required for acceptance of its validity.

SUMMARY

If designed well and administered adroitly, simulation games can be useful classroom tools. Developing new games or adapting existing games is largely a creative process and generalizations and guidelines are difficult to formulate. Further, games differ widely in internal structure and this again precludes generalization. Games can be distinguished from other classroom activities, however. In most instances they are used to involve students in active learning, moving them a step closer to an issue, topic or problem than would be the case in a lecture or discussion format. Some problems in game design which make this goal difficult to achieve have been raised in this paper. Problems arise with designs in which: not all students are active participants; outcomes of player decisions are not linked clearly to rewards or payoffs; assumptions are hidden; and context is not sufficient to warrant the necessary classroom time. There are few forums for discussing these problems and issues. Perhaps these comments will prompt some response so we all can improve our efforts in this area.

REFERENCES

Jakubs, John F. "Succeed." *Discussion Paper* no. 36 Department of Geography, Ohio State University, Columbus, Ohio, 1973.

Jakubs, John F.; King, Paul E.; Davis, George A.; and Adams, Lael S. "Port Sivad: A Locational Decision Game for a Noxious Public Facility." *Discussion Paper* no. 22. Department of Geography, Ohio State University, Columbus, Ohio, 1971.

Kibel, Barry M. *Simulation of the Urban Environment.* Washington, D.C.: Association of American Geographers, 1972.

Taylor, John L., and Walford, Rex. *Simulation in the Classroom.* Baltimore: Penguin, 1972.

Walford, Rex. *Games in Geography.* London: Longman, 1969.

The Introductory Course in Geography A Curricular Perspective

Gary Manson
Michigan State University

For more than ten years, a growing number of professional geographers have been working to improve geographic education in schools, colleges, and universities.[1] At the precollegiate level, one result of their efforts has been the High School Geography Project, a curriculum development venture that involved scores of geographers as well as numerous professional educators and classroom teachers.[2] In higher education, which is the focus of this paper, geographers' interest in the teaching and learning of geography has been evident in several ways. Their most visible activities have been sponsored by the Association of American Geographers and funded by the National Science Foundation and the U.S. Office of Education. These activities include the creation of: the Commission of College Geography, which published more than fifty resource papers and technical reports aimed at bridging the "content gap" between research and teaching; the Commission on Geographic Education, which conducted six regional conferences designed to help faculty become more reflective about the teaching/learning process; and the Teaching and Learning in Graduate Geography Project, which aimed at helping doctoral students prepare for their role as college teachers.[3]

Less conspicuous than national commission and projects but undoubtedly signficant in their impact on undergraduate education are the many geographers who have written introductory-level textbooks, developed innovative courses, experimented with new ways of teaching, and spoke out on behalf of pedagogical and curricular reforms.[4] Equally important are institutional policies related to the employment and tenure of faculty. In a recent survey, 358 chairpersons of geography departments claimed that teaching was weighted as heavily as research in hiring and promoting faculty.[5] Thus, there may be some justification for the assertion, "Professional geographers have awakened to the fact that they are key elements in any reform of geography teaching."[6]

What prompted this impressive, perhaps unprecedented, array of activities by scholars who were trained as researchers but who are now concerned and involved with geographic education? Several factors must be recognized. Federal funds may have stimulated, or at least facilitated, efforts at the national level., Declining enrollments and dwindling budgets may have prompted chairmen to posit a relationship between credit hour production and the quality of teaching. Political, economic, and social turmoil may have led some faculty to reconsider the means and ends of higher education, as well as the societal role and responsibility of geography. At present, however, these are only speculations because there has been no comprehensive study of this "decade of reform in geographic education."

THINKING ABOUT THE CURRICULUM OF TODAY

But the past is prologue. What is the present state of affairs with respect to teaching and learning geography at the undergraduate level? Quite frankly, we cannot answer that question because we have so little information. We lack data about students, curriculum, and teaching practices. We do not know which innovative courses, which experimental teaching methods, and which of the newer introductory textbooks are in use, not to mention which have proven effective. Beyond such rudiments lies the paucity of concepts, models, and theories necessary for analyzing, planning, conducting, and assessing undergraduate education in geography. Without such information and ideas we shall continue, no doubt, to do things differently, and on occasion we will even do things better. However, we probably will not be doing things as efficiently and effectively as we might, and we certainly will not know why things work as they do.

Joseph Schwab, an eminent scientist who often functions as a gadfly on the rump of higher education, has pointed out a curious anomaly:

> Scholars and scientists, trained, skilled, and committed to rational activity, do relatively little thinking about that which matters most to their collegiate community—the curriculum.
>
> As far as students are allowed to see, the curriculum is not a subject for thought; it merely is. In many cases, indeed, thought about curriculum is not merely invisible; it barely occurs. Single courses are sometimes the outcome of single happy thoughts but are rarely accorded the reflexive, critical scrutiny we give as a matter of duty and right to our "scholarly" productions.
>
> Their origins are left obscure and their outcomes only occasionally examined. Sequences and groupings of courses are sustained by tradition and inertia and only rarely examined. Least of all do we worry about our sins of omission, the courses we do not give, the outcomes we do not seek.[7]

Schwab is probably correct when he asserts that most curriculum in higher education just "happens." It is our need for thinking about geographic curriculum in higher education which prompted this paper. Certainly, geographic curricula should receive the same "reflexive, critical scrutiny we give as a matter of duty and right to our 'scholarly' productions."

This paper deals with only one facet of the undergraduate curriculum in geography—introductory courses. I have chosen introductory courses as my topic for several reasons. First, such courses enroll hundreds of students and generate thousands of credit hours in some departments. Second, many geographers believe that all is not well with introductory courses. Some have been dismayed by the diversity of their content:

> If any rational person were to undertake a survey of introductory geography offerings in the United States, he surely would be stunned and confused. In short order, he would discover that depending on the school, "Introductory Geography" is (1) physical geography, (2) geology and geomorphology, (3) meteorology, (4) economic geography, (5) cultural geography, (6) human geography, (7) world regional geography, (8) some combination of the above, or (9) some version of the individual's specialty.[8]

Others have been troubled by the quality of instruction:

> Teachers of most introductory or survey college geography courses give little attention to these courses because they have little interest in teaching them and few have had any training as teachers.[9]

Still others are troubled by the lack of intellectual rigor in introductory courses:

> At a time when geography has everything going for it, when students particularly are realizing that we exist in space as well as time, there is little need to teach watered-down general education courses, which end up so diluted and devoid of challenge that we drive the intelligent students away forever. Even today . . . we see too many beginning courses that are factual jumbles, bereft of the simplest and, therefore, the most powerful concepts of our discipline.[10]

A final reason for focusing on introductory courses is that faculty charged with teaching them cannot rely on profession-wide, or, in many cases, even a departmental consensus about the nature and conduct of these courses. Difficult and potentially controversial decisions must be made. What content shall be covered? What textbook shall be used? How shall the class be organized? What objectives shall be established for the course? Which evaluation procedures shall be employed? And how shall the results of each of these separate decisions be combined into an integrated and worthwhile educational experience, i.e., a good course? Perhaps the remainder of this paper will be helpful in dealing with such questions.

SOME PURPOSES FOR INTRODUCTORY COURSES

How should one begin thinking about introductory courses? In principle, and not surprisingly, one should begin with a consideration of their purposes. In practice, and also not surprisingly, course purposes are often afterthoughts which are neither intended nor taken seriously by faculty or students. That the principle is not practiced seems unfortunate because an explicit statement of purpose can guide such important activities as selecting content, designing mode(s) of instruction, and determining the criteria and forms of evaluation.

Introductory courses in geography usually are taught with one of three purposes in mind: "introduction to the discipline," "liberal education," and "self-fulfillment." Courses aimed at introducing students to the discipline of geography are intended to begin the process of training professional geographers. Such courses, if they were effective, would cause the student to know what a geographer knows and to think as a geographer thinks. An emphasis on knowing what a geographer knows produces a course that stresses essential facts, basic concepts and fundamental techniques necessary for success in advanced courses. Emphasis on thinking as a geographer produces a course which stresses geography as a research discipline and teaches students the ways of geographic inquiry. In such a course:

> [T]he student is taught to attack problems the way a geographer does. He is led to ask questions of a geographic nature, and to make use of appropriate methods to find answers. He is exposed to organizing themes which are the conceptual structures around which geographical data are grouped. He finds out about the research clusters that are currently popular. He is taught technique, methods, and terminology. He finds out what geography is all about.[11]

Gilbert White regards liberal education as education which cultivates reason and analysis, deals with the aims and modes of human life, and includes consideration for the esthetic, spiritual, and physical development of the student. With respect to the role of geography in liberal education, he says:

> A liberally educated person should know sufficient about the processes which shape the spatial distribution of selected landscape features so that with a minimum of memorization of basic facts and anomalous relationships he can state with a fair degree of accuracy the complex of landscape features he would expect to find on any given part of the earth's surface, expressly noting the amount of diversity at any given scale and the changes he would expect to result from any given shift in conditions affecting the processes.[12]

Thus, geography courses conceived as contributions to liberal education seek to develop the intellect of students by revealing to them the world as geographers see it. Through such courses students may learn to think and act like geographers, but that is not their purpose.

The most recent thrust in introductory courses seems to be associated with the "self-fulfillment" purpose. According to Pattison, advocates of this purpose believe education should foster the fully functioning individual through emphasis on self-reliance, self-fulfillment, and self-acceptance.[13] Most of the advocates of "self-fulfillment" geography believe that "we can't reduce the teaching of geography to a content model" and that "teaching people how to learn . . . is more important than telling them what to learn."[14] For these individuals, geographic education means human development or it means nothing. The task of the instructor is to devise ways of helping individuals learn more useful and satisfying ways of resolving the relationships between themselves and problematic situations.

THE CONTENT OF INTRODUCTORY COURSES

The second topic to be considered with respect to introductory courses is their content, or the facts, ideas, skills, and attitudes that make up the substance of a course. The content of a course is to be viewed as means to an end, not an end in itself. Content is taught and learned only so that students may become better geographers, more enlightened members of society, or more fully developed persons. No geographers that I know advocate learning geography for its own sake, although some occasionally permit their passion for the discipline to mislead the inattentive. Conversely, I know no geographers who would defend courses that are devoid of geographic content, although some do manage to convey that impression when they plead for more concern for students. Aside from these infrequent transgressions, geographers agree that content is important.

Before turning to the ways in which geographers have dealt with content in introductory courses, we shall make two assumptions. First, we shall assume that the content of a geography course should be drawn from the discipline of geography.[15] This assumption is made with an awareness of the nature of the discipline and a recognition of the contributions of geographers to interdisciplinary education. However, it seems reasonable to assume that the content of an introductory course in geography should be drawn from that field of inquiry. Second, we shall assume that the scope of geography, or any of its subdivisions, is greater than can be encompassed by any single course. This means that whatever the duration or intensity of a course, all the possible content cannot be learned. Thus, the so-called issues of content coverage and content mastery are really non-issues for one can neither cover nor master everything. The hard questions concern content selection and content organization.

There are at least three ways of approaching content for introductory courses: the "structure of the discipline" approach, the "findings of the discipline" approach, and the "needs of the students" approach. Each approach has received some attention from geographers writing about introductory courses and each affords an alternative method of dealing with content. We must emphasize once again that all three approaches are regarded by their respective proponents as instrumental to the achievement of their purposes. The content that is selected and the way in which it is organized are valued only insofar as they prove, or are thought, to be effective in the education of undergraduate students. While the question of which purpose is "best" is a normative question, the question of which approach to content is "best" is an empirical question.

The "structure of the discipline" approach rests on the premise that undergraduate education is best advanced by exposure to the fundamental concepts and methodology that make up particular fields of inquiry. One of the clearest examples of this approach to introductory courses in geography is offered by Rummage and Cummings:

> Such a course would make geographically significant facts more meaningful to students by helping them develop a conceptual structure of the discipline.[16]

Their syllabus leaves little doubt that the structure of the discipline approach is operative. Among the fundamental concepts emphasized are distributions, surfaces, sampling, the gravity model, and central place theory. With respect to methodology, Rummage and Cummings stress,

> 1. the collection and use of data in problem situations,
> 2. the employment of appropriate cartographic or mathematical-statistical techniques to facilitate comprehension of these problems,
> 3. the selection and use of suitable systems of measurement for verifying the validity of the selected hypothesis.[17]

The "findings of the discipline" approach derives from the belief that undergraduate education proceeds most effectively by having students learn the outcomes of scholarly inquiries rather than the processes and conceptual structures through which this knowledge was generated. This seems to be what Taaffe had in mind when, after reviewing the major geographic traditions, he proposed,

> Introductory courses should present the findings associated with each of these views. This could lead to three quite different introductory courses. One would be organized around the findings of essentially ecological hazards or pollution. A second might be organized around world regions, giving examples or areal interrelationships in different parts of the world. A third would be organized around concepts of spatial organization, such as diffusion and hierarchical systems. Still a fourth kind of introductory course would attempt to represent the three views as well as some of their overlaps. When asked by university administrators or curriculum committees what they have to contribute to general education, geographers make choices which reflect varying emphasis on the three views.[18]

The course most frequently mentioned in connection with the "findings of the discipline" is the world regional course. I have found no world regional textbook that rejects that approach. Typical is the statement found in the preface of one of the more enduring volumes:

> This book . . . seeks to assist college and university students in acquiring basic ideas and supporting facts about contemporary world geography which a person with a college education might reasonably be expected to know. Its aim, in short, is general education in world geography.[19]

Similar statements may be found in introductory level textbooks for physical geography, economic geography, cultural geography, and so forth.

The "needs of students" approach rests on the notion that geographic education proceeds in the most desirable fashion when teachers assume some responsibility for the development of the total human personality rather than confining their attention to the intellectual dimension:

> A student is not a passive digester of knowledge elegantly arranged for him by superior artists of curriculum design. He listens, reads, thinks, studies, and writes at the same time that he feels, worries, hopes, loves, and hates. He engages in all these activities not as an isolated individual but as a member of overlapping communities which greatly influence his reactions to classroom experience. To teach the subject matter and ignore the realities of the student's life and the social systems of college is hopelessly naive.[20]

To my knowledge, few introductory courses in geography have been developed, using the "needs of students" approach to content. One reason is that most proponents have been more concerned about the processes of teaching and learning than with courses and curriculum. However, the syllabus for one such course contains the following:

> There are several goals from the staff's point of view. You no doubt will have others which we hope you will share with us. Here are some of ours:
>
> 1. To help you learn basic ideas, skills, attitudes, and values of geography and see their relevance for you;
> 2. To have you take the major share of responsibility for your own learning and for teaching your fellow students;
> 3. To have you share in evaluating your own learning and that of your fellow students;
> 4. To help you see how effectively you can do tasks 2 and 3;
> 5. To explore new ways of conducting college classes;
> 6. To give us practice in serving more as "consultants to learners" than as "instructors."
>
> Of the above mentioned goals, please note that #1 is specific to geography; that is, it is concerned with the subject matter of the course. The others we think of as "process goals," that is, goals which could presumably be attained in any course no matter what the subject matter, since they relate to the "process"; of teaching and learning rather than to the "content" to be learned. A process of teaching and learning is present in every course, although one does not frequently find this explicitly stated in terms of course goals. We stress these process goals because of our strong belief that these are just as important, if not more important, than content goals. In fact, we believe that these are strongly interdependent; that the degree of subject matter learning will be strongly dependent on how well the process goals are attained.[21]

WAYS OF TEACHING INTRODUCTORY COURSES

As is the case with content, a way of teaching is a means not an end. Questions about "best" ways of teaching are not moral questions. In principle, they are questions that can be answered through persistent and clever research. Because these answers are not yet known does not mean that ways of teaching are most productively discussed with the emotional fervor of religious revivals, an appropriate metaphor for some of the recent literature on ways of teaching. Instead, ways of teaching should be discussed in terms of their relationship to the purposes of education.

Ways of teaching can be conceptualized at several different levels. For instance, a way of teaching can mean the image that a teacher has of "teachers-at-their-best." This is what Axelrod has in mind when he suggests that unless an observer has some understanding of the prototype governing a teacher's behavior, it is not possible to understand why the teacher is doing what he or she is doing, nor is it possible to determine how well it is being done.[22] A way of teaching can also mean the comprehensive, normative models of behavior and society which are used to make decisions about the ends and means of education. Such models derive from educational theories and philosophies. They include programmed instruction based on operant conditioning theory; nondirective teaching stemming from client-centered therapy; and group investigation, which derives from the democratic process philosophy.[23] In a third sense, ways of teaching can mean specific instructional techniques such as simulation, lecturing and questioning. This usage implies that teaching is a composite of teaching tactics. The more skillfully a teacher uses these tactics the more effective the teacher will be.[24]

For this paper, we shall regard a way of teaching as a pattern of treating people, objects and events that is directed purposely and recognizably toward the achievement of an educational goal.[25] This definition places our view of ways of teaching about midway between specific intructional techniques, on the one hand, and prototypes, models, theories and philosophies, on the other. We shall call these middle-range patterns "modes of teaching" and we shall examine three such modes—the discourse mode, the management of learning mode, and the inquiry mode.[26] Each of these modes has been represented in recent discussions of teaching and learning geography at the undergraduate level.

Discourse is probably the oldest and most common way of teaching an introductory course. Through discourse the teacher strives to transmit, as meaningfully as possible, the facts, concepts, skills, and attitudes that make up the content of the course. The principal means of discourse are lectures and textbooks, although laboratories and discussions may be used to clarify or reinforce certain parts of the content. Two points should be made regarding discourse before turning to its geographic applications and implications. First, as any other way of teaching, discourse may be done poorly or well. Most criticisms of the discourse mode assume that it is poorly done, yet we may be equally justified in assuming that it is well done, in which case it can be an extremely effective way of teaching. (Of course, the question remains, "An effective way of teaching *what*?") Second, discursive teaching should not be confused with didactic teaching. Didactic teaching, according to Axelrod, stresses cognitive knowledge acquired primarily by memorization or mastery of skills acquired by repetition and practice.[27] Discourse, on the other hand, aims at higher order learning.

McNee is not an advocate of the discourse mode, or lecture-demonstration as he calls it, but he suggests several advantages of that way of teaching.[28] First, discourse can have the elegance, order, and clarity not to be found in courses organized around other ways of teaching. Second, this way of teaching is efficient because many geographic generalizations can be covered in a limited amount of time. Third, a certain aesthetic value flows from the dramatic wall map, the grand design textbook and the spellbinding lecture, and these are the hallmarks of the discursive way of teaching geography.

The management of learning mode is similar to the discourse mode in that it also takes as its purpose the transmission of a fixed body of knowledge and skills to students. The principal difference lies in the systematic way in which instructional variables such as pacing, media, and materials are manipulated so as to accommodate individual differences among learners.

Davis, Alexander, and Yelon put the case for the management mode this way:

> . . . [T]here are alternate ways of presenting materials to different students—some of which are better than others . . . there are ways of dealing with subject matter that will encourage . . . future learning. A learning system can be designed to maximize student performance along prescribed dimensions. . . . There are ways to design a learning system that takes account of the fact that individuals differ from one another in their abilities, backgrounds, and styles of learning. . . . (The educator's) fundamental responsibility is the design, development, and evaluation of learning systems that will maximize student performance on specific criteria at a minimum cost in time, effort and money.[29]

Certain aspects of the management mode have been adopted by geographers. The audiovisual-tutorial method has attracted some attention because it offers the advantage of self-pacing.[30] Fielding has developed programmed instructional materials that are not only self-pacing, but that also provide immediate feedback.[31] Healy and Stephenson have opted for the repeated-tests feature of this mode.[32] However, to my knowledge, no geographer has developed a system whereby learning objectives, instructional methods, and learning styles are matched in the manner suggested by Davis, Alexander, and Yelon.

Both the discourse and management modes tend toward reception learning by the students. Reception learning means that students are expected to receive and assimilate the content messages that are transmitted from instructors, textbooks, films, instructional programs, and so forth. Reception learning is by no means a passive process nor does it suggest rote, or meaningless learning.[33] Indeed, students may receive and react to messages delivered through the discourse or management modes in extraordinarily impressive ways.[34] Nonetheless, the intellectual "posture" of the student engaged in reception learning and the "climate" of the classroom characterized by the discourse or management modes are quite different from the "posture" and "climate" found in conjunction with a third mode of teaching referred to as inquiry.[35]

Three versions of inquiry should be considered by teachers of introductory courses—academic inquiry, personal inquiry and reflective action.[36] Academic inquiry is a way of teaching based on the ways scholars and scientists generate and validate knowledge. Underlying the academic inquiry mode is the premise that students will comprehend, in the fullest sense of that word, a body of knowledge only to the degree that they understand how and why it was generated and verified. Teaching through academic inquiry means:

> . . . [T]o show students how knowledge arises from the interpretation of data. It means, second to show students that the interpretation of data—indeed, even the search for data—proceeds on the basis of concepts and assumptions that change as our knowledge grows. It means, third, to show students that as these principles and concepts change, knowledge changes, too. It means, fourth, to show students that though knowledge changes, it changes for a good reason—because we know better and more than we did before.[37]

Occasionally, the academic inquiry mode is called problem solving. This is an appropriate appellation, provided the problem arises from discipline-related questions, not personal or social problems. This is the meaning McNee has in mind when he discusses using problem solving as a means of teaching an introductory course:

> There are both the immediate satisfactions of doing something for oneself and the long-range satisfactions of self-confidence in probing the unknown. So it is with students in courses stressing problem-solving. There are both immediate and long-range satisfactions in creating your own map, in interpreting a complex pattern on an air photo, in analyzing store locations from primary data, in making a choice among alternative possibilities in the location of a factory, in finding the Corn Belt for yourself rather than having it thrust upon you, in discovering the diffusion process through plotting the spread of hula hoops or skyscrapers. . . . Such satisfactions are maximized in a course stressing problem-solving.[38]

Personal inquiry is a way of teaching that derives from each person's efforts to develop identity and autonomy through interactions with natural and social environments.[39] The task of the educator is to teach in ways that facilitate rather than retard those efforts. For instance, teachers should help students become more aware of the natural and social processes of which they are a part. Teachers should help students develop their abilities to choose from among alternative ways of behaving and evaluate the consequences of those choices.[40] According to Thelen:

> Probably the most distinctive feature of [teaching in the personal inquiry mode] will be the deep involvement of the student—his absorption, his concern, his releasing of a great deal of effort and energy into the activity; and there will be a commensurate emergence of insight from the well-spring of this meaningful experience.[41]

Thelen makes clear that natural inquiry is not enough and that an educative function is necessary. How might an introductory course in geography be conducted to facilitate students' capacity for personal inquiry? One example comes from a source quoted earlier, "Geography 200" which utilizes problem solving in a manner somewhat different from McNee:

> We have tried to design problems which engage you in relevant human concerns, which call upon how to utilize your own accumulated knowledge and to acquire new knowledge through research, and which help you develop core scientific attitudes of tentativeness, skepticism, and objectivity. The problems, if they are good ones, should lead to your recognition of multiple causes and alternative decisions in the spatial context of human behavior. As such, they are not "answer-oriented" but, rather, they should provide useful training for decision-making, not just in geography but in other contexts as well. We are not primarily interested in your acquisition and regurgitation of facts about the world's geography, although we believe you will learn and retain many such facts that are insinuated in the problem-solving contexts. Our major concern is that you grow in your capacity to solve problems. We think problem solving fulfills the need for you to learn to work with the scientific method as well as to use geography as a subject that is very directly concerned with important problems of mankind that necessitate skillful and informed decision-making.
>
> Staff members are to be both teachers and learners. We all want to help you learn by serving as your consultants or resource persons. We want to help you see yourself as able to learn on your own; to see yourself as a valuable resource for others' learning; and to see fellow students and staff as learning resources for you. We don't see ourselves primarily as "information-givers" but rather as "question-askers" and facilitators, since we think you will learn more effectively with this kind of student-staff relationship.[42]

Reflective action is a way of teaching in which students engage in efforts to transform (or maintain) their natural and social environments. If a more conservative view of reflective action were adopted, the particular environment in question, such as a neighborhood or a wilderness area, would be regarded as a laboratory for learning by students even though specific policies and practices may be changed as a result of student inquiries. In this case, the primary justification for reflective action is student learning. If a more radical view of reflective action were adopted, the distinction between learning as a goal and change as a goal would become blurred. Thelen rejects the latter view of reflective action because changing society is not the business of teachers except insofar as it produces enlightened citizens who will act intelligently.[43] At present, I am unaware of any introductory course in geography that is taught by means of the reflective action mode. There may be good reasons for avoiding this way of teaching but it seems to me that the conservative version offers some attractive pedagogical features.

EVALUATION AND INTRODUCTORY COURSES

Evaluation may be defined as the process of assembling evidence and making decisions about the status and performance of learners, teachers, or curriculum vis-a-vis the purpose of geographic education. Assembling evidence may include, but is not limited to, activities such as giving examinations, administering questionnaires, and observing teachers in their classrooms. Making decisions may include, but is not limited to, activities such as approving dissertations, selecting textbooks, and promoting faculty. Thus, evaluation is a more comprehensive and complex concept than merely evaluating students. However, because the evaluation of students is obviously an important part of teaching an introductory course, we shall focus our discussion on that topic.

Teachers evaluate students for three reasons. First, they may wish to ascertain the conditions prevailing before instruction begins. Students bring to any classroom knowledge, skills, attitudes, and aptitudes that may influence their learning. Having information about these pre-existing conditions should enable the teacher to design a more effective course. Gathering such information and using it to prescribe instruction is known as diagnostic evaluation.

Let us consider a simple example. Assume that three students enter your classroom. To begin studying the subject you are teaching, some minimum amount of knowledge is required. Since you cannot assume that all three students possess at least the necessary minimum knowledge, you administer a pretest to determine what they know. You find that Student 1 possesses more than the minimum, Student 2 has exactly that amount, and Student 3 has less than the necessary minimum knowledge. One major implication of this finding is that each student has different needs. Student 3 needs experiences enabling him to master the knowledge prerequisite to beginning the "regular" curriculum. Student 2 can probably follow the "regular" curriculum while Student 1 should be able to omit some of the early activities scheduled for Student 2 and move on to a level of difficulty more suitable for him. In some cases, only the amount of time needed to complete the curriculum will be different for each student. In other cases, each student will need substantially different learning experiences to achieve the objectives.

A second reason for evaluating students is to monitor the teaching-learning process in order to discern if changes are needed in instructional materials and procedures. This is called formative evaluation and, in Kurfman's opinion, is the most important form of evaluation:

> The most significant kind of educational decisions . . . are the decisions teachers make as a part of the instructional process to improve the learning of students.[44]

Suppose a teacher wishes to know whether students are learning what they should be learning. He may interview several of them part way through the series of lessons, ask questions about the material under study and find that many are having difficulty with a particular idea or technique. Or the teacher may listen to the class discuss a particularly controversial topic and find that students cannot differentiate between factual, inferential, and value statements despite lengthy instruction on that very point. Or the teacher may give a short quiz and find that most students do not understand a particular technique such as regression analysis after reading about it in a textbook. These findings probably mean that insufficient or inadequate instruction was provided, at least for some students. Therefore, it may be necessary to revise instructional materials, to provide more time for learning, or to modify instructional strategies.

Students should be informed about the nature of their errors, why they were made, and how they can be prevented. Information of this kind should also suggest what the student has yet to learn, thus converting an error into useful feedback. In fact, Bloom et al., argue this is the most important use of formative evaluation.[45] It should be emphasized that formative evaluation should play no part in any other evaluation activity such as grading. Rather, formative evaluation should be viewed as an effort to monitor and improve the teaching/learning process.

Summative evaluation, the third reason for grading, assesses the degree to which students have achieved the objectives of the course. Summative evaluation may be conducted in order to certify skills, predict success, provide feedback, compare outcomes, or assign grades. All these are important uses of summative evaluation but because grading seems to be one of the most troublesome issues facing teachers of introductory courses, our discussion will be confined to that problem.

To most people grading means sorting or classifying students into groups with similar test scores. Typically, fewer students are classified into "high" and "low" groups than in the "middle" or "average" groups; most teachers and students have learned to regard this pattern as a "normal" grade distribution. This grading practice derives from several assumptions. First, it is assumed that there should be fewer high and low grades and more average grades, that the distribution of grades should conform to a presumed normal distribution of ability. Second, it is assumed that the ideal examination should enable the teacher to classify his students according to the amount of knowledge they have. Third, it is assumed that the ideal test item should be missed by about one-half the students because such an item discriminates most efficiently. Fourth, and perhaps most important, it is assumed that the basis for grades should be the student's test score ranking relative to those of other students. That is why we call this the "ranking" model of grading.

An alternative to comparing students in order to derive a grade is the standards, or "mastery," model. Here the only question raised is "Has the student adequately mastered the understandings or skills that indicate successful completion of this unit of study and readiness to enter the next?" The decision is either "Yes" or "No;" A's, B's, C's, etc., are irrelevant. Grading is accomplished simply by indicating which instructional objectives have been achieved by the student. Teachers set the mastery level wherever they wish—100% correct answers, 90%, 80%, and so forth. If the mastery level were set at 80% and a student answered correctly four out of five questions pertaining to a particular principle, he has "mastered" that principle. Using the 100% mastery level means that all students should answer all test items correctly. Other models for grading have been proposed but ranking and mastery offer, in my opinion, the most viable alternatives for introductory courses.[46]

TOWARD A MODEL

This paper has sketched out a way of thinking about introductory courses by posing four questions:

1. What shall be the purpose of the course?
2. What shall be the content of the course and how shall it be organized?
3. Which mode of teaching shall be utilized?
4. What shall be the forms and functions of evaluations?[47]

Each question has been answered in several different ways:

1. Purposes
   - A. Professional education
   - B. Liberal Education
   - C. Self-fulfillment education
2. Content
   - A. Structure of the discipline
   - B. Findings of the discipline
   - C. Needs of the students
3. Modes
   - A. Discourse
   - B. Management
   - C. Inquiry
4. Evaluation
   - A. Prescribing instruction
   - B. Improving teaching and learning
   - C. Giving grades

Each answer has been presented as an ideal type but obviously such singularity does not prevail in the real teaching of introductory courses. Thus, we should not expect to find in most classrooms "pure" discourse, "pure" liberal education, and so forth, because most faculty, sometimes consciously and deliberately, range across purposes, contents, modes, and evaluations. Indeed, some maintain they are eclectic and pragmatic in their teaching, taking the best idea available at the time.

Nevertheless, it seems likely that certain patterns can be found in the behavior of individual faculty, within a department, or, for that matter, across the discipline of geography itself. Such patterns can be identified in several ways. I shall propose using four continua.[48] Continuum A focuses on purpose. The X suggests my notion that the majority of introductory courses in geography are taught as professional education or liberal education courses. Continuum B concerns content. It is suggested that most introductory courses emphasize disciplinary findings rather than the structure of the discipline or the needs of students' approaches to content. Continuum C involves ways of teaching. Even though much has been made of inquiry as a desirable way of teaching, the hypothesis offered here is that most introductory courses are taught through discourse. Finally, Continuum D suggests that most teachers of introductory courses continue to emphasize evaluation for the sole purpose of grading students.[49]

Continuum A
The Purposes Continuum

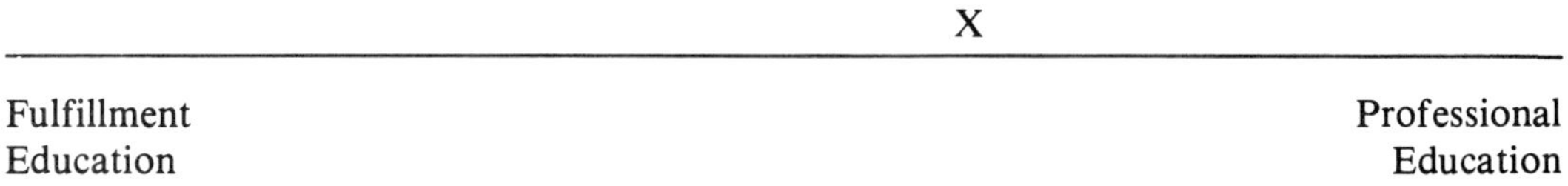

X

Fulfillment Education — Professional Education

Continuum B
The Content Continuum

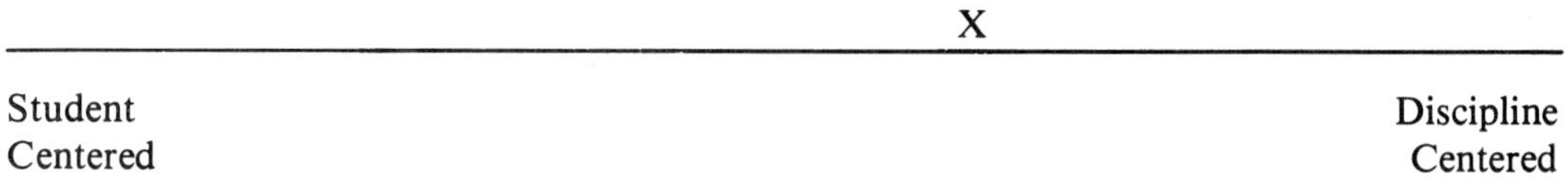

X

Student Centered — Discipline Centered

Continuum C
The Modes Continuum

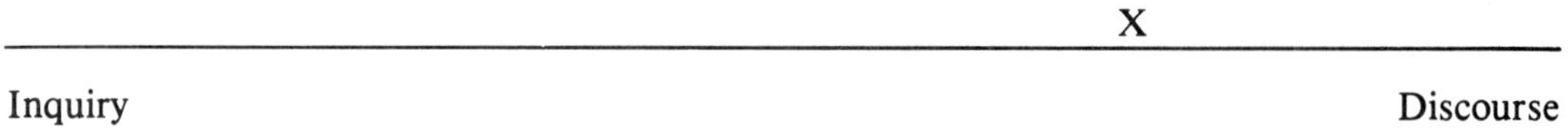

X

Inquiry — Discourse

Continuum D
The Evaluation Continuum

X

Evaluation for Learning — Evaluation for Grading

It is not difficult to recognize certain affinities among these continua. For example, would not fulfillment best combine with a student-centered approach to content, a personal inquiry mode of instruction, and an emphasis on evaluation for the improvement of teaching and learning? Would not professional education be best served by a discipline-centered approach to content, academic inquiry as the primary mode of teaching, and evaluation through grading? There is much to be said in support of such combinatorial thinking. Indeed, it is an essential requirement for curriculum development. However, these are not the only possibilities and, as this way of thinking is applied in actual situations, alternative combinations are likely to emerge. Developing and testing such combinations could well constitute an agenda for those interested in the development of geographic curriculum for the future.

NOTES

1. Geographers' involvement in school geography is an established tradition. Concerted, discipline-wide efforts to improve undergraduate education seems to be a relatively recent phenomenon. Preston James, "The Significance of Geography in American Education," *Journal of Geography* 68 (1969):473-483.
2. Published as *Geography in an Urban Age* (New York: The MacMillan Company, 1970). For a discussion of the origins and characteristics of this project, see Donald J. Patton, ed., *From Geographic Discipline to Inquiring Student* (Washington, D.C.: Association of American Geographers, 1970).
3. For further information, contact Dr. Salvatore Natoli, Educational Affairs Director, Association of American Geographers, 1710 Sixteenth Street, N.W., Washington, D.C.
4. Among these textbook authors are Peter Haggett, *Geography: A Modern Synthesis* (New York: Harper and Row, 1975); and Gordon Fielding, *Geography as Social Science* (New York: Harper and Row, 1974); Among the course developers are Janice J. Monk, and Charles S. Alexander, "Interactions Between Man and Environment: An Experimental College Course," *Journal of Geography* 74 (1975):212-222; and James Gardner, "Physical Geography and Environmental Problems: Relevance in a Course Content and Structure," *Journal of Geography* 70 (1971):163-168. Innovative ways of teaching have been exemplified by John D. Stephens, and Robert I. Wittick, "An Instructional Unit for Simulating Urban Residential Segregation," *The Professional Geographer* 37 (1975):340-346; and Michael E. Eliot Hurst, "Educational Environments: The Use of Media in the Classroom," *Journal of Geography* 72 (1973):41-48. Recent critiques of undergraduate education in geography have been offered by Peter Gould, "The Open Geographic Curriculum" (unpublished manuscript); and "Experiments in College Teaching: A Report to the Profession," *The Professional Geographer* 24 (1972):350-361.
5. L. Dee Fink, and David J. Morgan, "The Importance of Teaching in the Appointment and Promotion of Academic Geographers" (unpublished manuscript, 1975).
6. Clyde F. Kohn, "The 1960's: A Decade of Progress in Geographical Research and Instruction," *Annals of the Association of American Geographers* 60 (1970):219.
7. Joseph J. Schwab, *College Curriculum and Student Protest* (Chicago: University of Chicago Press, 1969), p. 246.
8. Melvin G. Marcus, "Introductory Physical Geography in the College Curriculum," *Introductory Geography: Viewpoints and Themes*, Commission on College Geography, Publication no. 5 (Washington, D.C.: Association of American Geographers, 1971), p. 3.
9. "Experiments in College Teaching," *op. cit.*, p. 351.
10. Gould, *op. cit.*, p. 16.
11. Preston E. James, "Introductory Geography: Topical or Regional?" *Journal of Geography* 66 (1967):52-53.
12. Gilbert White, "Geography in Liberal Education," *Geography in Undergraduate Liberal Education,* Commission on College Geography Publication no. 1 (Washington, D.C.: Association of American Geographers, 1967).
13. William Pattison, "The Educational Purpose of Geography," *Evaluation in Geographic Education.* In Dana Kurfman, ed., The 1971 Yearbook of the National Council for Geographic Education (Belmont, California: Fearon Publishers, 1970).
14. "Experiments," *op. cit.*, p. 352.
15. Those who wish to investigate the relationships between curricular structures and disciplinary structures might begin by reading Jerome Bruner, *The Process of Education* (New York: Vintage Books, 1960). Then they should continue with Arthur R. King, Jr., and John A. Brownell, *The Curriculum and the Disciplines of Knowledge: A Theory of Curriculum Practice* (New York: John Wiley & Sons, Inc., 1966); and Philip Phenix, *Realms of Meaning* (New York: McGraw-Hill Book Co., 1964).
16. Kennard W. Rummage, and Leslie P. Cummings, "Introduction to Geography—A Spatial Approach: A One-Semester Course Outline," *New Approaches in Introductory College Geography Courses*, Commission on College Geography Publication no. 4 (Washington, D.C.: Association of American Geographers, 1967), p. 114.
17. *Ibid.*, p. 115.
18. Edward J. Taaffe, "The Spatial View in Context," *Annals of the Association of American Geographers* 64 (1974):3.
19. Jesse H. Wheeler, Jr.; Trenton J. Kostbade; and Richard S. Thoman, *Regional Geography of the World,* 3rd ed. (New York: Holt, Rinehart & Winston, 1975), p. v.

20. *The Student in Higher Education* (New Haven, Conn.: The Hazen Foundation, 1968), p. 6.
21. "Syllabus: Geography 200"
22. Joseph Axelrod, *The University Teacher as Artist* (San Francisco: Jossey-Bass, 1973).
23. Bruce Joyce, and Marsha Weil, *Models of Teaching* (Englewood Cliffs, New Jersey: Prentice-Hall, Inc., 1972).
24. Dwight Allen, and Kevin Ryan, *Microteaching* (Reading, Mass.: Addison-Wesley Publishing Co., 1969).
25. Ronald T. Hyman, *Ways of Teaching* (Philadelphia: J.B. Lippincott, 1970), p. 25.
26. I am deeply indebted to L. Dee Fink, for sharing his manuscript "Approaches to Teaching," as well as personal communications showing how these approaches have been evident in geographic education.
27. Axelrod, *op. cit.*, p. 10.
28. Robert B. McNee, "A Proposal for a New Geography Course for Liberal Education: Introduction to Geographic Behavior," *New Approaches, op. cit.*, pp. 2-3.
29. Robert H. Davis; Lawrence T. Alexander; and Stephen L. Yelon, *Learning System Design: An Approach to the Improvement of Instruction* (New York: McGraw-Hill Book Co., 1974), p. 3. This may sound mechanistic and unrelated to the needs of students. However, Patricia Cross says that a curriculum is student-centered to the degree it provides for individual differences in learning. Patricia K. Cross, "Learner-Centered Curricula." In *Learner-Centered Reform: Current Issues in Higher Education* (San Francisco: Jossey-Bass, 1975).
30. See, for instance, Benjamin F. Richason, Jr., and Arthur G. Wilner, "The Audio-Visual-Tutorial Method in Geography." In *Methods of Geographic Instruction*, John W. Morris, ed. (Waltham, Massachusetts: Blaisdell Publishing Co., 1968), pp. 112-127; and Rainer R. Erhart, and David S. Mellander, "Experiences with an Audio-Visual-Tutorial Laboratory," *Journal of Geography* 68 (1969):82-92.
31. Gordon J. Fielding, *Programmed Case Studies in Geography* (New York: Harper & Row, 1974).
32. John R. Healy, and Larry K. Stephenson, "Unit Mastery Learning in an Introductory Geography Course," *Journal of Geography* 74 (1975):25-31.
33. For some differences between rote, or meaningless, learning and meaningful learning, see David P. Ausubel, *Educational Psychology: A Cognitive View* (New York: Holt, Rinehart and Winston, Inc., 1968), pp. 24-26.
34. G.H. Dury, "Points of Contact: Or, On Not Despising Lower Classmen," *Geographic Perspectives,* no. 33, 1974).
35. Inquiry is sometimes presented as a way of learning in which students engage in discovering rather than receiving meaningful ways of viewing and coping with reality. I distinguish ways of teaching from ways of learning; therefore, I shall refer to such learning as discovery learning and will use inquiry to refer to a way of teaching. For a discussion of discovery learning, see, Ausubel, *op. cit.,* Chapter 25.
36. Adapted from Herbert Thelen, *Education and the Human Quest* (New York: Harper & Bros., 1960), Chapters 5-9. Again, my appreciation to Fink for calling this book to my attention.
37. Biological Sciences Curriculum Study, Joseph J. Schwab, Supervisor, *Biology Teachers Handbook* (New York: John Wiley & Sons, 1965), pp. 39-40 as quoted in Joyce and Weil, *op. cit.,* p. 155.
38. McNee, *op. cit.,* p. 3.
39. Thelen was writing about precollegiate education. However, the models he proposed lend themselves equally well to higher education. Furthermore, the personal inquiry mode has received some attention in recent discussions of teaching and learning in colleges and universities.
40. For further discussion of how teachers might work in this mode, see Carl Rogers, "The Facilitation of Significant Learning." In *Instruction: Some Contemporary Viewpoints*, Lawrence Seigel, ed., pp. 37-54.
41. Thelen, *op. cit.,* p. 141.
42. "Geography 200," pp. 4-5.
43. Thelen, *op. cit.,* p. 84.
44. Dana G. Kurfman, "Evaluation Developments Useful in Geographic Education." In Dana G. Kurfman, ed., *Evaluation in Geographic Education* (Belmont, California: Fearon Publishers, 1970), p. 3. For a similar view, see David A. Hill, "Evaluation of College Geographic Learning," *op. cit.,* pp. 53-69.
45. Benjamin S. Bloom; Thomas J. Hastings; and George F. Madaus, *Handbook on Formative and Summative Evaluation of Student Learning* (New York: McGraw-Hill Book Co., 1971), p. 129.
46. For other approaches to grading, see *Improving Teaching and Learning at the University,* A Report to the President of the University of Colorado and the University Community (Boulder Colorado: Office of the President, University of Colorado, 1975), pp. 38-44.
47. This framework for curriculum analysis usually is attributed to Ralph Tyler, *Basic Principles of Curriculum and Instruction* (Chicago, Illinois: University of Chicago Press, 1950).
48. The idea of using continua for curriculum and analysis has been adopted from Paul Dressel, *College and University Curriculum* (Berkley, California: McCutchan Publishing Co., 1968), pp. 16-18.
49. At present there are no data to substantiate these notions. I am currently engaged in a pilot study to determine their validity and to develop ways of gathering information about them.